FLAMES, LASERS, AND REACTIVE SYSTEMS

Edited by
J.R. Bowen
University of Washington
Seattle, Washington

N. Manson
Université de Poitiers
Poitiers, France

A.K. Oppenheim
University of California
Berkeley, California

R.I. Soloukhin
Institute of Heat and Mass Transfer
BSSR Academy of Sciences
Minsk, USSR

Volume 88
PROGRESS IN
ASTRONAUTICS AND AERONAUTICS

Martin Summerfield, Series Editor-in-Chief
Princeton Combustion Research Laboratories, Inc.
Princeton, New Jersey

Technical papers selected from the Eighth International Colloquium on Gasdynamics of Explosions and Reactive Systems, Minsk, USSR, August 1981, and subsequently revised for this volume.

Published by the American Institute of Aeronautics and Astronautics, Inc.
1633 Broadway, New York, NY 10019

American Institute of Aeronautics and Astronautics, Inc.
New York, New York

Library of Congress Cataloging in Publication Data
Main entry under title:

International Colloquium on Gasdynamics of Explosions
   and Reactive Systems (8th: 1981: Minsk, Byelorussian S.S.R.)
   Flames, lasers, and reactive systems.

   (Progress in astronautics and aeronautics: v. 88)
   Includes index.
   1. Flame—Congresses.   2. Lasers—Congresses.   3. Explosions—
Congresses.   I. Bowen, J.R. (J. Ray)   II. Title.   III. Series.
TL507.P75   vol. 88   [QD516]   629.1s   [621.402'3]   83-15463
ISBN 0-915928-77-9;   Set: 0-915928-79-5

Copyright © 1983 by
American Institute of Aeronautics and Astronautics, Inc.

All rights reserved. No part of this book may be reproduced in any form or by any means, electronic or mechanical, including photocopying, recording, or by any information storage and retrieval system, without permission in writing from the publisher.

Progress in Astronautics and Aeronautics

Series Editor-in-Chief

Martin Summerfield
Princeton Combustion Research Laboratories, Inc.

Series Associate Editors

Burton I. Edelson
*National Aeronautics
  and Space Administration*

Leroy S. Fletcher
Texas A&M University

Allen E. Fuhs
Naval Postgraduate School

J. Leith Potter
Vanderbilt University

Norma J. Brennan
Director, Editorial Department
AIAA

Camille S. Koorey
Series Managing Editor
AIAA

Table of Contents

Preface xi

List of Series Volumes xv

Chapter I.   Laminar Flames 1

  Resonant Response of a Flat Flame Near a Flame-Holder 3
    A.C. McIntosh and J.F. Clarke, *Cranfield Institute of Technology, Bedford, England*

  Temperature and Pressure Effect in Cool and Blue Flames 38
    Y. Ohta and H. Takahashi, *Nagoya Institute of Technology, Nagoya, Japan*

  An Excess Enthalpy Flame Stabilized in Ceramic Tubes 57
    T. Hashimoto and S. Yamasaki, *Hitachi, Ltd., Tsuchiura, Japan*, and T. Takeno, *The University of Tokyo, Tokyo, Japan*

Chapter II.   Turbulent Flames 79

  Differential Diffusion Effects on Measurements in Turbulent Diffusion Flames by the Mie Scattering Technique 81
    S.H. Stårner and R.W. Bilger, *The University of Sydney, Sydney, Australia*

  Concentration and Velocity Measurements in a Turbulent Reacting Mixing Layer 105
    J.L. Bousgarbies and J. Nérault, *Laboratoire d'Études Aérodynamiques et Thermiques, Poitiers, France*

  Turbulent Combustion Zone in a Tubular Reactor 119
    Y. Chauveau, P. Cambray, E. Gengembre, M. Champion, and J.C. Bellet, *Université de Poitiers, Poitiers, France*

  Premixed Turbulent Flames—Interplay of Hydrodynamic and Chemical Phenomena 133
    A.M. Klimov, *Academy of Sciences, Moscow, USSR*

  Modification of Turbulent Flowfield by an Oblique Premixed Hydrogen-Air Flame 147
    D. Escudie, M. Trinite, and P. Paranthoen, *Laboratoire de Thermodynamique, Mont Saint Aignan, France*

Chapter III.   Combustion of Solids 165

  Combustion of Lithium Perchlorate, Ammonium Chloride-Ammonium Perchlorate Solid Mixtures 167
    M.S. Al Fakir, P. Joulain, and J.M. Most, *Université de Poitiers, Poitiers, France*

Radiation from Polyurethane Pile Fires 182
  J.M. Souil, H. Azov, and P. Joulain, *Université de Poitiers, St. Julien l'Ars, France*, and S. Galant, *Bertin et Cie, Plaisir, France*

Study of Condensed System Flames by Molecular Beam Mass Spectrometry 197
  O.P. Korobeinichev and L.V. Kuibida, *Academy of Sciences, Novosibirsk, USSR*

Unsteady Burning of Double-Base Propellants 208
  V.E. Zarko, V.N. Simonenko, and A.B. Kiskin, *Academy of Sciences, Novosibirsk, USSR*

Surface Layer Destruction during Combustion of Homogeneous Powders 220
  V.E. Zarko and V.Y. Zyryanov, *Academy of Sciences, Novosibirsk, USSR*

Heat Transfer Ahead of Flame Spreading over a Cured Epoxy Resin Surface in an Opposed Flow 228
  B.Y. Kolesnikov, V.L. Efremov, N.S. Umarbekov, and A.B. Kolesnikov *Kazakh State University, Alma-Ata, USSR*

Chapter IV. Ignition and Extinction 237

Self-Ignition of Atomized Liquid Fuel in Gaseous Medium 239
  A.A. Borisov, B.E. Gel'fand, E.I. Timofeev, S.A. Tsyganov, and S.V. Khomik, *Academy of Sciences, Moscow, USSR*

Hydrocarbon Induced Acceleration of Ignition of Methane-Air Ignition 252
  R. Zellner and K.J. Niemitz, *Institut für Physikalische Chemie der Universität Göttingen, Göttingen, W. Germany*, and J. Warnatz, *Institut für Physikalische Technischen Hochschule Darmstadt, Darmstadt, W. Germany*, and W.C. Gardiner Jr., C.S. Eubank, and J.M. Simmie, *University of Texas, Austin, Texas*

Extinction of In-Flight Engine Fuel-Leak Fires with Dry Chemicals 273
  R.L. Altman, *NASA Ames Research Center, Moffett Field, Calif.*

Chapter V. Nonequilibrium Systems 291

Linearized Kinetic Models for Polyatomic Gases and Mixtures of Gases: Application to Vibrationally Relaxing Flows 293
  R. Brun, G. Duran, P.C. Philippi, M.F. Dourieu, and R. Tosello, *Université de Provence, Marseille, France*

**Theoretical Model for Sound Output from a
Pulsating Arc Discharge** 305
    N.I. Kidin and V.B. Librovich, *Academy of Sciences, Moscow, USSR*,
    and M.L. Vuillermoz and J.P. Roberts, *Polytechnic of the
    South Bank, London, England*

Chapter VI.   Lasers 317

**Gain Coefficient and Vibrational Temperature Measurements
in Shock Tube Driven CO_2-GDL, $CO_2 + N_2 + He$, and
$CO_2 + CO(N_2) + H_2$ Mixtures** 319
    S.S. Novikov, V.M. Doroshenko, and N.N. Kudryavtsev,
    Academy of Sciences, Moscow, USSR

**Time-Dependent Nozzle and Base Flow/Cavity Model
for CW Chemical Laser Flowfields** 336
    N.L. Rapagnani and D.W. Lankford, *Air Force Weapons
    Laboratory, Kirtland Air Force Base, N. Mex.*

Modeling of a Chemically Driven H_2-HCl Transfer Laser 369
    V.K. Baev and V.I. Golovichev, *Academy of Sciences, Novosibirsk,
    USSR*, and H. Guenoche and C. Sedes, *Université de Provence,
    Marseille, France*

A Gasdynamic Laser Using Products of Acetylene Explosions 391
    A.B. Britan, A.N. Khmelevskii, V.A. Levin, S.A. Losev,
    V.V. Lugovskoi, G.D. Smekhov, and A.M. Starik,
    Moscow University, Moscow, USSR

**Influence of Flow Structure on Optical Gain
in Gasdynamic Lasers** 411
    M.G. Ktalkherman, V.M. Malkow, and N.A. Ruban,
    Academy of Sciences, Novosibirsk, USSR

**Operation of Arc-Heated Gasdynamic CO_2 Laser
at 16.4-18.6 μm** 425
    A.I. Demin, E.M. Kidriavtsev, and A.Y. Volkov, *Academy of Sciences,
    Moscow, USSR*, and D.G. Bakanov, A.I. Fedoseev, and A.I. Odintsov,
    Moscow State University, Moscow, USSR, and V.F. Sharkov,
    Kurchatov Institute of Atomic Energy, Moscow, USSR

Author Index for Volume 88 436

Table of Contents for Companion Volume 87

Introduction 1
    Y.B. Zel'dovich, *Academy of Sciences, Moscow, USSR*

Chapter I. Shock Waves Interactions 7

The Study of Shock-Induced Signals and Coherent Effects in Solids by Molecular Dynamics 9
    A.M. Karo, F.E. Walker, and W.G. Cunningham, *Lawrence Livermore National Laboratory, Livermore, Calif.,*
    and J.R. Hardy, *University of Nebraska, Lincoln, Neb.*

Oblique Shock Waves in Two-Phase Flow 22
    K. Hayashi, *Princeton University, Princeton, N.J.,* and M.C. Branch, *University of Colorado, Boulder, Colo.*

Equilibrium Shock Wave Properties in Dusty and Clean Air 41
    A.L. Kuhl, *R&D Associates, Marina del Rey, Calif.*

Shock Waves in Water Induced by Focused Laser Radiation 64
    O.G. Martynenko, N.N. Stolovich, G.I. Rudin, and S.A. Levchenko, *Academy of Sciences, Minsk, USSR*

Ignition of Small Particles Behind Shock Waves 71
    V.M. Boiko, A.V. Fedorov, V.M. Fomin, A.N. Papyrin, and R.I. Soloukhin, *Academy of Sciences,*
    Novosibirsk, USSR

Reflection of Shock Waves at Rigid Walls in Two-Phase Media 88
    A.A. Borisov, B.E. Gel'fand, R.I. Nigmatulin, K.A. Rakhmatulin, and E.I. Timofeev, *Academy of Sciences,*
    Moscow, USSR

Relaxation Phenomena in a Foamy Structure 96
    V.M. Kudinov, B.I. Palamarchuk, V.A. Vakhnenko, A.V. Cherkashin, S.G. Lebed, and A.T. Malakhov,
    Academy of Sciences, Kiev, USSR

Chapter II. Blast Waves 119

Self-Similar Blast Waves Incorporating Deflagrations of Variable Speed 121
    R.H. Guirguis, M.M. Kamel, and A.K. Oppenheim, *University of California, Berkeley, Calif.*

Analysis of Reactive Blast Waves Propagating Through Gaseous Mixtures
with Spatially Distributed Chemical Energy 157
    S. Ohyagi and A. Ohsawa, *Saitama University, Saitama, Japan*

On the Use of General Equations of State in Similarity Analysis
of Flame-Driven Blast Waves 175
    A.H. Kuhl, *R&D Associates, Marina del Rey, Calif.*

Optical Interferometry of Spherical Shock Waves 196
    V.F. Klimkin and V.V. Pickalov, *Academy of Sciences, Novosibirsk,*
    USSR, and R.I. Soloukhin, *Academy of Sciences, Minsk, USSR*

A Study of Explosive Shock Tubes 205
    G.A. Shvetsov, V.M. Titov, V.P. Chistyakov, and I.A. Stadnichenko, *Academy of Sciences,*
    Novosibirsk, USSR

The Taylor Instability of Contact Boundary Between Expanding Detonation Products
and a Surrounding Gas 218
    S.I. Anisimov, Y.B. Zel'dovich, N.A. Inogamov, and M.F. Ivanov, *L.D. Landau Institute*
    for Theoretical Physics, Moscow, USSR

Chapter III. Gaseous Detonations 229

Properties of Detonation Waves in Hydrocarbon-Oxygen-Nitrogen Mixtures
at High Initial Pressures 231
    P. Bauer and C. Brochet, *Laboratoire d'Energètique et de Dètonique, Poitiers, France*

Overdriven Gaseous Detonations 244
    T.P. Gavrilenko and E.S. Prokhorov, *Academy of Sciences, Novosibirsk, USSR*

Motion of Solid Bodies in Combustible Gas Mixtures 251
  M.M. Gilinsky, L.I. Zak, and T.S. Novikova, *Moscow State University, Moscow, USSR*

Direct Initiation of Detonation in LNG/Air Clouds 262
  J. Kurylo, J.M. Thomsen, and F.M. Sauer, *Physics International Company, San Leandro, Calif.*

Influence of Walls on Pressure Behind Self-Sustained Expanding Cylindrical
and Plane Detonations in Gases 302
  D. Desbordes and N. Manson, *Université de Poitiers, Poitiers, France,* and J. Brossard, *Université d'Orléans, Bourges, France*

Kinetic Modelling of Ethane/Air Detonability 318
  R. Atkinson and D.C. Bull, *Shell Research Ltd., Thornton Research Center, Chester, England*

Chapter IV. Heterogeneous Detonations 333

Detonations Supported by Physical Explosions of Liquefied Gases 335
  S. Tsugé and S. Kadowaki, *University of Tsukuba, Ibaraki, Japan*

Effect of Liquid Films on Detonation in a Gaseous Mixture 352
  J.P. Saint-Cloud, C. Guerraud, and N. Manson, *Université de Poitiers, Poitiers, France*

Ignition of Aluminum Particles in a Gaseous Detonation 362
  B. Veyssiere, *Université de Poitiers, France*

A Model of Blast Waves Propagating in Coal Mines 376
  V.P. Korobeinikov and I.S. Men'shov, *Academy of Sciences, Moscow, USSR*

Multifront Combustion of Two-Phase Media 394
  L.A. Afanasiva, V.A. Levin, and Y.V. Tunik, *Moscow State University, Moscow, USSR*

Flame Propagation in Dust-Air Mixtures at Minimum Explosive Concentration 414
  W. Buksowicz, *Fire Protection Research and Development Center, Józefów, Poland,* and P. Wolánski, *Technical University of Warsaw, Warsaw, Poland*

Chemical Kinetics of Detonation in Some Organic Liquids 426
  B.N. Kondrikov, *Mendeleev Institute of Chemical Technology, Moscow, USSR*

Chapter V. Explosions in Solids 443

Shock Induced Hot-Spot Formation and Subsequent Decomposition in Granular,
Porous HNS Explosive 445
  D.B. Hayes, *Sandia National Laboratories, Albuquerque, N. Mex.*

Initiation of Detonations 468
  C.L. Mader, *Los Alamos National Laboratory, Los Alamos, N. Mex.*

Shock Wave Predetonation Processes in Porous High Explosives 492
  B.A. Khasainov, A.A. Borisov, and B.S. Ermolayev, *Academy of Sciences, Moscow, USSR*

Preface

This and a companion volume include revised and edited papers that were presented at the Eighth International Colloquium on the Gasdynamics of Explosions and Reactive Systems held in Minsk, USSR, in August 1981. The International Gasdynamic Colloquia had their origin in 1966 as a consequence of revolutionary advances in the understanding of detonation wave structure. Leading researchers in this field concluded that a regular forum should be available for discussion of important findings in the gasdynamics of flows associated with the exothermic process—the essential feature of detonation waves. However, it was felt that a much broader scope of applications should be included.

The gasdynamics of explosions is a subject concerned principally with the interrelationship between the rate processes of energy deposition in a compressible medium and its concurrent nonsteady flow as it occurs typically in explosion phenomena. Gasdynamics of reactive systems is a broader term referring to the processes of coupling between the dynamics of fluid flow and molecular transformations in reactive media occurring in any combustion system. In this connection, in addition to the usual topics of explosions, detonations, shock phenomena, and reactive flow, the Eighth Colloquium included papers that dealt especially with the gasdynamic aspects of nonsteady flow in combustion systems, the fluid mechanic aspects of combustion with particular emphasis on the effects of turbulence, and the diagnostic techniques for the study of combustion phenomena. Of special interest were papers dealing with the radiative heat-transfer effects on the fluid dynamic features of reactive systems such as luminous flames, intense fires, and gasdynamic lasers.

The contributions have been assembled into two volumes: *Flames, Lasers, and Reactive Systems* and *Shock Waves, Explosions, and Detonations*. In this volume, the papers have been arranged into chapters on laminar flames, turbulent flames, combustion of solids, ignition and extinction, nonequilibrium systems, and lasers. The material should be particularly useful to the reader who wishes to be informed about the current directions and important findings of Soviet scientists. About half of the papers are of Russian origin and provide a valuable insight into the current status of Russian research.

Many of the 55 papers in these two volumes provoked interesting discussions during the Colloquium. While the brevity of this Preface does not permit the editors to do justice to all of the stimulating papers, the more noteworthy contributions among them will be highlighted in the following.

The range of topics covered in this volume is indicated by the chapter titles. In Chap. I, Laminar Flames, *McIntosh* and *Clarke* report an application of asymtotic theory applicable to the case of large activation energy, a characteristic feature of combustion reactions, to model the effects of inlet composition and mass flux oscillations on the behavior of a premixed flame stabilized by a coplanar flame holder. The analysis indicates that a resonant oscillatory frequency exists for which the amplitudes of flame position and temperature oscillations become very large. This frequency is shown to vary with the Lewis number, the initial flame position, and the rate of heat loss to the flame holder.

Mie scattering due to small particles in turbulent diffusion flames has been used to compute the time series of significant scalar quantities. In this technique a critical assumption is that the marker particles mix and diffuse in the same manner as fuel atoms. In Chap. II, Turbulent Flames, *Stårner* and *Bilger* develop an analysis that may be used with experimental data to test whether differential diffusion between the fuel atom and the marker particles is significant. Molecular beam mass spectrometry has been used to determine transient species concentrations and to deduce chemical kinetics in a variety of experimental systems. In Chap. III, Combustion of Solids, *Korobeinichev* and *Kuibida* report the application of a molecular beam system to determine the species produced in the flames of ammonium perchlorate and its mixtures. The data confirmed experimentally the hypothesized reaction mechanism.

An important parameter characterizing the combustion capability of liquid fuels is the self-ignition delay. Optimization of combustion performance requires precise knowledge of the dependence of ignition delay on temperature, pressure, and composition. In Chap. IV, Ignition and Extinction, *Borisov, Gel'fand, Timofeev, Tsyganov,* and *Khomik* reported results of an experimental investigation of ignition delay in atomized fuel over a temperature range of 800-2000 K and for a pressure range of 1.0-10.0 MPa. The effects of fuel additives on ignition delay were also determined.

In Chap. VI, Lasers, *Rapagnani* and *Lankford* report a detailed calculation of the time-dependent flow in a continuous-wave chemical laser flowfield. The calculations showed that pressure mismatch between primary and secondary nozzles controlled the locations of the initial mixing zone. In a study of arc-heated gasdynamic CO_2 lasers *Demin, Kidriavtsev, Volkov, Bakanov, Fedoseev, Odintsov,* and *Sharkov* observed laser transitions not previously reported.

The companion volume includes papers on shock wave interactions, blast waves, gaseous and heterogeneous detonations, and explosions in solids (Vol. 87 in the *AIAA Progress in Astronautics and Aeronautics* series).

The first Colloquium was held in 1967 in Brussels, and Colloquia have been held on a biennial basis since then (1969 in Novosibirsk, 1971 in Marseilles, 1973 in La Jolla, 1975 in Bourges, 1977 in Stockholm, 1979 in Göttingen, and 1981 in Minsk). They have now achieved the status of a prime international meeting on these topics and attract contributions from scientists and engineers throughout the world. The Proceedings of the First through the Sixth Colloquia have appeared as part of the journal, *Acta Astronautica*, or its predecessor, *Astronautica Acta*. With the publication of the Seventh Colloquium the Proceedings now appear as part of the AIAA *Progress in Astronautics and Aeronautics* series.

Acknowledgments

The Eighth Colloquium was held under the auspices of the Institute of Heat and Mass Transfer, Soviet Academy of Sciences, Minsk, BSSR, Aug. 24-28, 1981. Arrangements in Minsk were made by Prof. R.I. Soloukhin. The publication of the Proceedings has been made possible by grants from the National Science Foundation (USA) and the Army Research Office (USA).

Preparations for the Ninth Colloquium are under way. The meeting is scheduled to take place in July 1983 at the Ecole Nationale Supérieure de Mécanique et d'Aérotechnique in Poitiers, France.

J. Ray Bowen
Numa Manson
Antoni K. Oppenheim
R.I. Soloukhin
April 1983

R.I. Soloukhin and Y.B. Zel'dovich in discussion.

Participants in session.

Volume Titles

Volume Editors

*1. **Solid Propellant Rocket Research.** 1960

Martin Summerfield
Princeton University

*2. **Liquid Rockets and Propellants.** 1960

Loren E. Bollinger
The Ohio State University
Martin Goldsmith
The Rand Corporation
Alexis W. Lemmon Jr.
Battelle Memorial Institute

*3. **Energy Conversion for Space Power.** 1961

Nathan W. Snyder
Institute for Defense Analyses

*4. **Space Power Systems.** 1961

Nathan W. Snyder
Institute for Defense Analyses

*5. **Electrostatic Propulsion.** 1961

David B. Langmuir
Space Technology Laboratories, Inc.
Ernst Stuhlinger
NASA George C. Marshall Space Flight Center
J. M. Sellen Jr.
Space Technology Laboratories, Inc.

*6. **Detonation and Two-Phase Flow.** 1962

S. S. Penner
California Institute of Technology
F. A. Williams
Harvard University

*7. **Hypersonic Flow Research.** 1962

Frederick R. Riddell
AVCO Corporation

*8. **Guidance and Control.** 1962

Robert E. Roberson
Consultant
James S. Farrior
Lockheed Missiles and Space Company

*9. **Electric Propulsion Development.** 1963

Ernst Stuhlinger
NASA George C. Marshall Space Flight Center

*10. **Technology of Lunar Exploration.** 1963

Clifford I. Cummings and
Harold R. Lawrence
Jet Propulsion Laboratory

*11. **Power Systems for Space Flight.** 1963

Morris A. Zipkin and
Russell N. Edwards
General Electric Company

*12. **Ionization in High-Temperature Gases.** 1963

Kurt E. Shuler, Editor
National Bureau of Standards
John B. Fenn, Associate Editor
Princeton University

*13. **Guidance and Control—II.** 1964

Robert C. Langford
General Precision Inc.
Charles J. Mundo
Institute of Naval Studies

*14. **Celestial Mechanics and Astrodynamics.** 1964

Victor G. Szebehely
Yale University Observatory

*15. **Heterogeneous Combustion.** 1964

Hans G. Wolfhard
Institute for Defense Analyses
Irvin Glassman
Princeton University
Leon Green Jr.
Air Force Systems Command

*16. **Space Power Systems Engineering.** 1966

George C. Szego
Institute for Defense Analyses
J. Edward Taylor
TRW Inc.

*17. **Methods in Astrodynamics and Celestial Mechanics.** 1966

Raynor L. Duncombe
U. S. Naval Observatory
Victor G. Szebehely
Yale University Observatory

*18. **Thermophysics and Temperature Control of Spacecraft and Entry Vehicles.** 1966

Gerhard B. Heller
NASA George C. Marshall Space Flight Center

*19. **Communication Satellite Systems Technology.** 1966

Richard B. Marsten
Radio Corporation of America

*20. Thermophysics of Spacecraft and Planetary Bodies: Radiation Properties of Solids and the Electromagnetic Radiation Environment in Space. 1967 — Gerhard B. Heller, *NASA George C. Marshall Space Flight Center*

*21. Thermal Design Principles of Spacecraft and Entry Bodies. 1969 — Jerry T. Bevans, *TRW Systems*

*22. Stratospheric Circulation. 1969 — Willis L. Webb, *Atmospheric Sciences Laboratory, White Sands, and University of Texas at El Paso*

*23. Thermophysics: Applications to Thermal Design of Spacecraft. 1970 — Jerry T. Bevans, *TRW Systems*

24. Heat Transfer and Spacecraft Thermal Control. 1971 — John W. Lucas, *Jet Propulsion Laboratory*

25. Communications Satellites for the 70's: Technology. 1971 — Nathaniel E. Feldman, *The Rand Corporation*; Charles M. Kelly, *The Aerospace Corporation*

26. Communications Satellites for the 70's: Systems. 1971 — Nathaniel E. Feldman, *The Rand Corporation*; Charles M. Kelly, *The Aerospace Corporation*

27. Thermospheric Circulation. 1972 — Willis L. Webb, *Atmospheric Sciences Laboratory, White Sands, and University of Texas at El Paso*

28. Thermal Characteristics of the Moon. 1972 — John W. Lucas, *Jet Propulsion Laboratory*

29. Fundamentals of Spacecraft Thermal Design. 1972 — John W. Lucas, *Jet Propulsion Laboratory*

30. Solar Activity Observations and Predictions. 1972 — Patrick S. McIntosh and Murray Dryer, *Environmental Research Laboratories, National Oceanic and Atmospheric Administration*

31. **Thermal Control and Radiation.** 1973

Chang-Lin Tien
University of California, Berkeley

32. **Communications Satellite Systems.** 1974

P. L. Bargellini
COMSAT Laboratories

33. **Communications Satellite Technology.** 1974

P. L. Bargellini
COMSAT Laboratories

34. **Instrumentation for Airbreathing Propulsion.** 1974

Allen E. Fuhs
Naval Postgraduate School
Marshall Kingery
Arnold Engineering Development Center

35. **Thermophysics and Spacecraft Thermal Control.** 1974

Robert G. Hering
University of Iowa

36. **Thermal Pollution Analysis.** 1975

Joseph A. Schetz
Virginia Polytechnic Institute

37. **Aeroacoustics: Jet and Combustion Noise; Duct Acoustics.** 1975

Henry T. Nagamatsu, Editor
General Electric Research and Development Center
Jack V. O'Keefe, Associate Editor
The Boeing Company
Ira R. Schwartz, Associate Editor
NASA Ames Research Center

38. **Aeroacoustics: Fan, STOL, and Boundary Layer Noise; Sonic Boom; Aeroacoustics Instrumentation.** 1975

Henry T. Nagamatsu, Editor
General Electric Research and Development Center
Jack V. O'Keefe, Associate Editor
The Boeing Company
Ira R. Schwartz, Associate Editor
NASA Ames Research Center

39. **Heat Transfer with Thermal Control Applications.** 1975

M. Michael Yovanovich
University of Waterloo

40. **Aerodynamics of Base Combustion.** 1976

S. N. B. Murthy, Editor
Purdue University
J. R. Osborn, Associate Editor
Purdue University
A. W. Barrows and J. R. Ward, Associate Editors
Ballistics Research Laboratories

41. Communication Satellite
    Developments: Systems.
    1976

    Gilbert E. LaVean
    *Defense Communications
    Engineering Center*
    William G. Schmidt
    CML Satellite Corporation

42. Communication Satellite
    Developments: Technology.
    1976

    William G. Schmidt
    CML Satellite Corporation
    Gilbert E. LaVean
    *Defense Communications
    Engineering Center*

43. Aeroacoustics: Jet Noise,
    Combustion and Core Engine
    Noise. 1976

    Ira R. Schwartz, Editor
    NASA Ames Research Center
    Henry T. Nagamatsu,
    Associate Editor
    *General Electric Research
    and Development Center*
    Warren C. Strahle,
    Associate Editor
    Georgia Institute of Technology

44. Aeroacoustics: Fan Noise and
    Control; Duct Acoustics;
    Rotor Noise. 1976

    Ira R. Schwartz, Editor
    NASA Ames Research Center
    Henry T. Nagamatsu,
    Associate Editor
    *General Electric Research
    and Development Center*
    Warren C. Strahle,
    Associate Editor
    Georgia Institute of Technology

45. Aeroacoustics: STOL Noise;
    Airframe and Airfoil Noise.
    1976

    Ira R. Schwartz, Editor
    NASA Ames Research Center
    Henry T. Nagamatsu,
    Associate Editor
    *General Electric Research
    and Development Center*
    Warren C. Strahle,
    Associate Editor
    Georgia Institute of Technology

46. **Aeroacoustics: Acoustic Wave Propagation; Aircraft Noise Prediction; Aeroacoustic Instrumentation.** 1976

Ira R. Schwartz, Editor
NASA Ames Research Center
Henry T. Nagamatsu, Associate Editor
General Electric Research and Development Center
Warren C. Strahle, Associate Editor
Georgia Institute of Technology

47. **Spacecraft Charging by Magnetospheric Plasmas.** 1976

Alan Rosen
TRW Inc.

48. **Scientific Investigations on the Skylab Satellite.** 1976

Marion I. Kent and Ernst Stuhlinger
NASA George C. Marshall Space Flight Center
Shi-Tsan Wu
The University of Alabama

49. **Radiative Transfer and Thermal Control.** 1976

Allie M. Smith
ARO Inc.

50. **Exploration of the Outer Solar System.** 1977

Eugene W. Greenstadt
TRW Inc.
Murray Dryer
National Oceanic and Atmospheric Administration
Devrie S. Intriligator
University of Southern California

51. **Rarefied Gas Dynamics, Parts I and II (two volumes).** 1977

J. Leith Potter
ARO Inc.

52. **Materials Sciences in Space with Application to Space Processing.** 1977

Leo Steg
General Electric Company

53. **Experimental Diagnostics in Gas Phase Combustion Systems.** 1977

Ben T. Zinn, Editor
Georgia Institute of Technology
Craig T. Bowman,
Associate Editor
Stanford University
Daniel L. Hartley,
Associate Editor
Sandia Laboratories
Edward W. Price,
Associate Editor
Georgia Institute of Technology
James G. Skifstad,
Associate Editor
Purdue University

54. **Satellite Communications: Future Systems.** 1977

David Jarett
TRW Inc.

55. **Satellite Communications: Advanced Technologies.** 1977

David Jarett
TRW Inc.

56. **Thermophysics of Spacecraft and Outer Planet Entry Probes.** 1977

Allie M. Smith
ARO Inc.

57. **Space-Based Manufacturing from Nonterrestrial Materials.** 1977

Gerard K. O'Neill, Editor
Princeton University
Brian O'Leary, Assistant Editor
Princeton University

58. **Turbulent Combustion.** 1978

Lawrence A. Kennedy
State University of New York at Buffalo

59. **Aerodynamic Heating and Thermal Protection Systems.** 1978

Leroy S. Fletcher
University of Virginia

60. **Heat Transfer and Thermal Control Systems.** 1978

Leroy S. Fletcher
University of Virginia

61. **Radiation Energy Conversion in Space.** 1978

Kenneth W. Billman
NASA Ames Research Center

62. **Alternative Hydrocarbon Fuels: Combustion and Chemical Kinetics.** 1978

Craig T. Bowman
Stanford University
Jørgen Birkeland
Department of Energy

63. Experimental Diagnostics in Combustion of Solids. 1978
Thomas L. Boggs
Naval Weapons Center
Ben T. Zinn
Georgia Institute of Technology

64. Outer Planet Entry Heating and Thermal Protection. 1979
Raymond Viskanta
Purdue University

65. Thermophysics and Thermal Control. 1979
Raymond Viskanta
Purdue University

66. Interior Ballistics of Guns. 1979
Herman Krier
University of Illinois at Urbana-Champaign
Martin Summerfield
New York University

67. Remote Sensing of Earth from Space: Role of "Smart Sensors." 1979
Roger A. Breckenridge
NASA Langley Research Center

68. Injection and Mixing in Turbulent Flow. 1980
Joseph A. Schetz
Virginia Polytechnic Institute and State University

69. Entry Heating and Thermal Protection. 1980
Walter B. Olstad
NASA Headquarters

70. Heat Transfer, Thermal Control, and Heat Pipes. 1980
Walter B. Olstad
NASA Headquarters

71. Space Systems and Their Interactions with Earth's Space Environment. 1980
Henry B. Garrett and Charles P. Pike
Hanscom Air Force Base

72. Viscous Flow Drag Reduction. 1980
Gary R. Hough
Vought Advanced Technology Center

73. Combustion Experiments in a Zero-Gravity Laboratory. 1981
Thomas H. Cochran
NASA Lewis Research Center

74. **Rarefied Gas Dynamics, Parts I and II (two volumes).** 1981
Sam S. Fisher
University of Virginia at Charlottesville

75. **Gasdynamics of Detonations and Explosions.** 1981
J. R. Bowen
University of Wisconsin at Madison
N. Manson
Université de Poitiers
A. K. Oppenheim
University of California at Berkeley
R. I. Soloukhin
Institute of Heat and Mass Transfer, BSSR Academy of Sciences

76. **Combustion in Reactive Systems.** 1981
J. R. Bowen
University of Wisconsin at Madison
N. Manson
Université de Poitiers
A. K. Oppenheim
University of California at Berkeley
R. I. Soloukhin
Institute of Heat and Mass Transfer, BSSR Academy of Sciences

77. **Aerothermodynamics and Planetary Entry.** 1981
A. L. Crosbie
University of Missouri-Rolla

78. **Heat Transfer and Thermal Control.** 1981
A. L. Crosbie
University of Missouri-Rolla

79. **Electric Propulsion and Its Applications to Space Missions.** 1981
Robert C. Finke
NASA Lewis Research Center

80. **Aero-Optical Phenomena.** 1982
Keith G. Gilbert and Leonard J. Otten
Air Force Weapons Laboratory

81. **Transonic Aerodynamics.** 1982
David Nixon
Nielsen Engineering & Research, Inc.

82. **Thermophysics of Atmospheric Entry.** 1982

T. E. Horton
The University of Mississippi

83. **Spacecraft Radiative Transfer and Temperature Control.** 1982

T. E. Horton
The University of Mississippi

84. **Liquid-Metal Flows and Magnetohydrodynamics.** 1983

H. Branover
Ben-Gurion University of the Negev
P. S. Lykoudis
Purdue University
A. Yakhot
Ben-Gurion University of the Negev

85. **Entry Vehicle Heating and Thermal Protection Systems: Space Shuttle, Solar Starprobe, Jupiter Galileo Probe.** 1983

Paul E. Bauer
McDonnell Douglas Astronautics Company
Howard E. Collicott
The Boeing Company

86. **Spacecraft Thermal Control, Design, and Operation.** 1983

Howard E. Collicott
The Boeing Company
Paul E. Bauer
McDonnell Douglas Astronautics Company

87. **Shock Waves, Explosions, and Detonations.** 1983

J.R. Bowen
University of Washington
N. Manson
Université de Poitiers
A.K. Oppenheim
University of California at Berkeley
R.I. Soloukhin
Institute of Heat and Mass Transfer, BSSR Academy of Sciences

88. **Flames, Lasers, and Reactive Systems.** 1983

J.R. Bowen
University of Washington
N. Manson
Université de Poitiers
A.K. Oppenheim
University of California at Berkeley
R.I. Soloukhin
Institute of Heat and Mass Transfer, BSSR Academy of Sciences

(Other volumes are planned.)

Chapter I. Laminar Flames

Resonant Response of a Flat Flame Near a Flame-Holder

A.C. McIntosh* and J.F. Clarke†

Cranfield Institute of Technology, Cranfield, Bedford, England

Abstract

The response to oscillatory inputs in composition and inlet mass flux of a premixed flame held on a coplanar flame-holder is investigated using large activation energy asymptotic theory. $O(1)$ changes in the preheat zone are linked to $O(\theta_1^{-1})$ changes in the equilibrium zone (where θ_1 is the initial nondimensionalized activation energy). Lewis numbers are not restricted to values in the neighborhood of unity. On the assumption that perturbations are small, the problem can be linearized and solutions obtained for equilibrium temperature and flame position. Despite the linearization, it is still necessary to use numerical methods to evaluate certain complex integrals which arise. The behavior at large times can then be evaluated for any given frequency of oscillation and all transients arising from the given input are accounted for. When the response is stable, it is found that for any relevant combination of overall Lewis number and initial flame position there is a resonant frequency. In other words, in the neighborhood of this frequency, the amplitudes of flame position and temperature oscillations (normalized with respect to the input amplitude of oscillations) are very large. The value of the resonant frequency varies with Lewis number and initial flame position (and hence also with heat loss).

Paper presented at the 8th ICOGER, Minsk, USSR, Aug. 23-26, 1981. Copyright © American Institute of Aeronautics and Astronautics, Inc., 1982. All rights reserved.

*Post-doctoral Research Fellow, Department of Aerodynamics.
†Professor of Gas Dynamics, Department of Aerodynamics.

Introduction

A number of authors have considered the solution of the full time-dependent equations governing the behavior of premixed laminar flames by using the large activation energy asymtotic theory. In the recent work of Sivashinsky (1977, 1979, 1980), the unsteady equations are solved as small $O(\theta_1^{-1})$ perturbations from the steady solution (where θ_1 is the initial nondimensional activation energy). In those papers, the main concern has been with the development of two-dimensional effects and in particular the phenomenon of cellularity, but in this paper and in the paper by Margolis (1980) the one-dimensional problem of a flat flame with heat loss to a nearby coplanar flame-holder is considered. A Hirschfelder-type model of the flame-holder is used, the details of which have been reviewed in some detail in Clarke and McIntosh (1979). Margolis, using techniques similar to that of Sivashinsky, considers the stability of the flame in response to small $O(\theta_1^{-1})$ perturbations from the steady solution, but the origin and type of incoming disturbances is not the main point under consideration in his paper.

The main aim of this work is to consider the response of a one-dimensional flame-front to known inputs and in particular to oscillatory inputs in composition and inlet mass flux. Large activation energy asymptotic theory is used to link $O(1)$ changes in the preheat zone to $O(\theta_1^{-1})$ changes in the equilibrium zone [a different approach to considering all changes to be $o(\theta_1^{-1})$]. Strict order matching produces a Stéfan-type problem which can then be linearized on the assumption that perturbations are small. However, it is still necessary to use numerical methods to evaluate some of the complex integrals which arise, and by this means the behavior at large times can then be evaluated for a given frequency of oscillation.

The basic concepts of this theory were first put forward in McIntosh and Clarke (1979), but in that work the quite severe restrictions of unit Lewis number and constant inlet mass flux were imposed. In this paper both these restrictions are removed so that the response to known inputs in both composition and inlet mass flux for general Lewis numbers can be investigated. This then provides a useful tool for investigating the existence of a resonant frequency for the flame/flame-holder system.

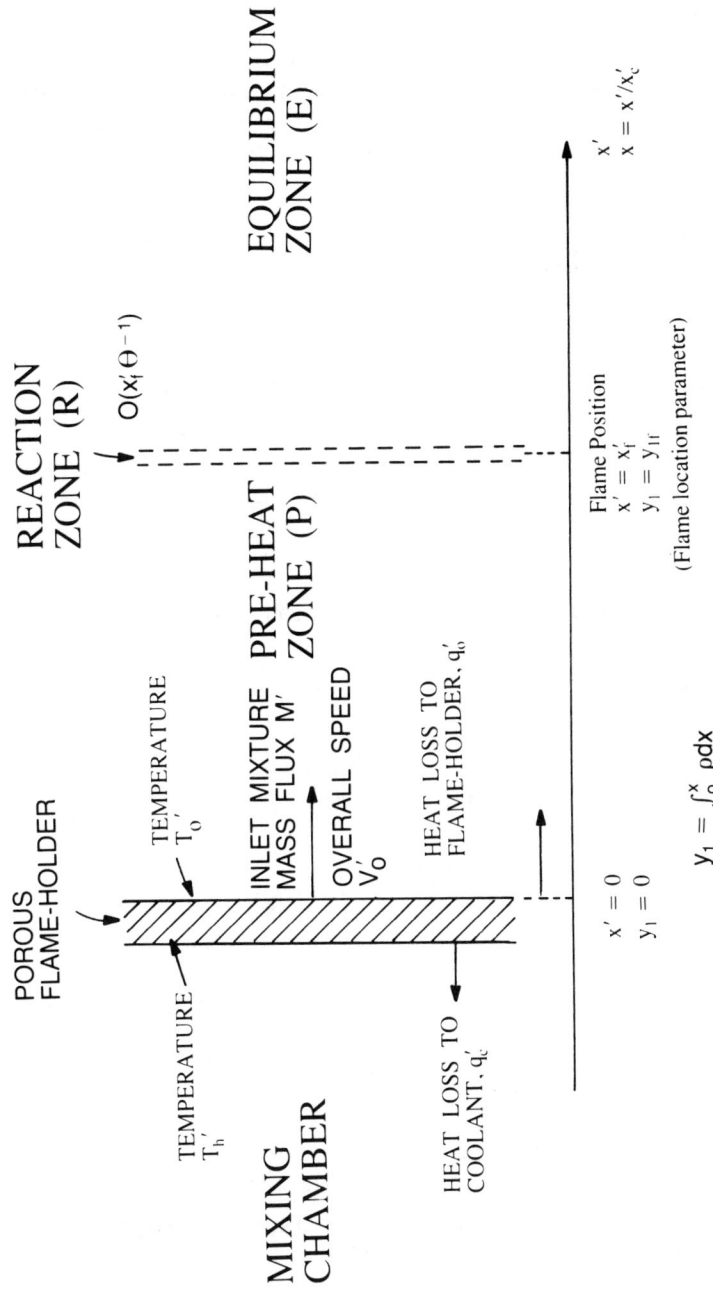

Fig. 1 Schematic of one-dimensional premixed flame with flame holder.

Formulation

Basic Equations

The main equations governing a simple one-step irreversible reaction in one-dimensional unsteady flow can be expressed as relations in temperature T and the stoichiometrically adjusted mass fractions C_ℓ (lean) and C_r (rich).

$$\frac{\partial C_\ell}{\partial t} + M_o \frac{\partial C_\ell}{\partial y_1} - \frac{\partial^2 C_\ell}{\partial y_1^2} = -\text{Le}\Lambda_1 C_\ell C_r \varepsilon^{\theta_1(1-1/T)} \qquad (1)$$

$$\frac{\partial T}{\partial t} + M_o \frac{\partial T}{\partial y_1} - \frac{1}{\text{Le}} \frac{\partial^2 T}{\partial y_1^2} = \frac{\text{Le}(1+\sigma)Q_1}{\sigma} \Lambda_1 C_\ell C_r \varepsilon^{\theta_1(1-1/T)} \qquad (2)$$

$$\frac{\partial \Delta}{\partial t} + M_o \frac{\partial \Delta}{\partial y_1} - \frac{\partial^2 \Delta}{\partial y_1^2} = 0 \qquad (3)$$

For simplicity, an order 2 reaction is assumed and Δ is the mixture strength variable

$$\Delta = C_X - C_F \qquad (4a)$$

so that

$$C_r - C_\ell = |\Delta| \qquad (4b)$$

The distance y_1 is defined as

$$y_1 \equiv \frac{M_1' C_p'}{(\rho'\lambda')\text{Le}} \int_0^{x'} \rho' dx' \qquad (5)$$

and the initial steady-state eigenvalue Λ_1 is defined as

$$\Lambda_1 \equiv W_F(\rho'\lambda') e^{-\theta_1}/\tau_o' C_p' M_1'^2 \qquad (6)$$

x' is the distance from the flame-holder (see Fig. 1) and $\sigma \equiv W_X/W_F$, where W_X and W_F are the molecular weights of oxidant and fuel, respectively. (In fact the adjusted mass fractions are given by $C_X = c_X$ and $C_F = \sigma c_f$, where c_X and c_F are the standard definitions for these quantities, respectively. Note also that $\ell = F,X$ and $r = X,F$ according to whether the mixture is fuel-lean or fuel-rich. The quantity τ_o' is the zero-activation energy reaction time, and C_p' is the overall specific heat of the mixture (assumed constant). The product of density ρ' and the thermal conductivity λ' is also assumed constant and writing $\mathcal{D}'$ for the coefficient of mass diffusion, the Lewis number Le is defined as

$$Le \equiv \rho' \mathcal{D}' C_p'/\lambda' \qquad (7)$$

The following constants form a suitable reference set and have been used to nondimensionalize the equations:

V_{01}' = initial inlet flow velocity (8a)

ρ_{01}' = initial inlet density of the mixture (8b)

M_1' = initial mass flux = $\rho_{01}' V_{01}'$ = constant (8c)

$\mathcal{D}_{01}'$ = initial inlet flow diffusion coefficient (8d)

T_{b1}' = initial temperature in the burnt zone (8e)

Any quantities with a prime (') denote dimensional quantities. The subscripts 01 and b1 mean that the parameters to which they are attached are initial values (1) and evaluated in the unburnt inlet flow stream (0) or the burnt stream (b), respectively. The definitions of the main nondimensional quantities are then as follows:

$$V \equiv V'/V_{01}' \quad \text{(velocity)} \qquad (9a)$$

$$\rho \equiv \rho'/\rho_{01}' \quad \text{(density)} \qquad (9b)$$

$$x \equiv x' V_{01}'/\mathcal{D}_{01}' \quad \text{(distance)} \qquad (9c)$$

$$T \equiv T'/T_{b1}' \quad \text{(temperature)} \qquad (9d)$$

$$t \equiv t' V_{01}'^2 / \mathcal{D}_{01}' \quad \text{(time)} \qquad (9e)$$

$$M \equiv M'/M_1' \quad \text{(mass flux)} \qquad (9f)$$

$$M_o \equiv M_o'/M_1' \quad \text{(inlet mass flux)} \qquad (9g)$$

(equation continued on next page)

$$\theta_1 \equiv E_A'/R'T_{b1}' \quad \text{(activation energy)} \tag{9h}$$

$$Q_1 \equiv Q'/C_p'T_{b1}' \quad \text{(heat of reaction)} \tag{9i}$$

Boundary Conditions

Figure 1 illustrates the geometric configuration of the present problem and shows schematically the porous-plug type of flame-holder first described by Hirschfelder et al. (1953). In this description the problem referred to as "the cold boundary difficulty" is effectively eliminated by stipulating that no product species is allowed to diffuse into the flame-holder. Mathematically this implies at the holder (O) that

$$G_{FO}/\sigma + G_{XO} + G_{NO} = 1 \tag{10a}$$

$$G_{PO} = 0 \tag{10b}$$

where G_{iO} are the fractional mass fluxes of species i (i = F, X, N, P, where F is the fuel, X is the oxidant, N is the diluent, and P is the product), and assuming an air-like situation where

$$G_{NO} = dG_{XO} \quad d = \text{const} \tag{11}$$

the mass fluxes are related to the mass fractions at the holder (see Clarke and McIntosh 1979) by

$$G_{iO} = C_{iO} - \frac{1}{M_o}\left(\frac{\partial C_{iO}}{\partial y_1}\right)_0 = \frac{1}{\hat{k}}(\sigma + P_i\Delta_{GO}) \quad i = F, X \tag{12}$$

$$\Delta_{GO} \equiv G_{XO} - G_{FO} = \Delta_O - \frac{1}{M_o}\left(\frac{\partial\Delta}{\partial y_1}\right)_0 = \frac{\sigma(1-\hat{\Phi})}{\hat{\Phi} + \sigma(1+d)} \tag{13}$$

where the inlet mass flux equivalence ratio $\hat{\Phi}$ is defined as

$$\hat{\Phi} = G_{FO}/G_{XO} \tag{14}$$

$$P_F \equiv -\sigma(1+d) \tag{15}$$

$$P_X \equiv 1 \tag{16}$$

$$\hat{k} \equiv 1 + \sigma(1+d) \tag{17}$$

In order to focus our attention on the main problem of the response of the flame to oscillatory inputs, the conductance of the flame-holder (Clarke and McIntosh 1979) is considered infinite. In this case the boundary condition for the temperature at the holder simplifies to

$$T_0 [\equiv T(0,t)] = T_{01} = \text{const} \tag{18}$$

This restriction does not have a serious effect on the main arguments presented here. The case where T_0 is not constant (and the conductance is not infinite) is considered by McIntosh (1981).

The dimensional heat loss (that is, heat lost per unit area per unit time) is defined by

$$q_0'(t) \equiv \lambda_0'(t)\left(\frac{\partial T'}{\partial x'}\right)_0 \tag{19}$$

and its nondimensional counterpart $q_0(t)$ is defined as

$$q_0(t) \equiv \left(\frac{\partial T}{\partial y_1}\right)_0 \left[= \frac{Le}{C_p' T_{b1}' M_1'} q_0'(t)\right] \tag{20}$$

Far downstream it is assumed that chemical equilibrium exists and that total consumption of either fuel or oxidant takes place. Thus in view of the fact that an initial value problem is being considered,

$$C_\ell(\infty,t) = 0 \tag{21}$$

$$\Delta(\infty,t) = \Delta_1 \tag{22}$$

$$T(\infty,t) = 1 \tag{23}$$

and the central problem amounts to the solution of the three second-order partial differential equations, Eqs. (1-3), subject to the six boundary conditions, Eqs. (12), (13), (18), and (21-23).

Asymptotic Analysis

We now seek to obtain series solutions to the full time-dependent equations, Eqs. (1-3) by exploiting the limit $\theta_1 \to \infty$. The analysis is similar to that of McIntosh and Clarke (1979). We therefore here only summarize the results.

The pre-exponential Λ_1 is given by

$$\Lambda_1 = \frac{\text{Le} B_1^2 \theta_1^2}{2|\Delta_1|} + \ldots \quad |\Delta_1| \sim O(1) \tag{24}$$

$$\Lambda_1 = \frac{\text{Le}^2 (1+\sigma) Q_1 B_1^2 \theta_1^3}{2\sigma[2 + \text{Le}(1+\sigma)Q_1 \theta_1 |\Delta_1|/\sigma]} + \ldots \quad |\Delta_1| \sim O(\theta_1^{-1}) \tag{25}$$

where

$$B_1 \equiv (1-T_{01})/(1-e^{-\text{Le} y_{1f1}}) \tag{26}$$

and

$$y_{1f1} \equiv y_1(x'_{f1}) \tag{27}$$

The subscript f1 denotes evaluation at the initial flame position (f on its own denotes evaluation at the flame at time $t \neq 0$). The asymptotic model is used in which a frozen preheat zone (P) is separated by a surface, $y_1 = y_{1f}(t)$, from an equilibrium zone (E). In this latter zone, the temperature is written as

$$T = 1 + (1/\theta_1) T_e^{(1)}(y_1, t) + \ldots \tag{28}$$

and in the outer limiting process ($\theta_1 \to \infty$, y_1, t fixed), C_ℓ and $T_e^{(1)}$ obey

$$C_\ell = 0 \tag{29}$$

$$\frac{\partial T_e^{(1)}}{\partial t} + M_o \frac{\partial T_e^{(1)}}{\partial y_1} - \frac{1}{\text{Le}} \frac{\partial^2 T_e^{(1)}}{\partial y_1^2} = 0 \quad y_1 > y_{1f} \tag{30}$$

In the preheat zone asymptotic solutions are sought of the form

$$T(y_1, t) = T^{(1)}(y_1, t) + \ldots \quad 0 \leq y_1 < y_{1f} \tag{31}$$

$$C_\ell(y_1, t) = C_\ell^{(1)}(y_1, t) + \ldots \quad 0 \leq y_1 < y_{1f} \tag{32}$$

and $T^{(1)}$, $C_\ell^{(1)}$ satisfy (in the region $0 \leq y_1 < y_{1f}$), in the outer limiting process.

$$\frac{\partial T^{(1)}}{\partial t} + M_o \frac{\partial T^{(1)}}{\partial y_1} - \frac{1}{Le} \frac{\partial^2 T^{(1)}}{\partial y_1^2} = 0 \tag{33}$$

$$\frac{\partial C_\ell^{(1)}}{\partial t} + M_o \frac{\partial C_\ell^{(1)}}{\partial y_1} - \frac{\partial^2 C_\ell^{(1)}}{\partial y_1^2} = 0 \tag{34}$$

In the reaction zone a new spatial variable y is defined as

$$y = \theta_1 [y_1 - y_{1f}(t)] \tag{35}$$

and T, C_ℓ have the asymptotic forms

$$T = 1 - (1/\theta_1) \, T(y,t) + \ldots \tag{36}$$

$$C_\ell = (1/\theta_1) \, C(y,t) + \ldots \tag{37}$$

whence under the inner limiting process ($\theta_1 \to \infty$, y,t fixed), T and C obey

$$\frac{\partial^2 T}{\partial y^2} = \frac{Le^2(1+\sigma)Q_1}{\sigma} \frac{\Lambda_1}{\theta_1^2} C \left(\frac{C}{\theta_1} + |\Delta_f|\right) \varepsilon^{-T} = \frac{Le(1+\sigma)Q_1}{\sigma} \frac{\partial^2 C}{\partial y^2} \tag{38}$$

These relations only hold if

$$\dot{y}_{1f} \equiv dy_{1f}/dt \sim 0(\theta_1) \tag{39}$$

a restriction which receives no mention in any of the current literature on this subject.

Equation (38) can be integrated once so that matching conditions can be found for T and C_ℓ and their derivatives between the inner zone solutions and those in the two outer zones. The following conditions are obtained:

$$T^{(1)}(y_{1f},t) = 1 \tag{40a}$$

$$C_\ell^{(1)}(y_{1f},t) = 0 \tag{40b}$$

$$T(-\infty,t) = \infty = C(-\infty,t) \tag{41}$$

$$-\frac{\partial T}{\partial y}(-\infty,t) = \frac{\partial T^{(1)}}{\partial y_1}(y_{1f},t) \tag{42a}$$

$$\frac{\partial C}{\partial y}(-\infty,t) = \frac{\partial C_\ell^{(1)}}{\partial y_1}(y_{1f},t) \quad (42b)$$

$$-T(\infty,t) = T_e^{(1)}(y_{1f},t) \equiv T_f^{(1)} \quad (43a)$$

$$C(\infty,t) = 0 \quad (43b)$$

$$\frac{\partial T}{\partial y}(\infty,t) = 0 = \frac{\partial C}{\partial y}(\infty,t) \quad (44)\ddagger$$

The downstream matching conditions now constitute flame boundary conditions for the preheat domain equations and the main task reduces to the solution of a Stéfan-like (moving-boundary) problem. Summarized, the equations to be solved [in the zone $0 \leq y_1 < y_{1f}(t)$] are found to be

$$\frac{\partial T^{(1)}}{\partial t} + M_o \frac{\partial T^{(1)}}{\partial y_1} - \frac{1}{Le} \frac{\partial^2 T^{(1)}}{\partial y_1^2} = 0 \quad (45)$$

$$\frac{\partial C_\ell^{(1)}}{\partial t} + M_o \frac{\partial C_\ell^{(1)}}{\partial y_1} - \frac{\partial^2 C_\ell^{(1)}}{\partial y_1^2} = 0 \quad (46)$$

$$\frac{\partial \Delta}{\partial t} + M_o \frac{\partial \Delta}{\partial y_1} - \frac{\partial^2 \Delta}{\partial y_1^2} = 0 \quad (47)$$

---

‡The conditions here derive from a strict application of order-matching of the relevant asymptotic series, including those that apply in the domain of intense chemical activity. The present asymptotic model of an unsteady flame is predicated on the assumption that Eq. (36) is valid and Eq. (44) is an inescapable corollary; flame models that describe the reaction zone as a sheet whose strength is proportional to $\exp(T_f^{(1)}/2)$ [see Eq. (54)], which is then replaced by $\exp[\frac{1}{2}\theta_1(T-1)]$, must be treated with great circumspection since they can provoke violations of the matching requirement (44). A particular case is discussed by McIntosh (1981) where it is shown that the sheet model applied to the one-dimensional burner stabilized flame only gives agreement with the strict order matching model used here when $\theta_1 = \infty$. This result is in accord with the comparison of the same two approaches to free flames. The earlier sheet model as posed by Sivashinsky (1977) is again only consistent with strict order matching models (e.g., Matkowsky and Olagunju 1980; Sivashinsky 1980) when $\theta_1 = \infty$.

and effectively link $O(1)$ changes in the preheat zone with $O(\theta_1^{-1})$ changes in equilibrium temperature and $O(1)$ changes in flame position. The conditions on these equations are

$$T^{(1)}(0,t) = T_{01} \tag{48}$$

$$c_{\ell 0}^{(1)} - \frac{1}{M_0}\left(\frac{\partial c_\ell^{(1)}}{\partial y_1}\right)_0 = \frac{1}{\hat{k}}(\sigma + p_\ell \Delta_{G0}) \tag{49}$$

$$\Delta_0 - \frac{1}{M_0}\left(\frac{\partial \Delta}{\partial y_1}\right)_0 = \Delta_{G0} \tag{50}$$

$$T^{(1)}(y_{1f},t) = 1 \tag{51}$$

$$c_\ell^{(1)}(y_{1f},t) = 0 \tag{52}$$

$$\Delta(\infty,t) = \Delta_1 \tag{53}$$

$$\left(\frac{\partial T^{(1)}}{\partial y_1}\right)_{y_1=y_{1f}} = \text{LeB}_1 F(|\Delta_f|) \, e^{T_f^{(1)}/2} \tag{54}$$

$$\left(\frac{\partial c_\ell^{(1)}}{\partial y_1}\right)_{y_1=y_{1f}} = \frac{-\sigma B_1}{(1+\sigma)Q_1} F(|\Delta_f|) e^{T_f^{(1)}/2} \tag{55}$$

where

$$F(|\Delta_f|) \equiv \frac{|\Delta_f|}{|\Delta_1|} \qquad |\Delta_1|, |\Delta_f| \sim O(1) \tag{56}$$

$$F(|\Delta_f|) \equiv \left|\frac{2+\text{Le}(1+\sigma)Q_1\theta_1|\Delta_f|/\sigma}{2+\text{Le}(1+\sigma)Q_1\theta_1|\Delta_1|/\sigma}\right|^{1/2} \qquad |\Delta_1|, |\Delta_f| \sim O(\theta_1^{-1}) \tag{57}$$

These later conditions [Eqs. (54-57)] arise from the integration of Eq. (38) across the inner zone.

Approximate Solution for Small Perturbations

An approximate solution to Eqs. (45-57) is now sought by considering the inputs at the flame-holder to be small, i.e.,

$$M_o(t) = 1 + \varepsilon M_{uo}(t) \tag{58}$$

$$\Delta_{GO}(t) = \Delta_1[1 + \varepsilon \Delta_{Guo}(t)] \tag{59}$$

The method for linearizing these equations follows that of our earlier work (McIntosh and Clarke 1979). Regarding flame movements as small, one writes

$$y_{1f}(t) = y_{1f1} + \varepsilon y_{fu}(t) + \ldots \tag{60}$$

$$\Delta(y_1, t) = \Delta_1[1 + \varepsilon \Delta_u(y_1, t) + \ldots \tag{61}$$

$$T^{(1)} = T_s^{(1)}(y_1) + \varepsilon T_u^{(1)}(y_1, t) + \ldots \tag{62}$$

$$C_\ell^{(1)} = C_{\ell s}^{(1)}(y_1) + \varepsilon C_{\ell u}^{(1)}(y_1, t) + \ldots \tag{63}$$

$$T_e^{(1)} = \varepsilon T_e^{*(1)} + \ldots \tag{64a}$$

$$T_f^{(1)} = \varepsilon T_e^{*(1)}(y_{1f1}, t) + \ldots \;(\equiv \varepsilon T_{f1}^* + \ldots) \tag{64b}$$

where $T_s^{(1)}$ and $C_{\ell s}^{(1)}$ are the steady solutions (Clarke and McIntosh 1979),

$$T_s^{(1)} = 1 - B_1 + B_1 e^{Le(y_1 - y_{1f1})} \tag{65}$$

$$C_{\ell s}^{(1)} = \frac{B_1}{(1+\sigma)Q_1}(1 - e^{y_1 - y_{1f1}}) \tag{66}$$

Substituting these relations into Eqs. (45-57) and linerizing, one obtains to leading order the following set of equations:

$$\frac{\partial T_u^{(1)}}{\partial t} + \frac{\partial T_u^{(1)}}{\partial y_1} - \frac{1}{Le}\frac{\partial^2 T_u^{(1)}}{\partial y_1^2} = -LeB_1 M_{u0} e^{Le(y_1 - y_{1f1})}$$

$$0 \leq y_1 < y_{1f1} \qquad (67)$$

$$\frac{\partial C_{\ell u}^{(1)}}{\partial t} + \frac{\partial C_{\ell u}^{(1)}}{\partial y_1} - \frac{\partial^2 C_{\ell u}^{(1)}}{\partial y_1^2} = \frac{\sigma B_1 M_{u0}}{(1+\sigma)Q_1} e^{y_1 - y_{1f1}}$$

$$0 \leq y_1 < y_{1f1} \qquad (68)$$

$$\frac{\partial \Delta_u}{\partial t} + \frac{\partial \Delta_u}{\partial y_1} - \frac{\partial^2 \Delta_u}{\partial y_1^2} = 0 \qquad 0 \leq y_1 < \infty \qquad (69)$$

$$T_u^{(1)}(0,t) \equiv T_{u0}(t) = 0 \qquad (70)$$

$$C_{\ell u0}^{(1)} - \left(\frac{\partial C_{\ell u}^{(1)}}{\partial y_1}\right)_{y_1=0} = \frac{P_\ell \Delta_1}{k} \Delta_{G u0} + \frac{\sigma B_1 M_{u0} e^{-y_{1f1}}}{(1+\sigma)Q_1} \qquad \Delta_1 \sim O(1)$$

$$= \frac{\sigma B_1 M_{u0} e^{-y_{1f1}}}{(1+\sigma)Q_1} \qquad \Delta_1 \sim O(\theta_1^{-1}) \qquad (71)$$

$$\Delta_{u0} - \left(\frac{\partial \Delta_u}{\partial y_1}\right)_{y_1=0} = \Delta_{G u0} \qquad (72)$$

$$T_u^{(1)}(y_{1f1},t) \equiv T_{uf1}^{(1)} = -Le B_1 y_{fu} \qquad (73)$$

$$C_{\ell u}^{(1)}(y_{1f1},t) \equiv C_{\ell uf1}^{(1)} = \sigma B_1 y_{fu}/(1+\sigma)Q_1 \qquad (74)$$

$$\Delta_u(\infty,t) = 0 \qquad (75)$$

$$Le^2 B_1 y_{fu} + \left(\frac{\partial T_u^{(1)}}{\partial y_1}\right)_{y_1=y_{1f1}} = \frac{LeB_1 \gamma_1}{2} \Delta_{uf1} + \frac{LeB_1}{2} T_{f1}^* \quad (76)$$

$$\frac{-\sigma B_1 y_{fu}}{(1+\sigma)Q_1} + \left(\frac{\partial C_{\ell u}^{(1)}}{\partial y_1}\right)_{y_1=y_{1f1}} = \frac{-\sigma B_1 \gamma_1 \Delta_{uf1}}{2(1+\sigma)Q_1} - \frac{\sigma B_1 T_{f1}^*}{2(1+\sigma)Q_1} \quad (77)$$

where

$$\Delta_{uf1} \equiv \Delta_u(y_{1f1}, t) \quad (78)$$

$$\gamma_1 \equiv 1 \quad \Delta_1 \sim O(1)$$

$$\equiv \frac{Le(1+\sigma)Q_1 \theta_1 |\Delta_1|/2\sigma}{1 + Le(1+\sigma)Q_1 \theta_1 |\Delta_1|/2\sigma} \quad \Delta_1 \sim O(\theta_1^{-1}) \quad (79)$$

Equations (67-69) can now be reduced to second-order ordinary differential equations by the method of Laplace transforms. The details involve lengthy algebra and are not repeated here. Using the following notations for the Laplace transform and its inverse,

$$\overline{X}(\omega) \equiv \int_0^\infty e^{\omega t} X(t) dt \quad (80)$$

$$L^{-1}[V(\omega)] \equiv (1/2\pi i) \int_{c-i\infty}^{c+i\infty} V(\omega) e^{\omega t} d\omega \quad (81)$$

the expressions for the displacements in flame position and temperature are found to be

$$y_{fu}(t) = L^{-1}\left(\frac{B \overline{M_{u0}}}{\omega A}\right) + \frac{\xi_1}{Le} L^{-1}\left(\frac{r^{y_{1f1}/2} \overline{\Delta_{Gu0}}}{A \operatorname{shr} y_{1f1}}\right) \quad (82)$$

$$\tfrac{1}{2} T_{f1}^*(t) = L^{-1}\left[\frac{1}{A}\left(\frac{r \exp(-y_{1f1}/2)}{\operatorname{shr} y_{1f1}}J - \frac{s \exp(-Le y_{1f1}/2)}{\operatorname{shLes} y_{1f1}}\right)\overline{M_{u0}}\right.$$

(equation continued on next page)

$$+ L^{-1}[(\frac{\xi_1 r J \exp(r y_{1f1})}{A \, shr y_{1f1}} - \frac{\gamma_1}{1+2r}) \overline{\Delta_{Gu0}} \exp((\frac{1}{2}-r)y_{1f1})]$$

where (83)

$$A(\omega) \equiv \frac{\omega}{Le} + (\frac{1}{2} + \frac{rchry_{1f1}}{shry_{1f1}})(\frac{1}{2} - \frac{schLesy_{1f1}}{shLesy_{1f1}}) \quad (84)$$

$$B(\omega) \equiv A(\omega) + \frac{\omega r \exp(-y_{1f1}/2)}{Leshry_{1f1}} + (\frac{1}{2} + \frac{\omega rchry_{1f1}}{shry_{1f1}}) \frac{s \exp(-Ley_{1f1}/2)}{shLesy_{1f1}}$$

(85)

$$J(\omega) \equiv \frac{1}{2} - \frac{schLesy_{1f1}}{shLesy_{1f1}} \quad (86a)$$

$$sh(x) \equiv \tfrac{1}{2}(e^x - e^{-x}) \qquad ch(x) \equiv \tfrac{1}{2}(e^x + e^{-x}) \quad (86b)$$

$$\xi_1 \equiv \frac{\ell|\Delta_1|}{\ell|\Delta_1| - 1} \equiv \frac{1}{n_1} \qquad \Delta_1 \sim O(1)$$

$$\equiv 0 \qquad \Delta_1 \sim O(\theta_1^{-1}) \quad (87)$$

$$\ell \equiv (1+d) \quad \text{(fuel-lean)}$$
$$\equiv 1/\sigma \quad \text{(fuel-rich)} \quad (88)$$

$$r \equiv \sqrt{\omega + \tfrac{1}{4}} \quad (89a)$$

$$s \equiv \sqrt{\omega/Le + \tfrac{1}{4}} \quad (89b)$$

and the symbol $\sqrt{}$ is used strictly to mean that root with its real part ≥ 0 (i.e. phase always in range $-\pi \leq$ phase $\leq \pi$).

Note that Eqs. (82) and (83) carry explicitly the functions Δ_{Gu0} and M_{u0} so that responses in flame positions and temperature can be theoretically described by these equations for determinate inputs in inlet composition and mass flux. When Le = 1 and M_{u0} = 0 (i.e., inlet mass flux is constant) the two solutions, Eqs. (82) and (83) collapse down to the simple results derived previously (McIntosh and

Clarke 1979):

$$y_{fu}(t) = -\xi_1 L^{-1}[\exp(y_{1f1}/2)\frac{\overline{\Delta_{Gu0}}}{r} \text{shry}_{1f1}] \quad (90)$$

$$\tfrac{1}{2}T^*_{f1}(t) = -L^{-1}[(\xi_1\frac{\text{shry}_{1f1}}{r}\exp(-y_{1f1})(\tfrac{1}{2}+\frac{r\text{chry}_{1f1}}{\text{shry}_{1f1}})$$
$$+\frac{\gamma_1\exp((\tfrac{1}{2}-r)y_{1f1})}{1+2r}\overline{\Delta_{Gu0}}] \quad (91)$$

Response at Large Times to Specific Inputs

As stated in the introduction, the main aim of this work is to consider the response of the flame to known inputs in both inlet composition and mass flux.

The full evaluation of the response of the flame-sheet from Eqs. (82) and (83) clearly requires detailed numerical calculations and involves the evaluation of contour integrals of the form of Eq. (81). However, a valuable contribution to the understanding of the behavior of the flame can be obtained by considering the singularities in the complex integrals represented by the inverse transforms in the above equations. In particular, if any poles exist in the ω plane, then the sign of the real part of ω at these poles will have a strong bearing on the stability of the flame. The response, of course, depends on the nature of the inputs and these are considered in the following two subsections.

Asymptotically Steady Inputs

The singularities in the complex integrals represented by the inverse transforms in Eqs. (82) and (83) arise through the zeros of $A(\omega)$, which can be shown always to cause poles of order unity. The position of these poles is critical to the stability of the flame-sheet response to changes in composition and mass flux at the holder. Writing

$$\omega = k + iz \quad (92)$$

one needs to find the position of the poles in the ω plane and solve the complex equation

$$(k + iz : Le, y_{1f1}) = 0 \quad (93)$$

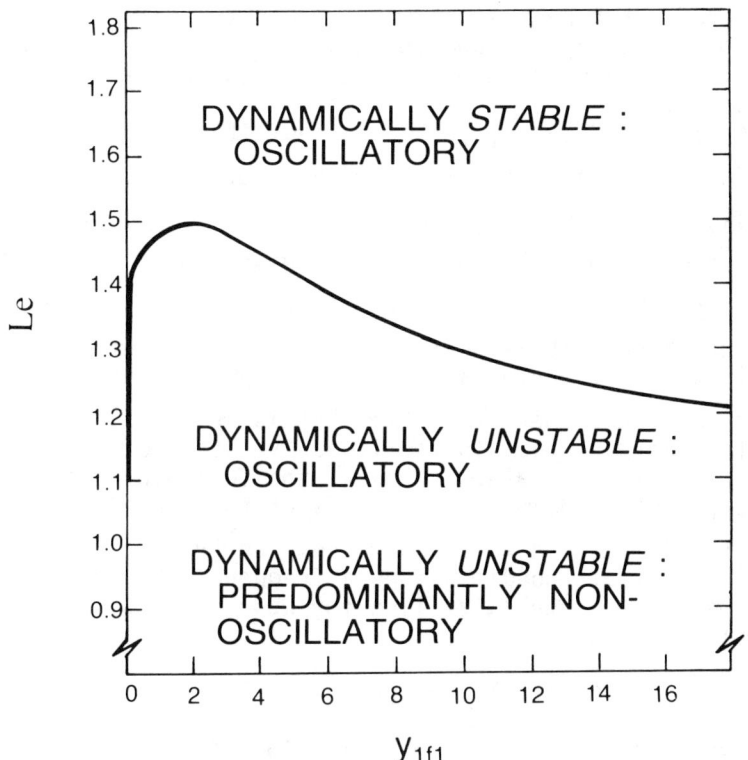

Fig. 2 Domains in the Le, y_{1f1} space shere a stable or unstable response of a plane flame to small disturbances is to be expected. For Le Le* (y_{1f1}), solutions will bifurcate. Heat loss q_0' = $(1+\sigma)Q'gM_1'\exp(-Ley_{1f1})$, where g is a constant for any fixed composition.

Let the formal solution to Eq. (93) be $k = k_0$, $z = z_0$ for a particular combination of Le and y_{1f1}. Recalling that the inverse transforms in Eqs. (82) and (83) are in fact complex integrals along the Bromwich contour like Eq. (81), the pole at $\omega_0 = k_0 + iz_0$ will either cause an initially stable or unstable response according to whether $k_0 < 0$ (stable) or $k_0 > 0$ (unstable). There is a neutral stability curve Le = Le*(y_{1f1}) where $k_0 = 0$ (see Fig. 2). Recall, however, that these equations are limited by the present linear theory, so that any predictions can only be considered accurate for small inputs at the flame-holder.

A computer program was written to "search" for the zeros of A numerically [note the highly transcendental nature of the terms in Eq. (84)]. The results of this work are summarized in Fig. 2, which shows the neutral stability curve and the domains where certain types of behavior are to be expected. For Le < 1, it was found that there is always a zero of A with $k_o > 0$ and for which $z_o = 0$ (i.e., ω_o is real). This indicates a predominantly monoscillatory instability.

For $1 < Le < Le^*(y_{1f1})$ the instabilities indicated for a particular combination of Le and y_{1f1} are oscillatory [that is, for $1 < Le < Le^*(y_{1f1})$, when $\omega_o = 0$ with $k_o > 0$, then $z_o \neq 0$]. Note the small region given approximately by $1.0 < Le < 1.5$, $0 < y_{1f1} < 2$, within which a stable response can be obtained. It is in fact very easy to assume that the neutral stability curve carries on its upward trend as y_{1f1} becomes small and thereby to "miss" this region. Below the neutral stability line, in the linearly unstable region, the solutions will bifurcate. However, the following sections of this paper are devoted to a study of the forced response within the linearly stable region $Le > Le^*(y_{1f1})$ (> 1).

The critical (nondimensional) frequency can be found along the neutral stability line for $Le = Le^*(y_{1f1})$. This is simply given by

$$\text{Critical frequency} = (1/2\pi)z_o \qquad (94)$$

and is shown in Fig. 3, where Le is plotted against z_o. Shown against the curves are the values of y_{1f1} at particular points. The high-heat-loss neutrally stable points are generally linked to high critical frequencies. Noting from Eq. (9e) that time is measured in units of $\mathcal{D}'_{01}/V'^2_{01}$, a typical unit of time is about 1 ms, so that from Fig. 3 the frequencies are (approximately) in the range 0-2 kHz).

Oscillatory Inputs

Much emphasis is laid in the literature on the stability of laminar flames. However, even if the flame is stable, its final response will depend on the type of inputs through Δ_{Gu0} and M_{uo} at the holder. Because in this work these inputs have been kept in the anlaysis, one can consider the response of the flame to different inlet conditions. The large time response will be dominated by

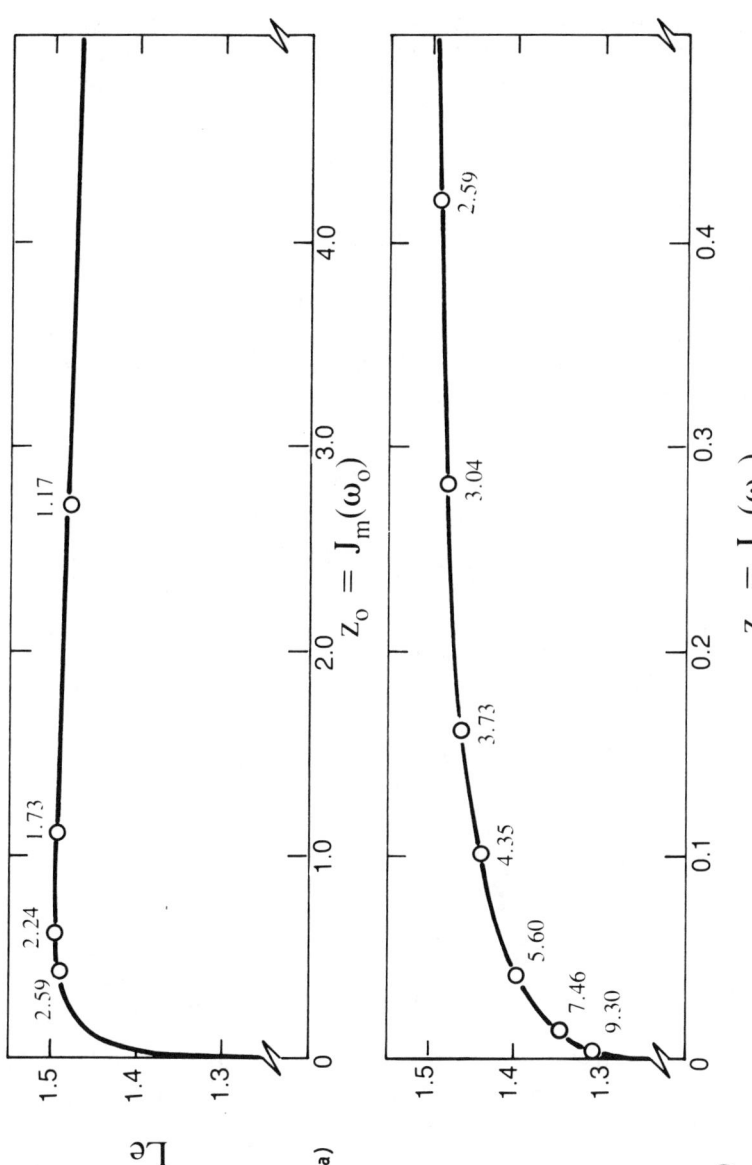

Fig. 3 Critical frequency for neutral stability as a function of the Lewis number Le and flame location y1f1.

any extra singularities occurring through the input functions. If the inputs in inlet mass flux (M_{uo}) and composition (Δ_{Gu0}) at the holder are oscillatory in nature, then the response will be governed not only by the stability criteria discussed in the previous subsection, but also by the extra poles on the imaginary axis (in the ω plane) due to the oscillatory inputs

$$M_{uo} = M^* \sin at \qquad (95)$$

and

$$\Delta_{Gu0} = \Delta^* \sin at \qquad (96)$$

If Le and y_{1f1} are such that at any poles of A the real part of ω is negative (that is, $Le > Le^*$), then the complex integrals at large times [Eqs. (82) and (83)] are simply $2\pi i$ poles $\omega = \pm ia$. One obtains

$$y_{fu}(t) = \frac{-M^*}{2a}[e^{iat}(\frac{B}{A})_+ + e^{-iat}(\frac{B}{A})_-]$$

$$+ \frac{\xi_1 e^{y_{1f1}/2}}{2iLe}\Delta^* [e^{iat}(\frac{r}{A shr y_{1f1}})_+ - e^{-iat}(\frac{r}{A shr y_{1f1}})_-] \qquad (97)$$

$$T^*_{f1}(t) = \frac{M^*}{i}[e^{iat}(\frac{X}{A})_+ - e^{-iat}(\frac{X}{A})_-]$$

$$+ \frac{\Delta^* e}{i} y_{1f1}/2 [e^{iat}(\frac{\xi_1 rJ}{A shr y_{1f1}} - \frac{\gamma_1 e^{-ry_{1f1}}}{1+2r})_+$$

$$- e^{-iat}(\frac{\xi_1 rJ}{A shry1f1} - \frac{\gamma_1 e^{-ry_{1f1}}}{1+2r})_-] \qquad (98)$$

where

$$X(\omega) \equiv \frac{re^{-y_{1f1}/2}}{shr y_{1f1}}(\frac{1}{2} - \frac{sch Le sy_{1f1}}{sh Le sy_{1f1}}) - \frac{se^{-Le y_{1f1}/2}}{sh Le sy_{1f1}} \qquad (99)$$

and A, B, J are defined by Eqs. (84-86), respectively. The subscripts + and − mean the evaluation of the quantity at $\omega = +ia$ and $\omega = -ia$, respectively.

Oscillations in Inlet Mass Flux. The response at large times to oscillations in inlet mass flux alone is given by

$$y_{fu} = \frac{-M^*}{2a}[e^{iat}(\frac{B}{A})_+ + e^{-iat}(\frac{B}{A})_-] \qquad (100)$$

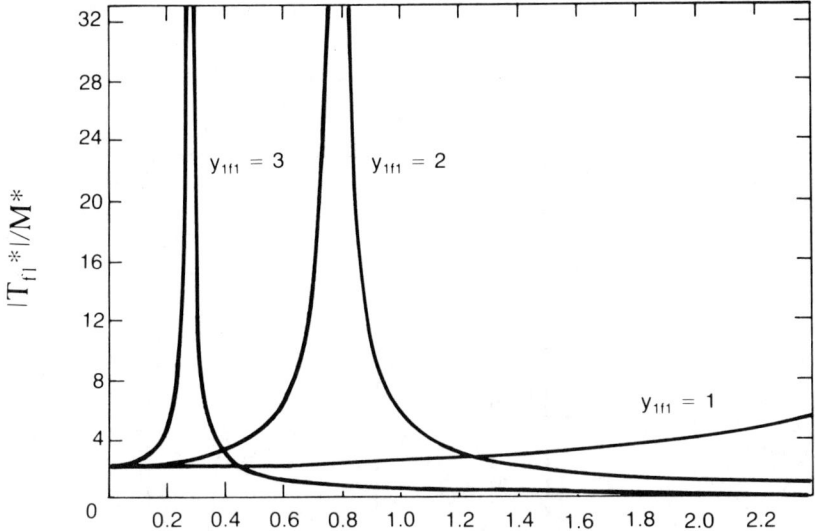

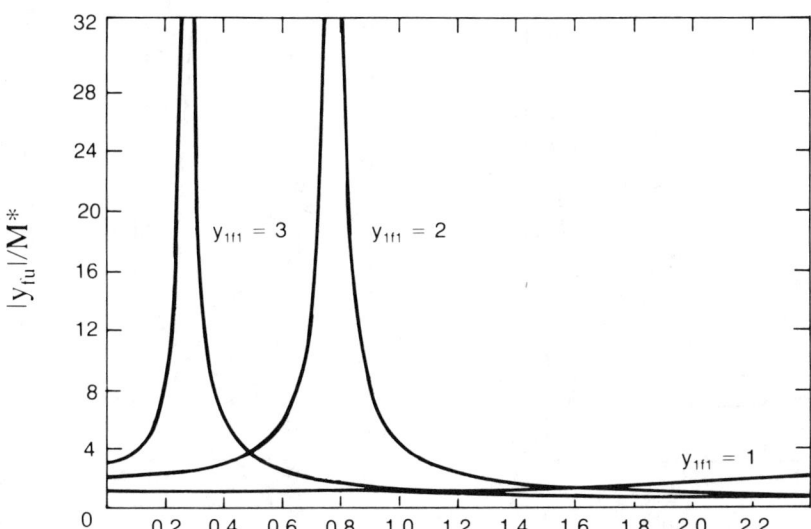

Fig. 4 Positional and temperature response at large times to oscillations in inlet mass flux; i.e., $\bar{M}_{uo} = M^* \sin at$. Lewis number Le = 1.5.

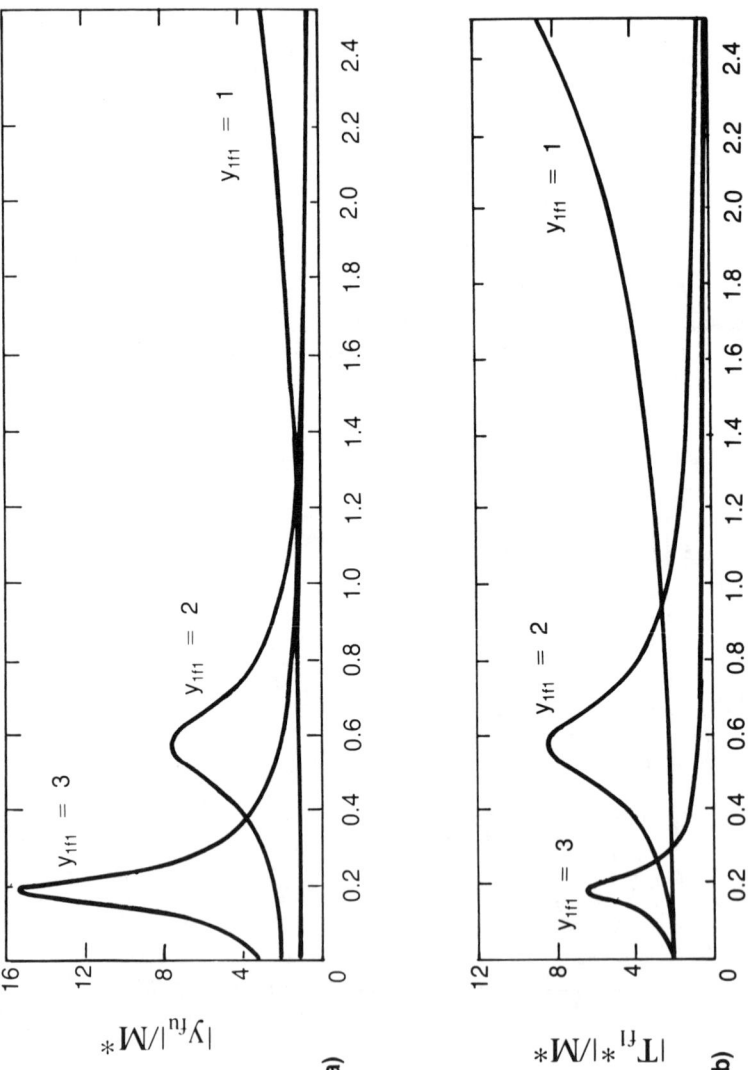

Fig. 5 Positional and temperature response at large times to oscillations in inlet mass flux; i.e., $M_{uo} = MI \sin\alpha t$. Lewis number Le - 1.75.

$$T^*_{f1}(t) = \frac{M^*}{i}[e^{iat}(\frac{X}{A})_+ - e^{-iat}(\frac{X}{A})_-] \quad (101)$$

The evaluation of

$$(B/A)_\pm \equiv u_1 \pm iv_1 \quad (102)$$

$$(X/A)_\pm \equiv u_2 \pm iv_2 \quad (103)$$

is in principle simple, but in practice for general Le, involves much algebra. Once u_1, v_1, u_2, v_2 are calculated, then

$$y_{fu} = \frac{-M^*}{a}(u_1 \cos at - v_1 \sin at) \equiv -R_1 \sin(at + \phi_1) \quad (104)$$

$$T^*_{f1} = 2M^*(u_2 \sin at + v_2 \cos at) \equiv R_2 \sin(at + \phi_2) \quad (105)$$

and

$$\frac{|y_{fu}|}{M^*} \equiv \frac{R_1}{M^*} = \frac{1}{a}(u_1^2 + v_1^2)^{1/2} \qquad \frac{|T^*_{f1}|}{M^*} \equiv \frac{R_2}{M^*} = 2(u_2^2 + v_2^2)^{1/2} \quad (106)$$

$$\tan \phi_1 = -u_1/v_1 \qquad \tan \phi_2 = v_2/u_2 \quad (107)$$

Typical results are plotted in Fig. 4 for Le = 1.5 and Fig. 5 for Le = 1.75, where the relative amplitudes $|y_{fu}|/M^*$ and $|T^*_{f1}|/M^*$ are plotted against frequency for different values of y_{1f1}. Because of the instabilities due to the zeros of A, (see the section entitled Asymptotically Steady Inputs and Fig. 2), Eqs. (100) and (101) are only valid for Le > Le^* (y_{1f1}), where it is recalled that Le^* (> 1) is the neutral stability curve in Fig. 2.

As Figs. 4 and 5 show, there emerges a "peak frequency" for which the oscillations are at a maximum, and a resonant frequency is indicated. Clearly, when the predicted peak amplitudes are very large, the theory must break down, but the graphs serve to show the important effect of the Lewis number and heat loss on laminar flame longitudinal oscillations.

<u>Oscillations in Composition</u>. The method follows a very similar pattern to that for the oscillations in mass flux. The positional and equilibrium temperature fluctuations (with inlet mass flux constant) are given by

$$y_{fu} = \frac{\xi_1 \exp(y_{1f1}/2)\Delta^*}{2iLe}[e^{iat}(\frac{r}{Ashry_{1f1}})_+ - e^{-iat}(\frac{r}{Ashry_{1f1}})_-] \quad (108)$$

$$T^*_{f1} = \frac{\Delta^* \exp(y_{1f1}/2)}{i} [e^{iat}(\frac{\xi_1 rJ}{\Delta shry_{1f1}} - \frac{\gamma_1 e^{-ry_{1f1}}}{1+2r})_+$$

$$- e^{-iat}(\frac{\xi_1 rJ}{\Delta shry_{1f1}} - \frac{\gamma_1 e^{-ry_{1f1}}}{1+2r})_-] \qquad (109)$$

For near-stoichiometric conditions $[\Delta_1 \sim 0(\theta_1^{-1})]$ these become

$$y_{fu} = 0 \qquad (110)$$

$$T^*_{f1} = \frac{-\Delta^* \gamma_1 e^{y_{1f1}/2}}{2i} [e^{iat}(\frac{e^{-ry_{1f1}}}{\frac{1}{2}+r})_+ - e^{-iat}(\frac{e^{-ry_{1f1}}}{\frac{1}{2}+r})_-] \qquad (111)$$

where, recalling from Eq. (79) that for $\Delta_1 \sim 0(\theta_1^{-1})$

$$\gamma_1 \equiv \frac{Le(1+\sigma)Q_1\theta_1 |\Delta_1|/2\sigma}{1 + Le(1+\sigma)Q_1\theta_1|\Delta_1|/2\sigma} \qquad (112)$$

Writing

$$(\frac{e^{-ry_{1f1}}}{\frac{1}{2}+r})_\pm \equiv u_3 \pm iv_3 \qquad (113)$$

Eq. (111) is then

$$T^*_{f1} = -\Delta^* \gamma_1 e^{y_{1f1}/2}(u_3 \sin at + v_3 \cos at) \equiv R_3 \sin(at + \phi_3)$$

so that $\qquad (114)$

$$\frac{T_{f1}}{\gamma_1\Delta^*} \equiv \frac{R_3}{\gamma_1\Delta^*} = e^{y_{1f1}/2}(u_3^2 + v_3^2)^{1/2} \qquad (115)$$

$$\tan \phi_3 = v_3/u_3 \qquad (116)$$

In fact, it can be shown that Eq. (113) gives

$$u_3 = \frac{\exp(-py_{1f1})}{(\frac{1}{2}+p)^2+q^2} [q\sin qy - (\frac{1}{2} + p)\cos qy] \qquad (117a)$$

$$v_3 = \frac{\exp(-py_{1f1})}{(\frac{1}{2}+p)^2+q^2} [q\cos qy + (\frac{1}{2} + p)\sin qy] \qquad (117b)$$

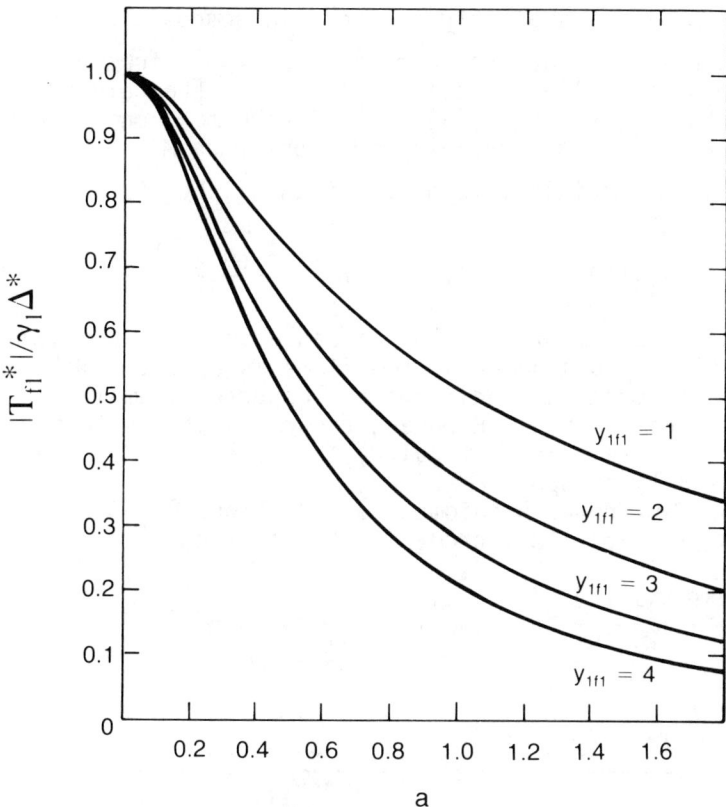

Fig. 6 Temperature response at large times to oscillations in near-stoichiometric composition; i.e., $\Delta_{GuO} = \Delta^*$ sinαt.

where

$$p \pm iq \equiv \sqrt{\tfrac{1}{4} \pm ia} \qquad (118)$$

and

$$p = \Theta/2\sqrt{2} \quad q = \sqrt{2}\, a/\Theta \quad \Theta \equiv (1 + (16a^2 + 1)^{1/2})^{1/2} \qquad (119)$$

It should be recalled from Eqs. (59) and (96) that Δ^* represents the fractional change in composition, so that even at near-stoichiometric conditions $[\Delta_1 \sim O(\theta_1^{-1})]$, Δ^* is an $O(1)$ quantity. In Fig. 6 is plotted the ratio $|T_{f1}^*|/(\gamma_1 \Delta^*)$ as a function of frequency and initial flame location

y_{1f1}. For these small fluctuations in composition, the temperature oscillations damp down at higher frequencies. The plots are valid for all Lewis numbers [Le only affects the factor γ_1; see Eq. (79)], and furthermore one should note that for these special conditions when $M_{uo} \equiv 0$ and $\Delta_1 \sim O(\theta_1^{-1})$, the original response integral, Eq. (83) is

$$T_{f1}^*(t) = -\gamma_1 \, e^{y_{1f1}/2} \, L^{-1}[\frac{\Delta_{Gu0} \, e^{-ry_{1f1}}}{\frac{1}{2} + r}] \quad (120)$$

so that the zeros of A no longer cause poles in the complex integration. Consequently, the flame is always stable under near-stoichiometric conditions in response to small compositional changes. However, the restrictions in this case are severe. In particular, the condition $M_{uo} \equiv 0$ is most unlikely to apply.

For far-from-stoichiometric conditions [$\Delta_1 \sim O(1)$], one obtains from Eqs. (83) and (84) the relations

$$y_{fu} = \frac{\exp(y_{1f1}/2)}{2iLen_1} \Delta^* [e^{iat}(\frac{r}{shry_{1f1}})_+ - e^{-iat}(\frac{r}{A \, shry_{1f1}})_-] \quad (121)$$

$$T_{f1}^* = \frac{\Delta^* \exp(y_{1f1}/2)}{i}[e^{iat}(\frac{r}{n_1 A \, shry_{1f1}} - \frac{\exp(-ry_{1f1})}{1+2r})$$

$$- e^{-iat}(\frac{r}{n_1 A \, shry_{1f1}} - \frac{\exp(-ry_{1f1})}{1+2r})_-] \quad (122)$$

and recall from Eq. (87) that

$$n_1 \equiv 1 - \frac{1}{\ell|\Delta_1|} = 1 - 1/(1+d)C_{Xb1} \quad \text{(fuel-lean)}$$

$$= 1 - 1/C_{Fb1} \quad \text{(fuel-rich)} \quad (123)$$

Writing

$$(\frac{r}{A \, shry_{1f1}})_\pm \equiv u_4 \pm iv_4 \quad (124)$$

$$(\frac{rJ}{A \, shry_{1f1}})_\pm \equiv u_5 \pm iv_5 \quad (125)$$

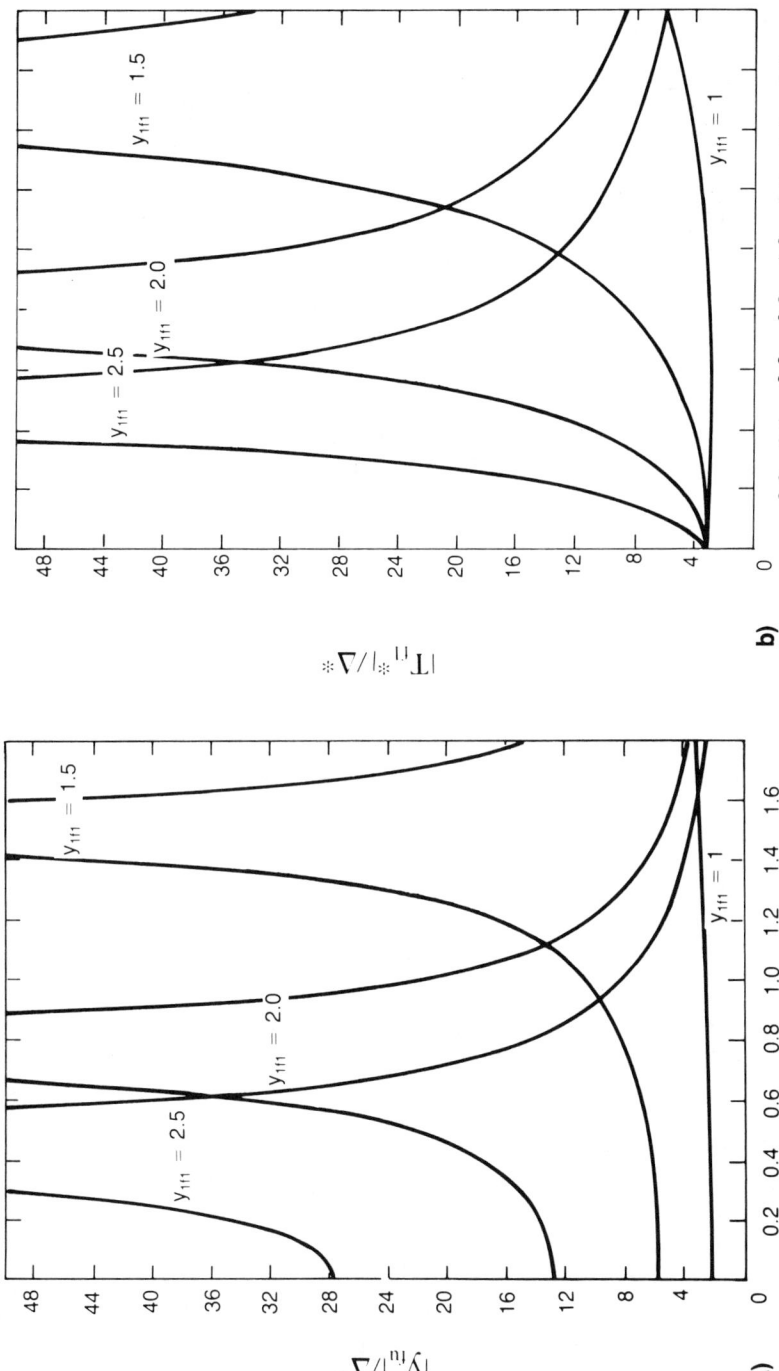

Fig. 7 Positional and temperature response at large times to oscillations in far-from-stoichiometric composition; i.e., $\Delta_{Gu0} = \Delta^* \sin\alpha t$, the Lewis number Le = 1.5 and C_{Fb1} = 0.5.

one can then express y_{fu}^* and T_{f1}^* as

$$y_{fu}^* = \frac{\Delta^* \exp(y_{1f1}/2)}{\text{Len}_1}(u_4 \sin at + v_4 \cos at) \qquad R_4 \sin(at + \phi_4) \quad (126)$$

$$T_{f1}^* = \Delta^* \exp(y_{1f1}/2)(u_6 \sin at + v_6 \cos at) \qquad R_5 \sin(at + \phi_5) \quad (127)$$

where

$$u_6 \equiv \frac{2u_5}{n_1} + \frac{u_3 \exp(-py_{1f1})}{(\frac{1}{2}+p)^2 + q^2} \quad (128)$$

$$v_6 \equiv \frac{2v_5}{n_1} + \frac{v_3 \exp(-py_{1f1})}{(\frac{1}{2}+p)^2 + q^2} \quad (129)$$

Consequently

$$\frac{|y_{fu}^*|}{\Delta^*} \equiv \frac{R_4}{\Delta^*} = \frac{(u_4^2 + v_4^2)^{1/2} \exp(y_{1f1})}{\text{Len}_1} \quad (130a)$$

$$\frac{|T_{f1}^*|}{\Delta^*} \equiv \frac{R_5}{\Delta^*} = \frac{(u_4^2 + v_4^2)^{1/2} \exp(y_{1f1})}{\text{Len}_1} \quad (130b)$$

$$\tan \phi_4 = v_4/u_4 \quad (131a)$$

$$\tan \phi_5 = v_6/u_6 \quad (131b)$$

Equations (121) and (122) are again only valid in the stable regions, so that oscillations for Le < 1 cannot be considered. However, for Le > Le* (> 1), one can derive the relative amplitudes $|y_{fu}^*|/\Delta^*$, $|T_{f1}^*|/\Delta^*$ of the large time response to forced oscillations. Figures 7 and 8 are plots, at Le = 1.5 and 1.75, respectively, of the amplitude of the fluctuations in position and temperature, when C_{Fb1}(fuel-rich) or $(1+d)C_{Xb1}$(fuel-lean) is 0.5 and for a range of y_{1f1} values. In a manner similar to that for the mass flux oscillations, peak frequencies emerge. For some

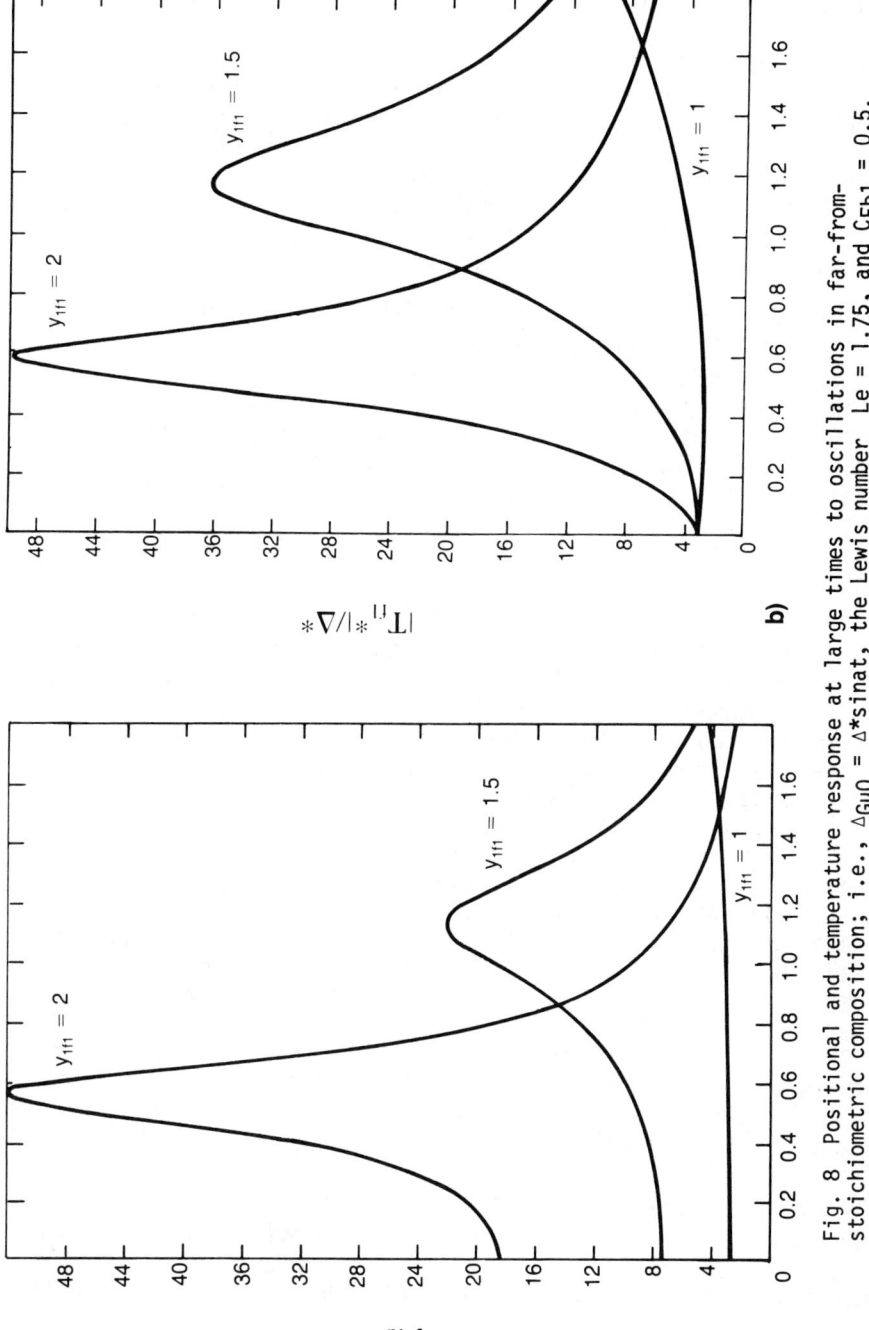

Fig. 8 Positional and temperature response at large times to oscillations in far-from-stoichiometric composition; i.e., $\Delta_{Gu0} = \Delta^* \sin at$, the Lewis number Le = 1.75, and $C_{Fb1} = 0.5$.

cases the associated peak amplitudes are very large and the theory would need to be examined very carefully in such cases. However, of great interest is the fact that these resonant frequencies are very nearly the same for both forced compositional and mass flux oscillations (at any composition). In other words, this frequency is in fact a "system resonance." Mathematically it arises through the function being common to both sets of Eqs. (100) and (101) and (121) and (122). There is a frequency which is a function of y_{1f1} and Le at which the combined contribution from $(1/A)_+$ and $(1/A)_-$ [see the notation defined after Eq. (99)] is at a maximum. The other terms in the two sets of equations do have a slight effect on the position of the peak frequency and, strictly, there is therefore difference between the peak frequency for compositional oscillations and that for mass flux oscillations. However, in most practical cases it was found that this difference is negligible.

Further investigations revealed the interesting fact that near these peaks the phase of the oscillations (that is ϕ_1, ϕ_2, ϕ_3, ϕ_4, ϕ_5) is changing rapidly with frequency, and the peak frequency lies close to the frequnces where the phase of the positional and temperature oscillations is $\pi/2$. This perhaps is as one would intuitively expect. At the resonant frequency the displacements in both equilibrium temperature and position are almost fully out of phase with the changes at the holder.

The most plausible explanation for this behavior is that it arises from an imbalance of the effects of convection, diffusion, and burner losses. Any changes in input conditions at the holder will take a certain time to reach the flame - this will be of the order of the characteristic time defined in Eq. (9e). For example, a Lewis number of 1.5 and an inlet flow velocity of approximately 20 cm/s indicate from the steady theory (Clarke and McIntosh 1979) that the equilibrium distance of the flame from the holder will be approximately 0.14 cm, and the associated time for the flow to reach the flame will be of the order of 7 ms. This time will consequently define a natural frequency of the flame/flame-holder system which, for the general case, will be dependent on the flame location y_{1f1} and Lewis number Le. Note that the reaction time only plays a part in determining the equilibrium balance of initial inlet mass flux M_1', with flame temperature T_{b1}' [see Clarke and McIntosh (1979) and note the explicit dependence on τ_0'

and $\exp(\theta_1)$ in Eqs. (6), (24), and (25)]. Once determined, the reaction time plays no further part in the unsteady moving boundary problem. This would not of course be the case if very fast flame movements were allowed. The assumption that y_{1f} is $o(\theta_1)$ would then no longer be true, so that the diffusion-reaction balance of the inner zone would not be maintained and a much more complicated problem would emerge.

To illustrate the predicted amplitudes of oscillation of the flame, note first that for Le = 1.75 and $y_{1f1} = 3$, the flame standoff distance is roughly 1 mm (Clarke and McIntosh 1979). Under these conditions, Fig. 5a gives a resonant (dimensionless) amplitude of oscillation of approximately 16; for forced oscillations of 0.1% of inlet flow speed (i.e., $\varepsilon = 0.001$), the actual amplitude of oscillation would be approximately from 0.02 to 0.08 mm, depending on the value of v'_{01} (initial inlet flow speed).

The unit of frequency is equivalent to about 1 kHz, [although this can vary according to the values of $\mathcal{D}'_{01}$ and V'_{01} used in Eq. (9e)]. Thus the resonant frequency can in fact be as low as $a = 0.1$; that is, therefore about 100 Hz, (for low inlet flow speeds). Of interest is that some experimenters (Blackshear and Rayle 1957) have shown that resonant frequencies do occur at quite low frequencies for plane flames. The phenomenon of flame "buzz," as this is generally referred to, is mainly attributed to acoustic excitation and of course cannot then be related directly to this essentially isobaric theory. However, El Banhawy et al. (1978) report that low-frequency ($\sim$ 500 Hz) high-amplitude oscillations have been observed in their experiments with flames in small tubes and that the driving mechanism involves fluctuations in the rate of heat release. It is possible that the resonant frequencies predicted by the present theory can be linked to low-frequency nonacoustic oscillations observed in such experiments.

In Fig. 9 the predicted resonant frequency is plotted as a function of initial flame location y_{1f1} (and thus heat loss). Each curve is for constant Lewis number. For branches of the curve where y_{1f1} lies within the instability region (Le < Le*; see Fig. 2), the curve is dotted to indicate that these conditions will in fact not be achieved. Next to the curves are indicated the associated peak amplitudes of $|y_{fu}|/M^*$. The amplitudes of $|T_{f1}|/M^*$, $|y_{fu}|/\Delta^*$

and $|T^*_{f1}|/\Delta^*$ are different in value but follow the same trend.

Further investigation confirms that the resonant nondimensional frequency a for a particular set of Le, y_{1f1} values is given by the value of z_0 associated with the location of the pair of conjugate poles from the zero of for that particular combination of Le and y_{1f1}, and the trends in the peak amplitudes (marked in Fig. 9) also confirm this observation. The peaks get larger nearer the neutral stability line. Physically one would expect this. As the natural frequency (equal to the resonant frequency) gets closer and closer to the critical frequency for neutral stability, so the "resistance" dies away to forced oscillations and the large time oscillations have larger and larger amplitudes.

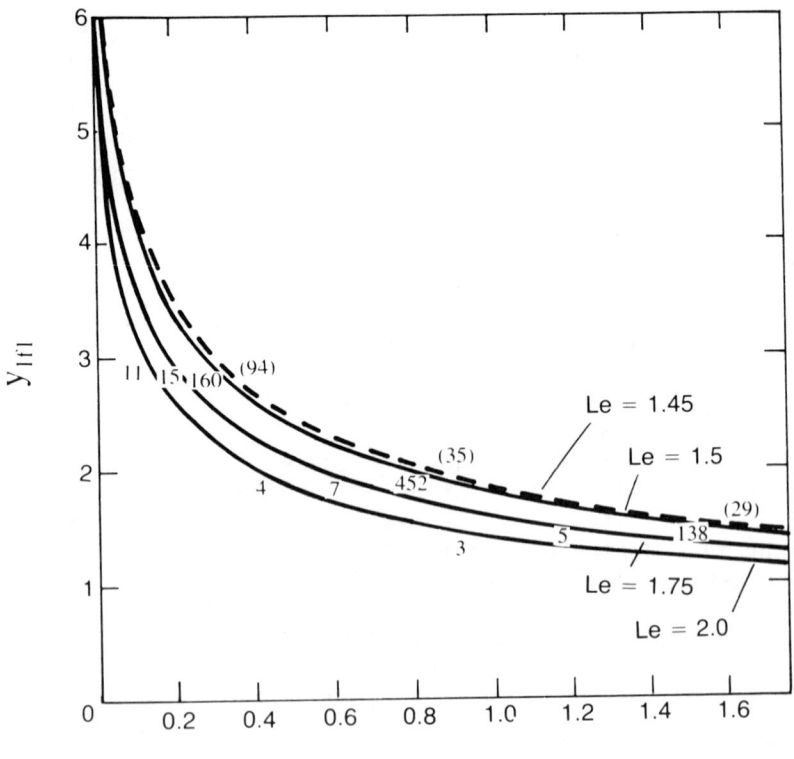

Fig. 9 Resonant frequency as a function of the Lewis number Le and initial nondimensional flame location y_{1f1}.

One should bear in mind that in this entire section only large time oscillations have been considered; detailed conclusions can only be drawn from examining the behavior over the whole time domain. However, the curves in Fig. 9 serve to demonstrate that resonant frequencies as low as 100 Hz are quite within the bounds of possibility for plane flames next to large conductance flame-holders. The phenomenon of resonance described in this paper indicates that if there is unsteadiness in the oncoming flow covering a wide range of frequencies, the plane flame is likely to oscillate at a single frequency which is a function of heat loss and Lewis number.

Conclusions

Using large activation energy asymptotics, a theory has been developed to describe the unsteady response to longitudinal disturbances of a one-dimensional premixed laminar flame next to a porous-plug-type flame-holder. A one-step irreversible reaction has been assumed with an Arrhenius form for the reaction rate, and the pressure variations have been assumed to be negligible. The flame-holder has been modeled using the approach of Hirschfelder, where it is assumed that no combustion products can filter upstream. Considering the reaction to take place in a thin zone of $O(\theta_1^{-1})$ thickness, $O(1)$ changes in inlet mass flux and composition are linked to $O(1)$ changes in flame position and $O(\theta_1^{-1})$ changes in equilibrium temperature. The rate of change of the flame position is restricted to being $o(\theta_1)$ in order to maintain the diffusion-reaction balance in the governing equations; 1 is the nondimensional activation energy of the initial steady-state reaction. Note that the Lewis number is unrestricted in magnitude.

In that the resulting matching conditions are applied on the moving flame boundary, the full task of finding a complete solution reduces to solving a Stefan (moving boundary) problem. The derived equations describe the response of the flame to determinate, well-defined inputs where both the response and inputs take place in real finite time intervals. In that the inputs start at a definite time (t = 0), we have taken full account of the transient behavior of the flame/flame-holder system. The whole formulation of the unsteady problem used in this paper has been firmly founded on well-established solutions of the steady-state equations (Clarke and McIntosh 1979).

An approximate solution for the response of the flame to changes in inlet mass flux and composition has been obtained on the assumption that perturbations are small. This effectively linearizes the boundary conditions and the differential equations, which have then been formally solved using Laplace transform techniques. These solutions have then been examined in order to extract information as to the response at large times to both asymptotically steady and oscillatory inputs.

In investigating the response to oscillatory inputs at large times, the theory predicts not only the anticipated resonant frequency but also the actual amplitudes of oscillations in flame position and temperature. It has been demonstrated that the resonant frequency is a function of both Lewis number and heat loss to the flame-holder. This interesting phenomenon has not previously been demonstrated analytically and may well afford an explaination for some heat-loss-associated flame oscillations observed by some experimenters.

References

Blackshear, P. L. and Rayle, N. D. (1957) Oscillations in combustors. NACA Rept. 1300, p. 229.

Clarke, J. F. and McIntosh, A. C. (1979) The influence of a flame-holder on a plane flame, including its static stability. Proc. R. Soc. London Ser. A 372, 367-392.

El Banhawy, Y., Melling, A., and Whitelaw, J. H. (1978) Combustion driven oscillations in a small tube. Combust. Flame 33, 281-290.

Hirschfelder, J. O., Curtiss, C. F. and Campbell, D. E. (1953) The theory of flames and detonations. Fourth Symposium (International) on Combustion, Baltimore, Md., p. 190.

Margolis, S. B. (1980) Bifurcation phenomena in burner-stabilized pre-mixed laminar flames. Combust. Sci. Technol. 22, 143-169.

Matkowsky, B. J. and Olagunju, D. O. (1980) Propagation of a pulsating flame front in a gaseous combustible mixture. SIAM J. Appl. Math. 39, 290-300.

McIntosh, A. C. (1981) Theoretical studies of unsteady pre-mixed flames. Ph.D. Thesis, Cranfield Institute of Technology, Cranfield, Bedford, England.

McIntosh, A. C. and Clarke, J. F. (1979) The time response of a pre-mixed flame to changes in mixture strength. Combustion Reactive Systems: AIAA Progress in Astronautics and

Aeronautics (edited by Bowen, Manson, Oppenheim, and Soloukhin), Vol. 76, pp. 443-462. AIAA, New York.

Sivashinsky, G. I. (1977) Diffusional-thermal theory of cellular flames. Combust. Sci. Technol. 15, 137-146.

Sivashinsky, G. I. (1979) On self-turbulization of a laminar flame. Acta Astron. 6, 569-592.

Sivashinsky, G. I. (1980) On flame propagation under conditions of stoichiometry. SIAM J. Appl. Math. 39, 67-82.

Temperature and Pressure Effects in Cool and Blue Flames

Yasuhiko Ohta* and Hitoshi Takahashi†
Nagoya Institute of Technology, Nagoya, Japan

Abstract

By using the stabilized flat two-stage low-temperature flames of rich diethyl ether-air mixtures, the heat release rates and the temperature coefficients (activation energies) of these flames were investigated. Heat release in a cool flame has little temperature sensitivity $E \cong 5$ kcal/mole (20 kJmole^{-1}), while the temperature coefficient of a blue flame is 20-30 kcal/mole (84-126 kJmole^{-1}), as generally found in a hot flame. Another experiment has been carried out in a special single-cylinder engine with lean diethyl ether-air mixture to determine the effects of pressure on the heat releases of low-temperature flames. A cool flame gives a pressure index (effective reaction order) of 1.48. In the region of τ_2 induction period, the pressure index reaches a level of 2.26. Blue flame ignition occurs even before the cool flame subsides, and the blue flame cannot propagate through the whole cylinder charge before red flames occur. The coupling of higher temperature flames with lower ones, especially hot flames after blue flames, is more noticeable at higher pressure conditions.

Introduction

It is generally believed that for hydrocarbon derived fuels, the preignition phenomena hold the key to various combustion mechanisms.

Presented at the 8th ICOGER, Minsk, USSR, Aug. 23-26, 1981. Copyright © American Institute of Aeronautics and Astronautics, Inc., 1982. All rights reserved.
*Lecturer, Department of Mechanical Engineering.
†Professor, Department of Mechanical Engineering.

TEMPERATURE AND PRESSURE EFFECTS IN COOL FLAMES

The self-ignition process of hydrocarbon fuels by which initially slow low-temperature reactions develop into rapid high-temperature flames is one aspect of the problem which is of particular interest in the field of internal combustion engines. The energy crisis has rekindled interest in the low-temperature flames which are a part of this process.

Self-ignition phenomena can generally be divided into several areas: slow oxidation, cool flames, blue flames and intermediate dark zones. Many different experimental approaches have been used to study the events leading up to and including hot ignition. There are many papers dealing with the early stages of spontaneous ignition, especially cool flames. Few investigations of blue flames, however, have been reported. The thermochemistry of hydrocarbon low-temperature oxidation, particularly heat release in low-temperature flames, is described with emphasis on the pre-blue flame regime during which active species are produced in cool flame reactions.

Flames stabilized in the Powling burner (Powling 1949) or vertical tube reactor (Williams and Sheinson 1973) are stable and spatially resolved, with favorable two-stage separation between cool and blue flames. These systems cannot simulate the low-temperature flames conditions, such as pressure effects encountered in engine cylinders. The laboratory-burner system was used to examine the temperature effects on the heat release rates of low-temperature flames, and a special single-cylinder engine was employed for the study of the pressure effects.

Theoretical Considerations

Two-Stage Flat Flames

Analysis based on the thermal theory of flame propagation was used to determine the heat release rates and temperature coefficients of these flames. The temperature coefficients are apparent activation energies of single-step reaction models of the low-temperature combustion reactions and the dark reactions occurring before and after the luminous flame reactions.

The energy conservation equation for steady one-dimensional laminar flame propagation is

$$\frac{d[\lambda(dT/dx)]}{dx} - \frac{d(\overline{C}_p T \rho u)}{dx} + qw = 0 \qquad (1)$$

where T is the absolute temperature; u is the velocity; λ is the thermal conductivity; ρ is the density; $\bar{C}_p$ is the mean specific heat

$$\bar{C}_p = \int_0^T C_p \, dT/T$$

of gas at the position of x; and q and w are the heat of reaction and the reaction rate, respectively.

The temperature coefficient E can be deduced from the relation between the gas temperature T and the heat release rate qw, which is assumed to be expressed by the empirical equation with an Arrhenius temperature dependence:

$$qw = q A P^n \exp(-E/RT) \qquad (2)$$

where A, R, P, and n denote a constant pre-exponential factor, universal gas constant, pressure, and reaction order, respectively.

Engine Combustion

The heat release rate $\dot{Q}_{lib}$ is inferred as the total amount of effective net heat release deduced by an energy balance from measurements of the cylinder pressure development and the heat loss from the cylinder walls $\dot{Q}_{elim}$

$$\dot{Q}_{lib} = \dot{U} + A P \dot{V} + \dot{Q}_{elim}$$

$$U = n C_v T \qquad C_v = AR/(k-1) \qquad PV = nRT$$

$$\dot{Q}_{lib} = A k P \dot{V}/(k-1) + A \dot{P} V/(k-1) - A P V \dot{k}/(k-1)^2 + \dot{Q}_{elim}$$
$$(3)$$

where U is the internal energy of the charge, A the thermal equivalent of work, V the combustion chamber volume, n the number of molecules, C_v the molar heat capacity at constant volume, and k the ratio of specific heats.

Estimates of the heat loss from the chamber walls were deduced from the experimental temperature histories of the wall surfaces and the equation for wall heat transfer for the unsteady-state heat conduction into a one-dimensional plate of infinite thickness (Carslaw and Jaeger 1959).

TEMPERATURE AND PRESSURE EFFECTS IN COOL FLAMES

Experimental Apparatus

Two-Stage Flat Flames

A modified Powling burner at atmospheric pressure was used to stabilize flat two-stage low-temperature flames of rich diethyl ether-air mixtures in vertically flowing streams. A sketch of this flat-flame burner is shown in Fig. 1. A fuel-air mixture flowed with almost uniform velocity distribution from an inner burner tube. An outer glass tube supported the stabilizing screen. When the mixture was ignited and the fuel-air ratio made quite rich, two distinct flame fronts were established in the positions shown in Fig. 1.

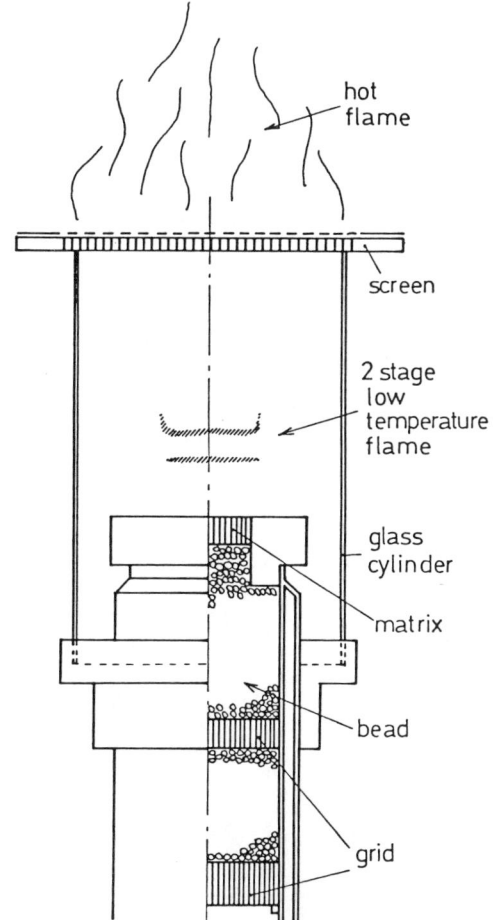

Fig. 1 Schematic of the flat-flame burner for low-temperature flames.

Table 1 Chromatograph specifications

Devices	Packing	Diameter, mm	Length, cm	Temperature, C	Monitored species	Carrier gas	Flow rate, cc/m	Response, s
GCG-550T (Yanagimoto) TCD	Cilicagel	3.0	20	70	$C_2H_4 + CO_2$	H_2	10	0.5
	Molecular sieve 5A	3.0	200	70	$O_2 + Ar$, N_2, CH_4, CO	H_2		
GCG-220 (Yanagimoto) TCD	Molecular sieve 5A	3.8	320	60	H_2	Ar	10	0.5
					N_2			
GC-4BT (Shimadsu) TCD +FID	Porapak T	3.0	180	90	HCHO, H_2O, CH_3CHO, $(C_2H_3)_2O$	H_2	10	0.5
	Alumina+Sq.	3.0	180	20	C_2H_4	H_2		

Temperature profiles in the two-stage flames were measured with a type R (platinum vs platinum-13% rhodium) 25-μm-diam wire, coated with SiO_2 to eliminate catalytic effects on the surface of the wire. A gas sample is removed continuously through a stainless steel probe into gas chromatographs (indicated in Table 1 for analysis).

Engine Combustion

Engine combustion experiments were carried out in a special single-cylinder (ϕ 85 x 100) one-cycle compression engine with lean diethyl ether-air mixture (equivalence ratio, ϕ = 0.5) at an engine speed of 300 rpm (5 s^{-1}), to determine the effects of pressure on the heat release rates in low-temperature flames. With this system a single homogeneous fuel-air charge may be ignited by compression in the absence of residual gases and cylinder hot spots. A compression ratio of 9.64:1 was used in conjunction with the preheat condition to give temperatures of the compressed charge in the range 600-670 K.

The surface temperature sensing junction, made of copper plating of 3 μm thickness on the surface, which connects the type T (copper vs constantan) thermocouple leads and is located on the cylinder head inner-wall surface, is shown in Fig. 2.

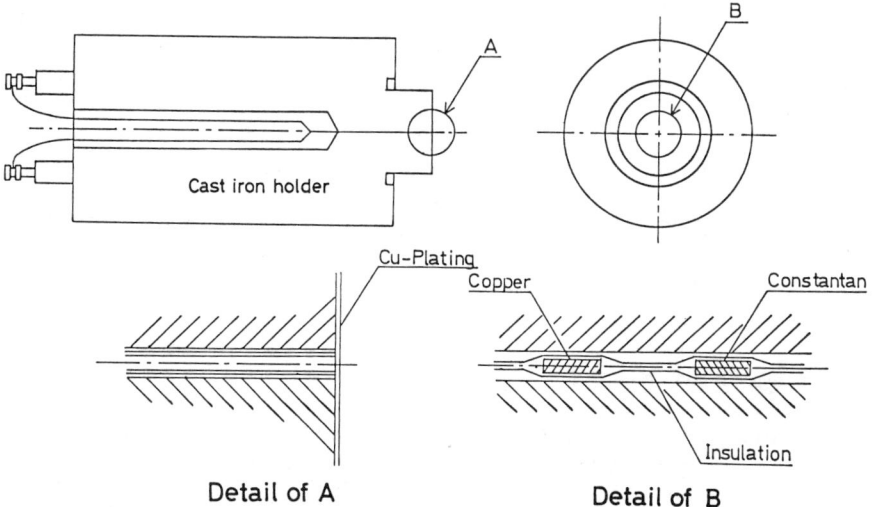

Fig. 2. Schematic of the thermocouple for surface temperature measurement in the combustion chamber.

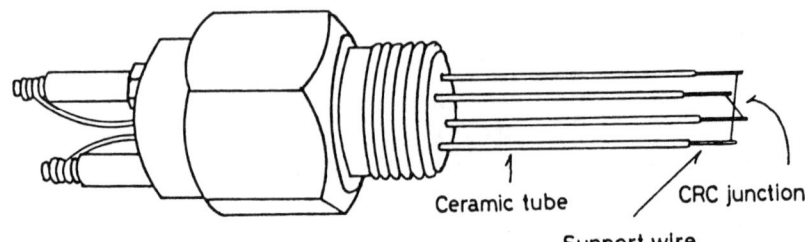

Fig. 3 Schematic of the fine wire thermocouple gas temperature measurement in the single-cylinder engine.

The measurements of the gas temperature in flames in the cylinder were made with an uncoated type E (chromel vs constantan) thermocouple, 25 μm in diameter and 10 mm in length, illustrated in Fig. 3. Compensation for the thermal inertia of the thermocouple wires was accomplished by combining the thermocouple voltage at any instant with the rate of change of voltage. An electrical circuit shown in Fig. 4 amplified the high-frequency components of the thermocouple signal in inverse proportion to their attenuation by thermal inertia. In this system the electronic time constant of the circuit must be adjusted so that the proper temperature is deduced from the circuits output. In the present measurements the time constant was selected so that just before the appearance of cool flames (at the time of 12 deg before top dead center) the compensated temperature is identical to that calculated from the measured pressure. The level of the compensation was not large and was only 20% of the uncompensated voltage at this time.

Pressures in the cylinder were measured dynamically with a strain-gage pressure transducer (Kyowa: PE-70KJ, response time: 40 μs). Emitted light was monitored by photomultiplier tubes (Hamamatsu TV: 1P28, response time: 50 ns) through glass filters [Toshiba: V-42 (blue, 320-510 nm), R-62 (red, >620 nm)].

Results

Two-Stage Flat Flames

Typical temperature traverses obtained by thermocouple measurements are shown in Fig. 5. Gas compositions of the stage species in the reaction zones are compared with the experimental data of Agnew and Agnew (1965) and of Foresti (1955) in Fig. 6. Despite the differences in burner design

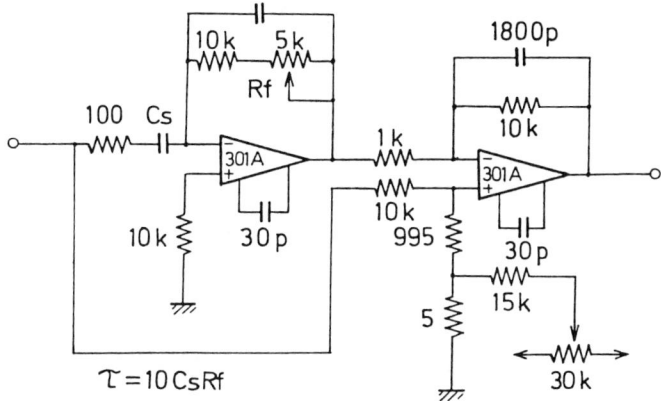

Fig. 4 Circuit diagram for thermal inertia compensation for fine wire thermocouple.

and flow conditions, it is evident that a common homogeneous chemical oxidation occurs. Since the thermochemical and transport property estimates are available for the major constituents, heat release calculations were made at intervals in the flame. All of the chemical species monitored in the gas chromatograph were included in the thermochemical calculations. Thermochemical and transport calculations were taken from standard literature sources: thermal conductivity (Eucken 1913), viscosity (Bromeley and Wilke 1951; Licht and Stechert 1944), heat capacity (Hougen et al. 1959).

Heat release rates through two-stage low-temperature flames are shown in Fig. 7. Heat release rates as a function of gas temperatures are shown in Figs. 8. The temperature coefficients are deduced from the slopes of these heat releases. Cool flames, or first appearances of low-temperature flames, have little temperature sensitivity [$E \simeq 5$ kcal/mole (20 kJmole^{-1})], while blue flames, secondary or the last stage of low-temperature flames have temperature coefficients of 20-40 kcal/mole (84-126 kJmole^{-1}), as generally observed for hot flames.

Engine Combustion

Oscilloscope records of pressure (P), compensated gas temperature (T), and wall surface temperature (T_w) are shown in Fig. 9a for subatmospheric initial pressure and in Fig. 9b for atmospheric initial pressure. Pressures of the former during the compression stroke were half of those of

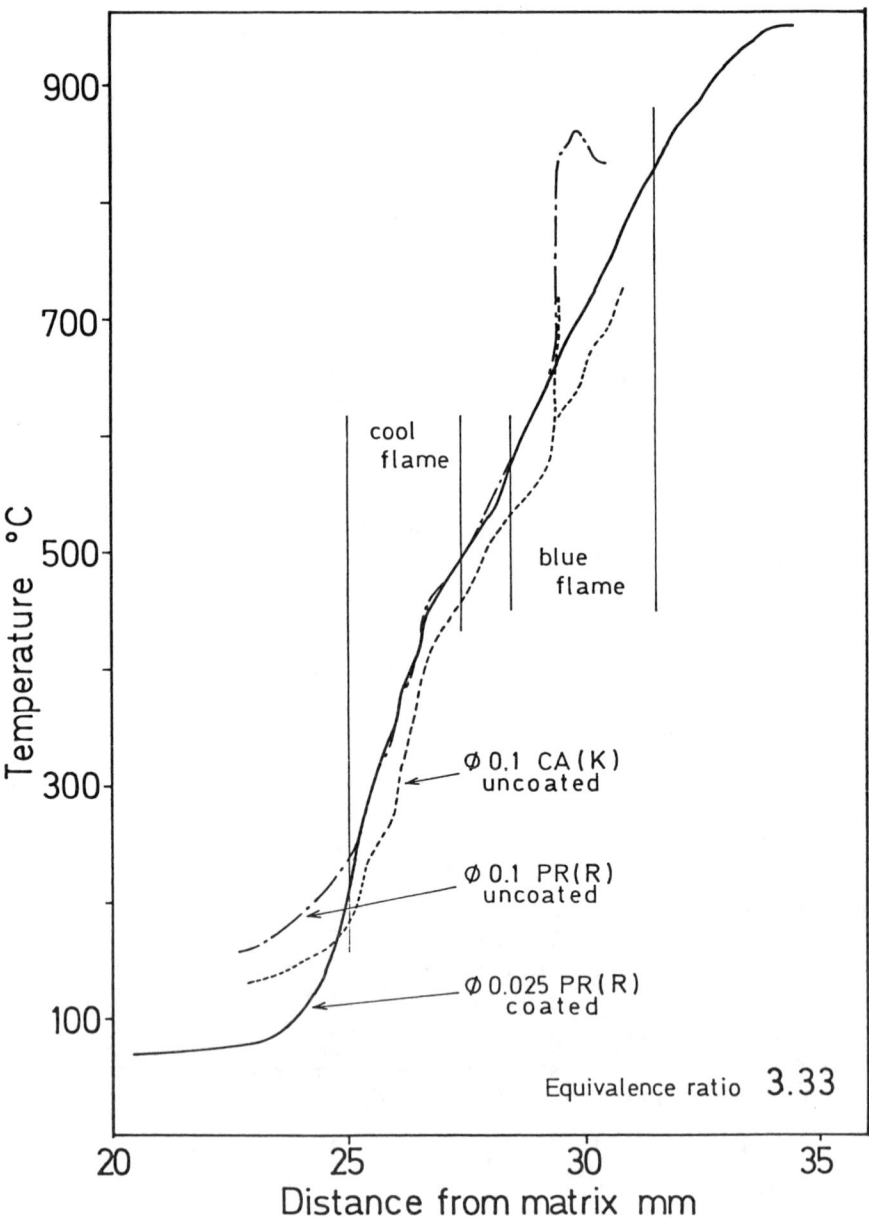

Fig. 5 Reaction zone temperature profile in the flat-flame burner with effects of catalytic heating and conduction errors.

TEMPERATURE AND PRESSURE EFFECTS IN COOL FLAMES

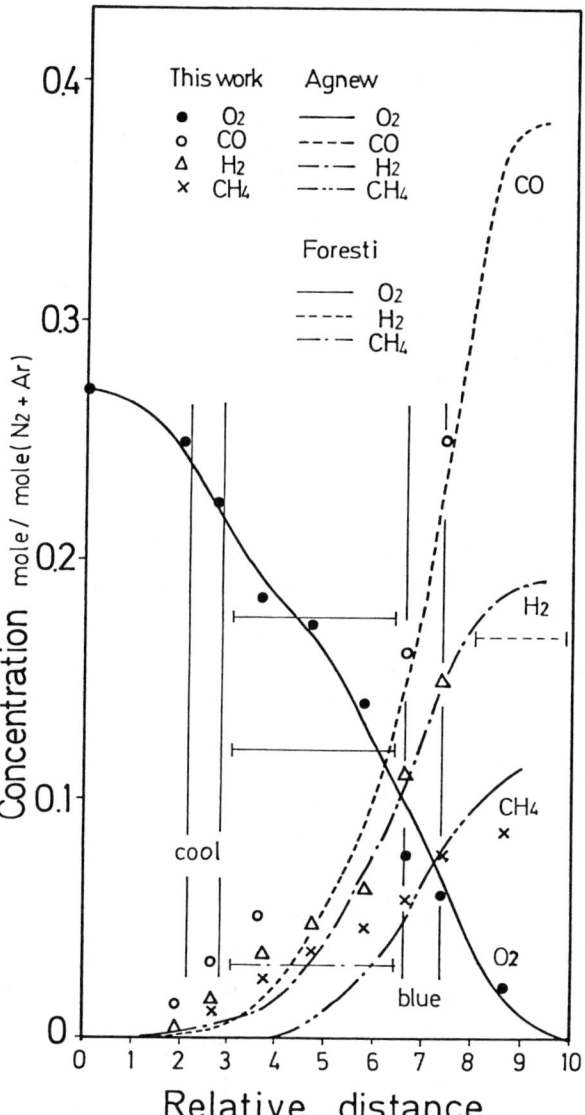

Fig. 6 Reaction zone composition profiles for stable species in the flat-flame burner.

the latter (the pressure scale of Fig. 9a is 1/2.5 to that of Fig. 9b). There is scarcely any difference in the gas temperature histories because the initial mixture temperatures were the same. The first inflection points of the compensated temperature correspond to those of pressure and indicate the points of departure of cool flames. At subsequent times, the temperature profiles were partially overcompensated because of flame generated gas motion.

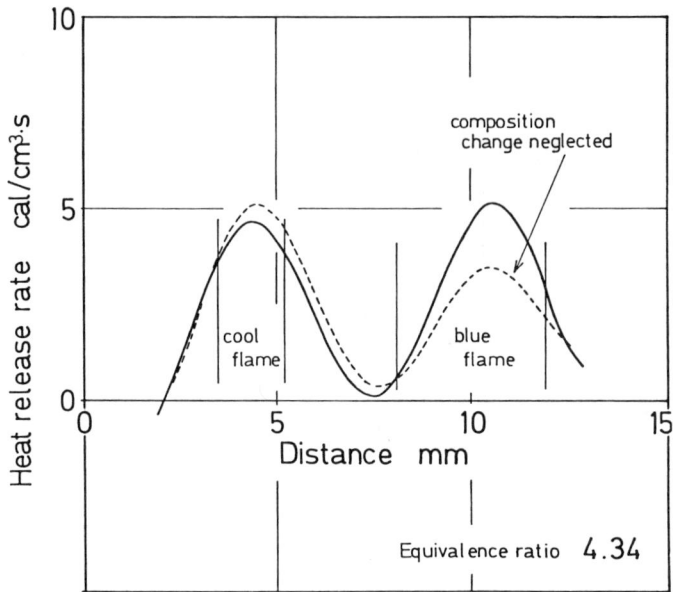

Fig. 7 Heat release rates in low-temperature flat-flame burner.

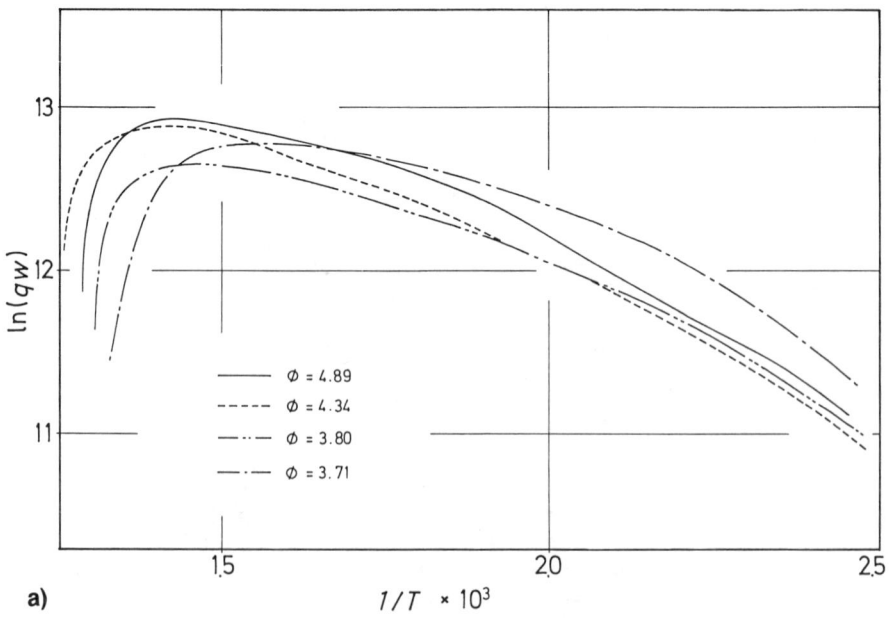

a)

Fig. 8 Heat release rates in low-temperature burner flames. a) Cool flame region. b) Blue flame region.

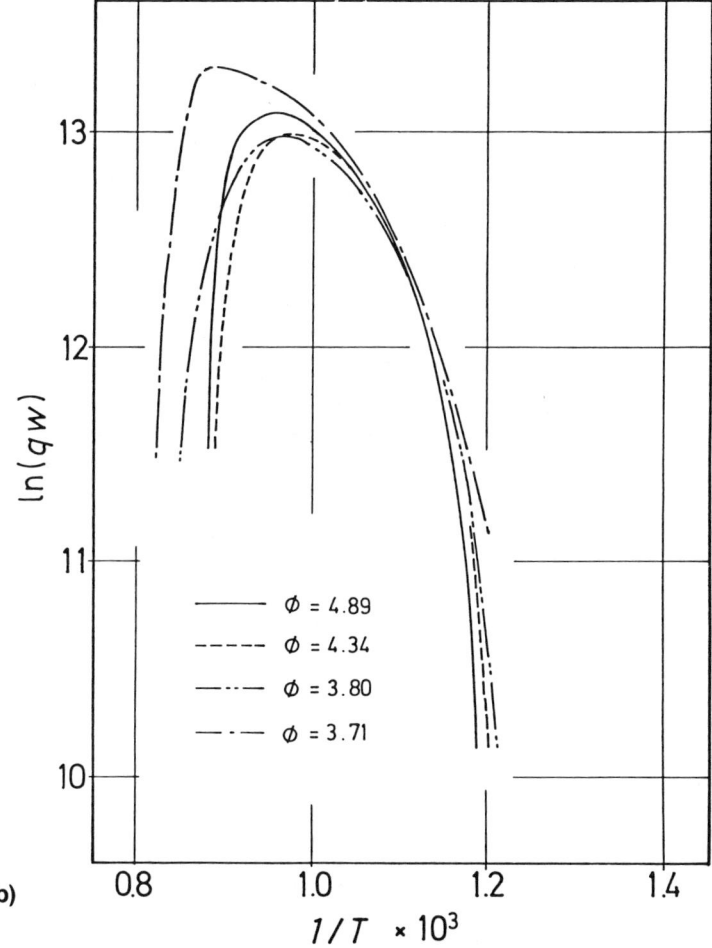

Fig. 8 (cont.) Heat release rates in low-temperature burner flames
a) Cool flame region. b) Blue flame region.

The next inflection points (or rises) of temperature denote the onset of a low-temperature flame of a different kind. Emitted light viewed through blue and red filters, shown in Fig. 10, suggests the interpretation of the pressure and temperature records. The first blue peak is associated with the cool flame, and the second peak, the second low-temperature flame. The latter should not be referred to as a second cool flame but as a blue flame

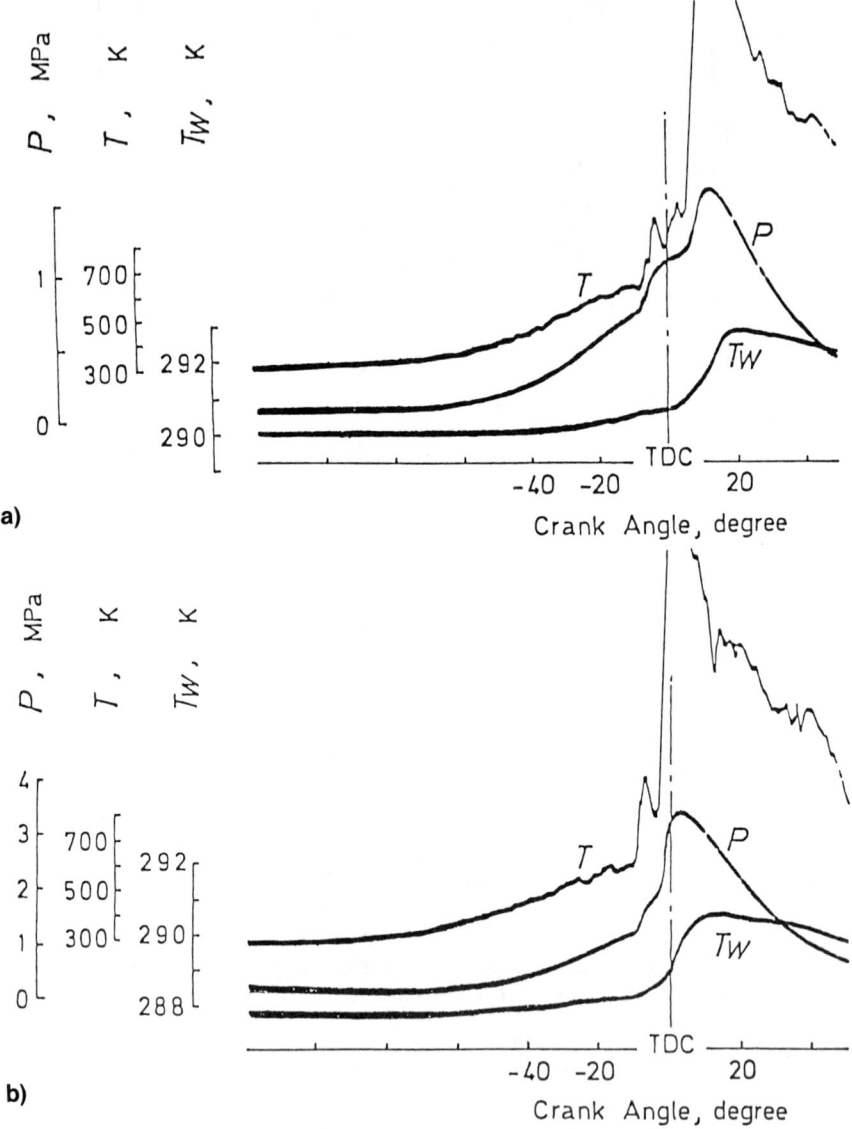

Fig. 9 Typical oscilloscope records of single-cylinder engine combustion. Pressure (P), gas temperature indication (T) and wall surface temperature (T_W). a) Subatmospheric initial pressure ϕ = 0.5, 300 rpm; initial conditions: 47.0 kPa, 293 K. b) Atmospheric initial pressure ϕ = 0.5, 300 rpm; initial conditions: 80.1 kPa, 293 K.

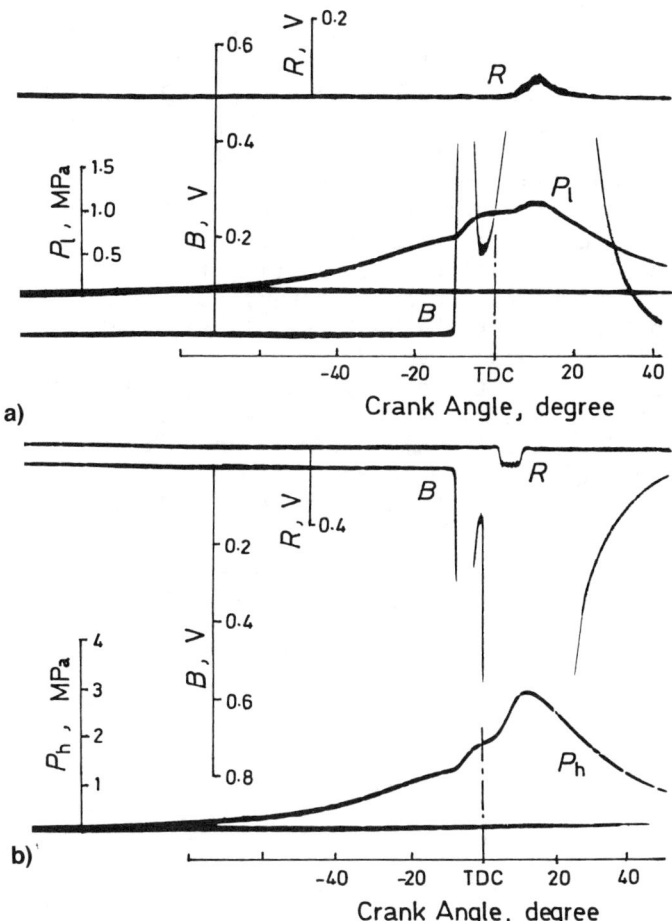

Fig. 10 Light emission records through blue (B) and red (R) filters and pressure profiles for single-cylinder combustion. a) Subatmospheric initial pressure φ = 0.5, 300 rpm; initial conditions: 46.2 kPa, 291 K. b) Atmospheric initial pressure φ = 0.5, 300 rpm; initial conditions: 78.6 kPa, 291 K.

because of its radiation intensity and profile. The last rise of pressure and temperature is associated with a hot flame which produces substantial red emissions.

Pressure and heat release rate vs crank angle under the lower and higher initial pressures are shown in Fig. 11. However, qw is the expression of heat release rate per 1 cm^3 of reaction volume, on the other hand, Q is the expression per whole combustion chamber volume. Low-temperature flames occur near the top dead center as shown in Fig. 11, so that the combustion chamber volume is held nearly constant.

Over a wide range of pressure, the cool flame first appears at the same temperature and same crank angle position. However, the time required for the reaction process at the lower-pressure conditions is almost double that at the higher pressure. The first peak of the heat release rate curves has a tailing edge; and the second peak, a preceding front. The peak separation is slight even under the lower pressure conditions. Moreover, each of the two peaks do not represent the heat release rates of the cool and blue flames, respectively.

The profiles of these curves and light emission records show that blue flame ignitions occur even before the cool flames subsides, and the blue flames cannot propagate through the whole cylinder charge before red hot flames occur. The coupling of higher-temperature flames with lower

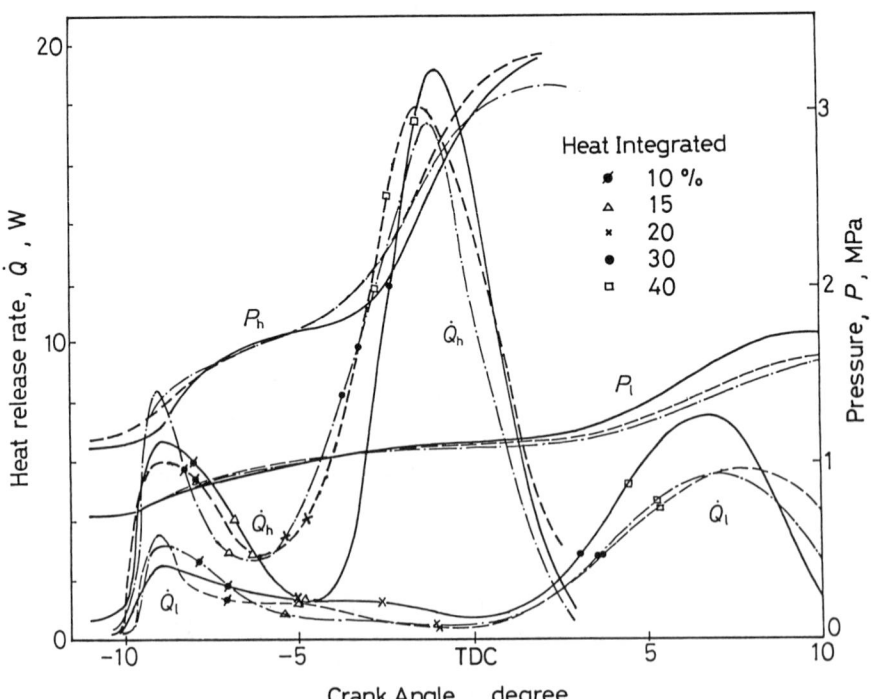

Fig. 11 Pressure heat and release rate for single-cylinder combustion. Pressure (P), heat release rate ($\dot{Q}$). P_l: ϕ = 0.5, 300 rpm; initial conditions 47.0 kPa, 293 K. P_h: ϕ = 0.5, 300 rpm; initial conditions 80.1 kPa, 293 K.

ones, especially hot flames after blue flames, is more noticeable at higher initial pressure conditions.

The effects of pressure on the heat release rates are indicated in Fig. 12 by the gradients of estimated lines,

$$\dot{Q}_h/\dot{Q}_L = \left(P_h/P_L\right)^n \qquad n = \tan \alpha$$

in which the heat release rates $\dot{Q}$ are compared for two sets of conditions at the same ratios of cumulative amount of heat released up to that time, $\int_0^t \dot{Q} dt$, to the whole heat of combustion of mixture H:

$$\frac{\int_0^{t_h} \dot{Q}_h dt}{H_h} = \frac{\int_0^{t_L} \dot{Q}_L dt}{H_L}$$

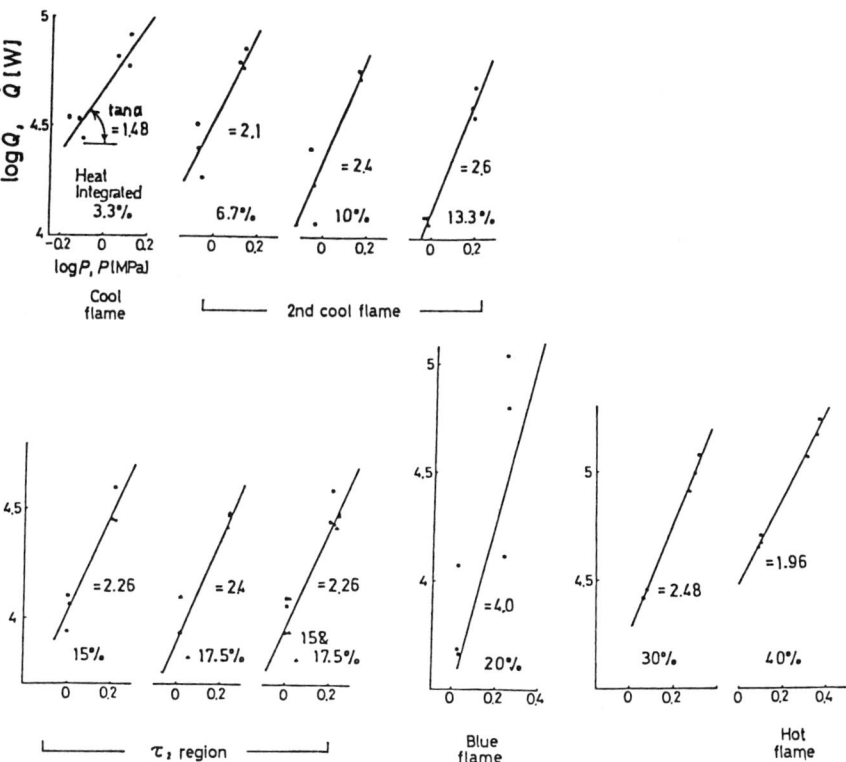

Fig. 12 Effects of pressure on heat release rates in low-temperature engine flames.

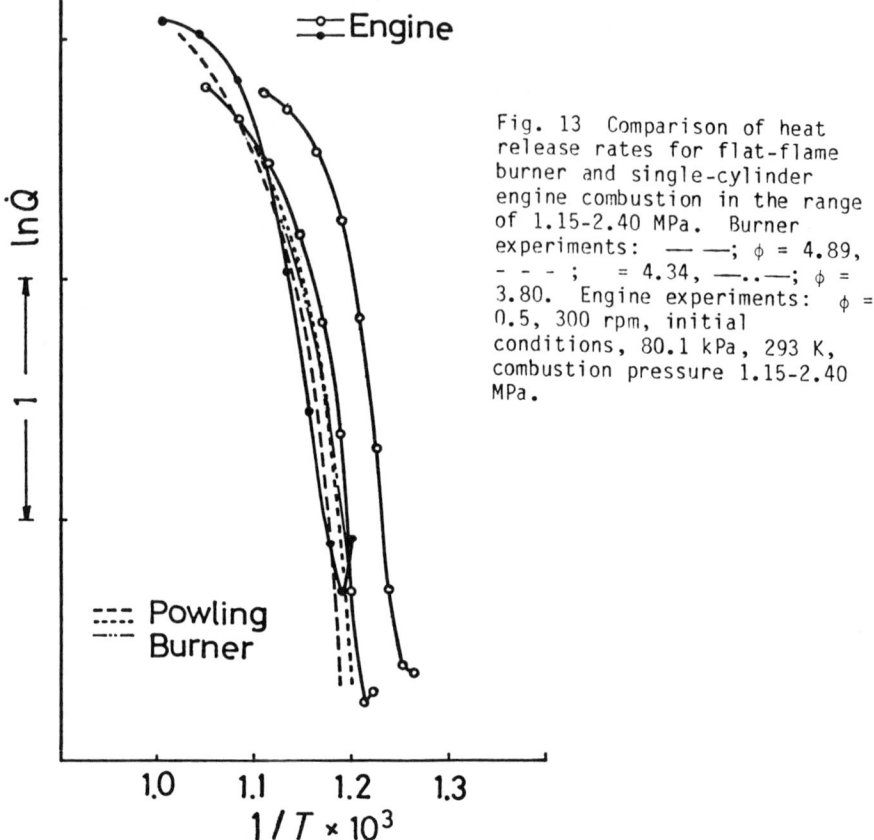

Fig. 13 Comparison of heat release rates for flat-flame burner and single-cylinder engine combustion in the range of 1.15-2.40 MPa. Burner experiments: ———; ϕ = 4.89, - - -; = 4.34, —··—; ϕ = 3.80. Engine experiments: ϕ = 0.5, 300 rpm, initial conditions, 80.1 kPa, 293 K, combustion pressure 1.15-2.40 MPa.

This ratio is termed the ratio of integrated heat. As the ratio of the pressure rise up to the time $\Delta P(t)$ to the total pressure rise during the whole combustion process $(\Delta P)_{whole}$ is close to the ratio of the cumulative heat to the heat of combustion of the mixture:

$$\frac{\Delta P(t)}{(\Delta P)_{whole}} \cong \frac{\int_0^t \dot{Q}\, dt}{H}$$

The ratio of integrated heat of 3.3% is associated with the region of cool flame appearances, for which the pressure index is 1.48.

This value may be compared to the reaction order for single-step reaction model cool flame explosions, and is in close agreement with the value of $\cong$ 1.5 reported by Sokolik

(1963) as the index of ΔP, the pressure rise during the cool flame of n-heptane. The pressure index increases as the amount of the integrated heat increases, probably as the consequence of a second cool flame. Subsequently, during the τ_2 induction period the cool flame changes into a blue or hot flame (typically seen in the rapid compression machine). This period corresponds to an integrated heat release of 15-17.5% and is accompanied by a decrease in the pressure index to 2.26.

In the engine experiments the lack of information on the area of the flame fronts or the reaction volume of the flames in the cylinder makes data interpretations difficult. Therefore only the values which have subsided or settled down to certain levels can give the quality data, which are considered to be obtained in the situations of nearly unchanged area or volume.

The temperature dependence for the heat release rates in the blue flame region under the engine cylinder conditions is compared with those of the Powling burner in Fig. 13. There is little to choose from between the two conditions of pressure. The behavior of blue flames at high pressures in the range 1.15-2.40 MPa does not differ greatly from that observed for flames in the low-pressure Powling burner. The temperature coefficients for the single-step reaction model at the high-pressure conditions presented in combustion chambers of engines are essentially the same as those for the single-step reaction model in the Powling burner. Hence it may be inferred that the reaction processes are similar for the two combustion systems.

Conclusions

Temperature coefficients (effective activation energies) of about 5 kcal/mole (20 kJmole^{-1}) in the cool flame region and of 20-30 kcal/mole (84-126 kJmole^{-1}) in the blue flame region for a single-step reaction model were deduced from experimental data on flat flames. Similar activation energies were observed for combustion at atmospheric- and higher-pressure conditions of engine cylinders. In the engine combustion experiments the pressure indexes (effective reaction orders) of the values of 1.48 in the cool flame region and of 2.26 in the τ_2 induction period region.

The distinction between cool and blue flames becomes less apparent as the initial pressure of the system is increased.

References

Agnew, A. G. and Agnew, J. T. (1965) Composition profiles of the diethyl ether-air two-stage reaction stabilized in a flat-flame burner. 10th International Symposium on Combustion, p. 123. The Combustion Institute, Pittsburgh, Pa.

Bromeley, L. A. and Wilke, C. R. (1951) Viscosity behavior of gases. Ind. Eng. Chem. 43, 1941.

Carslaw, H. S. and Jaeger, J. C. (1959) Conduction of Heat in Solids, 2nd ed., p. 62, Oxford University Press, Oxford England.

Eucken E. (1913) Über das Wärmeleitvermögen, die spezifische Wärme und die innere Reibung der Gase. Physik. Z. 14, 324.

Foresti, R. J. (1955) Stabilization and temperature measurement of flat cool flame. 5th International Symposium on Combustion, p. 582. Reinhold Publishing, New York.

Hougen, O. A., Watson, K. M., and Ragatz, R. A. (1959) Chemical Process Principles, Part II, 2nd ed., Appendix. John Wiley & Sons, New York.

Licht, W. and Stechert, D. G. Jr. (1944) Variation of the viscosity of gases and vapors with temperature, J. Phys. Chem. 48, 23.

Powling, J. (1949) A new burner method for the determination of low burning velocities and limits of inflammability. Fuel, 28, 25.

Sokolik, A. S. (1963) Self-Ignition, Flame and Detonation in Gases, p. 78. Israel Program for Scientific Translations, Jerusalem, Israel.

Williams, F. W. and Sheinson, R. S. (1973) Manipulation of cool and blue flames in the winged vertical tube reactor. Combust. Sci. Technol. 7, 85.

An Excess Enthalpy Flame Stabilized in Ceramic Tubes

T. Hashimoto* and S. Yamasaki†
Hitachi Ltd., Tsuchiura-shi, Japan
and
T. Takeno‡
The University of Tokyo, Tokyo, Japan

Abstract

An experimental study on the excess enthalpy flame system was made by using a ceramic burner. The burner was designed in the light of previous experimental studies to reduce heat loss to a minimum and to replace the metal block by a bundle of narrow ceramic tubes in order to recirculate heat inside the flame zone. The stability and combustion characteristics of the burner were studied to explore to what extent the range of the flame stability was extended below the normal lean limit. It was found that there were two distinct types of flame, and the flames exhibited a pronounced hysteresis behavior. The first was a flame stabilized inside the narrow tubes, while the second flame stabilized downstream of the tube exit. The emission characteristics of the first flame were excellent. The essential characteristics of the flames were found to correlate with those predicted in the previous, simplified one-dimensional theory.

Introduction

On the basis of the concept of excess enthalpy burning, Weinberg (1971), Hardesty and Weinberg (1974), and Takeno

---

Presented at the 8th ICOGER, Minsk, USSR, Aug. 23-26, 1981. Copyright © American Institute of Aeronautics and Astronautics, Inc., 1982. All rights reserved.

*Researcher, 6th Department, Mechanical Engineering Research Laboratory.

†Senior Researcher, Planning Office, Mechanical Engineering Research Laboratory.

‡Professor, Institute of Interdisciplinary Research, Faculty of Engineering.

and Sato (1979) recently proposed a simple way of producing an excess enthalpy flame by inserting a high-conductivity porous solid into a one-dimensional flame zone to recirculate heat through the solid from the downstream high-temperature region to the upstream low-temperature region. The potential of the proposed artificially modified flame system was analyzed on the basis of the simplified one-dimensional flame theory to reveal that the idea is a promising one for burning mixtures of low heat content in a simple system without pollutant formation (Takeno and Sato 1979, 1981b; Takeno, Sato, and Hase 1981; Buckmaster and Takeno 1981; Takeno and Hase 1981). In subsequent experimental studies, Takeno and co-workers studied the

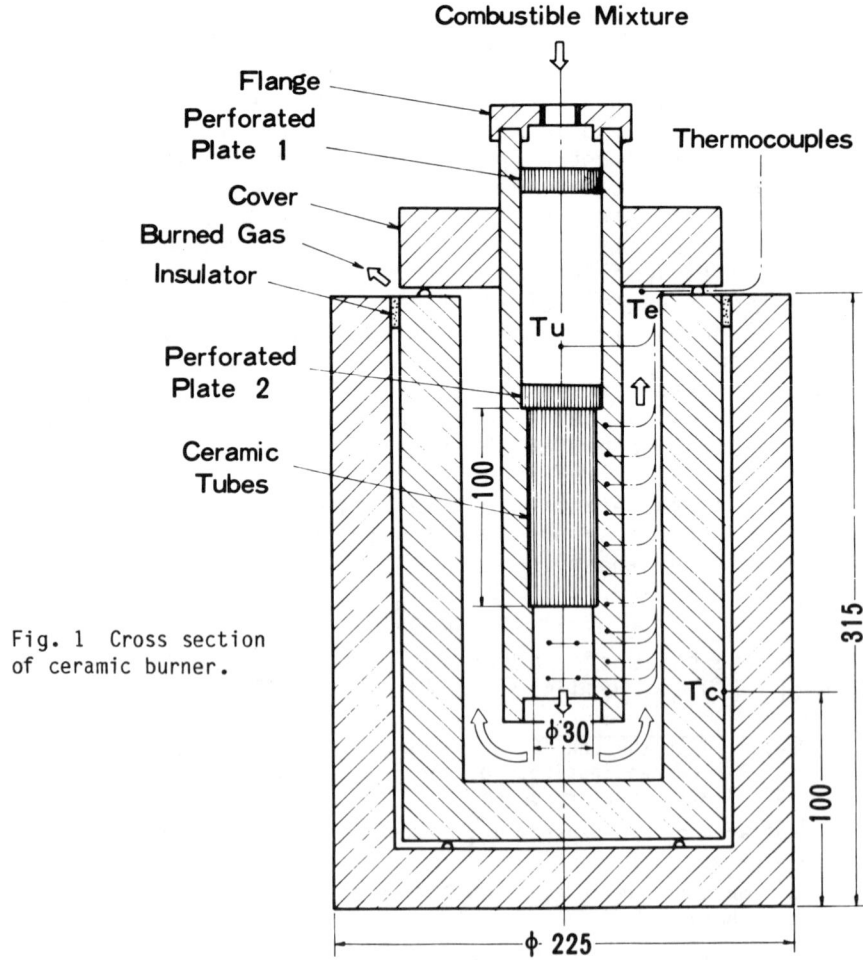

Fig. 1 Cross section of ceramic burner.

stability characteristics of the flame system by using methane-air mixtures and a specially designed burner equipped with a metal block to recirculate heat internally within the flame zone (Takeno and Sato 1980, 1981a, 1981b; Sato, Hase, and Takeno 1981). The block was made of iron at first and then of stainless steel, and had a number of straight holes through which the mixture flowed while undergoing chemical reactions. The diameters of the holes were made smaller than the thickness of the usual flame, so as to prevent the flame from anchoring and forming a curved surface along the heated wall, thus interrupting the intended heat transfer between solid and gas. It was found that the burner could extend the range of flammability below the normal lean limit, and the extension was produced as a result of the combined effects of heat recirculation inside and outside the metal block. The extension, however, was not as large as had been expected because the intended internal heat recirculation was not well realized inside the block. It is believed that formation of an oxide film on the metal surface, consequently preventing efficient heat transfer between solid and gas, is a major cause of the difficulty.

In view of these findings, in the present study, a ceramic burner was newly designed to reduce the heat loss to a minimum. Furthermore, the metal block was replaced by a bundle of narrow ceramic tubes so as to avoid formation of oxide films. Although the heat conduction through the solid

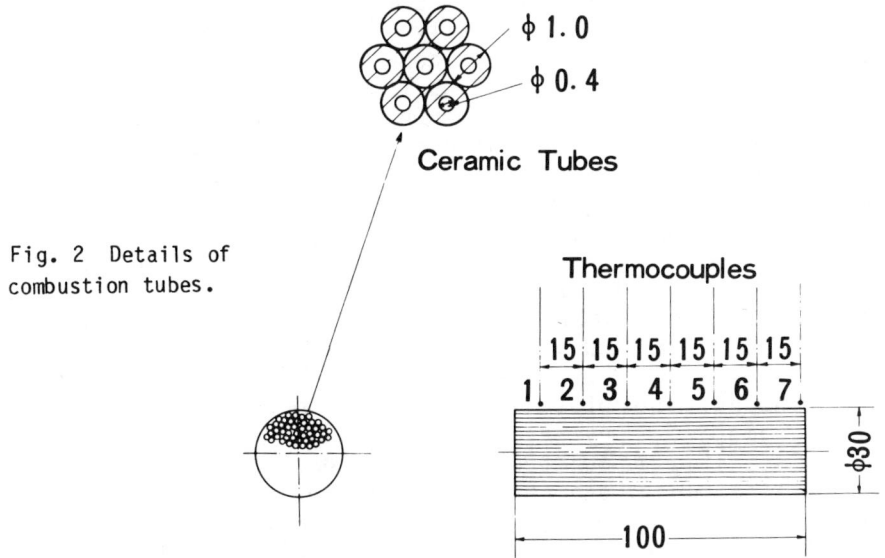

Fig. 2 Details of combustion tubes.

may be reduced, as compared to the metal block, it will still be effective in producing the internal heat recirculation. The objective of the present study is to investigate the feasibility of stabilizing such a flame by using this ceramic burner system.

Experimental

In the experiment, methane-air mixtures were used. Figure 1 shows the cross section of the axisymmetric burner made of ceramics. It was composed of an inner tube, an outer container, and a top cover. The inner tube was further equipped with a flange to introduce the mixture, two perforated plates, and a bundle of narrow ceramic tubes in which the combustion was expected to occur. The perforated plates were also made of ceramics and had coarser pores than those of the combustion tubes. They were intended to make the flow uniform and to reduce the radiative heat loss from the heated combustion tubes downstream. In the outer container, a narrow hollow space helped to reduce the heat escaping from the outer surface of the burner. In addition, the whole system was wrapped in Kaowool during the experiment.

The mixture was introduced into the inner tube at the top and descended through the plates and then through the combustion tubes where the chemical reactions occur. The combustion products emerged from the tubes and reversed their flow direction at the bottom of the container to flow up through an annular passage formed between the inner tube and the container. They finally left the burner through a narrow gap formed between the cover and the top of the container. Figure 2 shows details of the combustion tubes, which are key parts of the burner. Each tube was a straight alumina thermocouple insulator of 0.4 and 1.0 mm inside and outside diameter, respectively. The length was 100 mm, which was more than five times longer than that of the metal block used in previous work. The longer tube was expected to compensate for the inefficient heat conduction through the ceramics. Over 700 tubes were inserted within the 30-mm-i.d. inner tube.

In order to study the development of the flame, the wall temperature distribution along the flow direction was measured by means of ten Pt/Pt-Rh thermocouples of 0.1 mm wire diameter. They were spaced regularly at intervals of 15 mm starting at 10 mm downstream from the inlet of the combustion tubes, and were designated by numbers 1, 2, 3,... from upstream to downstream. Thus thermcouple 7 was located right at the tube exit, while thermocouple 10 was located 45 mm farther downstream. The burned gas temperatures 20 and 35 mm downstream of the exit were measured by using similar thermocouples. The unburned gas temperature T_u was measured on the axis 30 mm upstream of the tube inlet, while the final burned gas temperature T_e was measured at the burner exit. The temperature at the hollow space in the container was monitored by a chromel/alumel thermocouple. The burned gas composition was studied by sampling at the burner exit. The analysis was by an automatic analyzer (Yanagimoto Model EGA-37N), and the species studied were CO, CO_2, O_2, NO_x, and total hydrocarbons. The sampling rate was 6 cc/s, and the calibration for each component was made before every experiment.

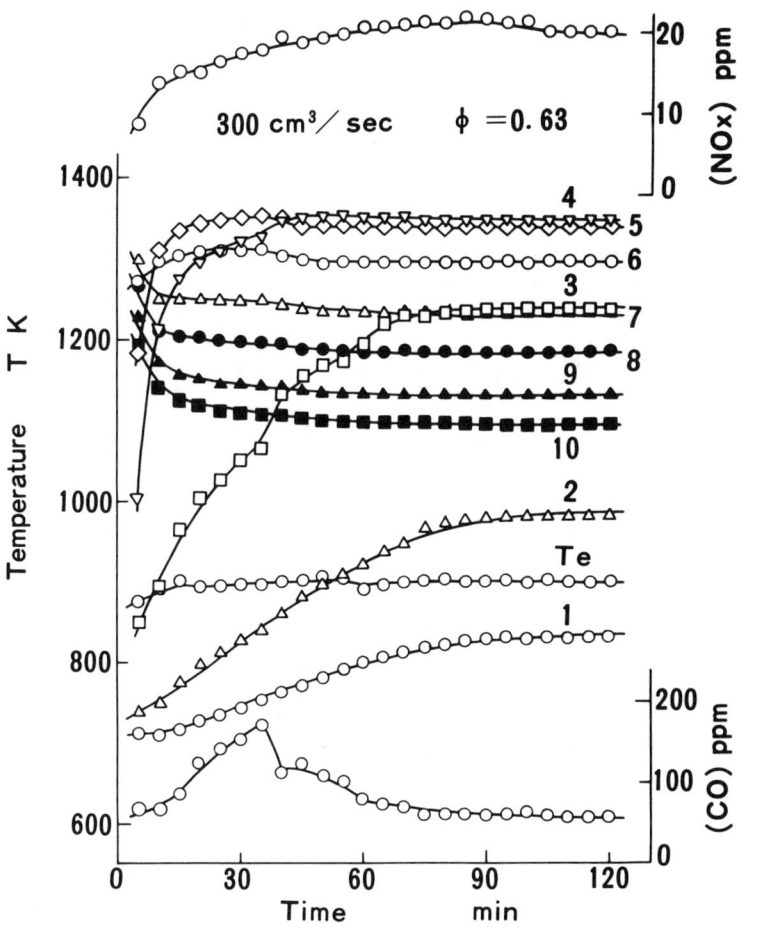

Fig. 3 Transient behavior of burner system.

The experiment was started by supplying the burner with the mixture at a flow rate and equivalence ratio favorable for ignition. The mixture flowing up the annular passage was ignited by taking out the cover and inserting a pilot flame into the passage. The flame then flashed back into the inner tube. It took more than 6 h to heat up the large heat capacity of the burner system to attain a steady state. The flow rate and equivalence ratio were then adjusted to a prescribed value, and it often took another 2 h before the final steady state was established. An example of this transient process is shown in Fig. 3, in which the temperatures and concentrations of NO_x and CO are plotted against time. The process was accompanied by the flame transition, which will be described later, and some 2 h were required for attainment of the steady state.

Experimental Results

Flame Stability

Figure 4 shows the observed flame stability diagram in the flow rate (Q)/equivalance ratio (ϕ) plane. Two different types of stable flames were observed; and the temperature distributions revealed that the first flame was stabilized inside the combustion tubes, while the second flame was stabilized downstream of the tubes. They are represented in Fig. 4 by the solid and the open circles, respectively. The stability limit for the second flame is extended below the normal lean limit, and leaner than the limit is observed for the burner equipped with the metal block (Takeno and Sato 1981a). The limit of the first flame, on the other hand, is located around the normal lean limit, and there also exists a critical maximum flow rate above which the flame is not sustained in the combustion tubes. Another important feature of the diagram is the coexistence of the two flames in the shaded region at lower right. What happened in this region can be seen in Fig. 5, in which the wall temperature around the center of the combustion tubes is plotted against ϕ for a constant flow rate. The flame exhibited a pronounced hysteresis behavior. As ϕ was increased from a low value, the temperature increased gradually along the lower branch (open circles in the figure), indicating that the flame remained downstream of the tube exit. At a critical value it increased very sharply and the flame flashed back into the combustion tubes to be stabilized there. The flame remained in the tubes for a further increase of ϕ. On the other hand, as ϕ was decreased from this state, the temperature decreased along

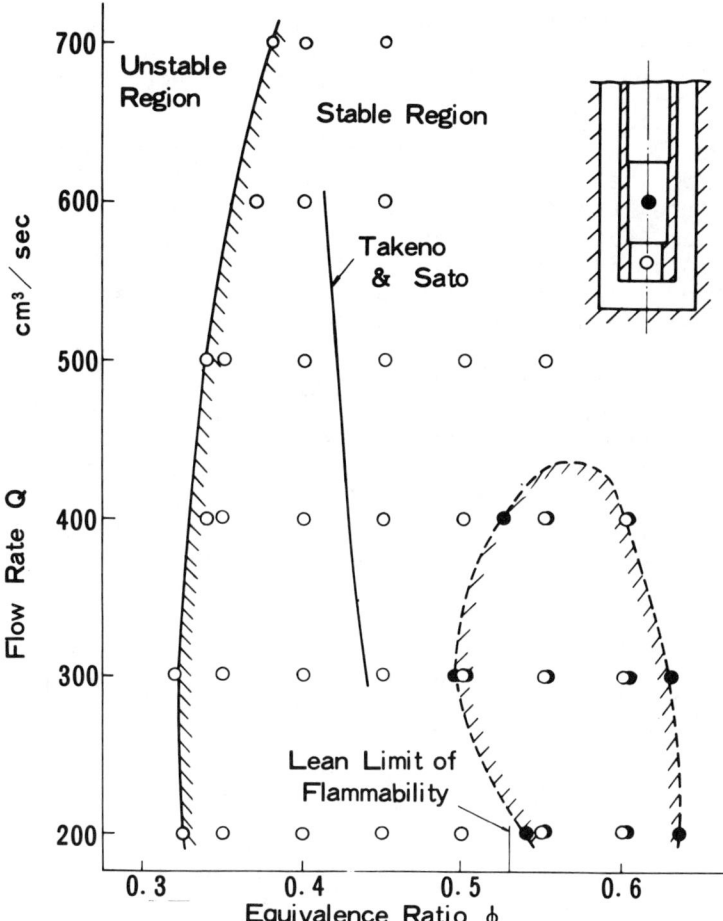

Fig. 4 Flame stability diagram in flow rate/equivalence ratio plane.

the upper curve (solid circles) and the flame remained in the tubes until, at another critical value, a sudden transition to the lower branch occurred and the flame was forced to emerge from the tubes to be stabilized downstream of the exit.

Wall Temperature Distribution

The wall temperature distribution for the two flames are compared in Fig. 6. The solid circles represent the temperatures for the first flame stabilized in the combustion tubes, while the open circles represent those of the second flame downstream of the tube exit. T_u indicates the unburned gas temperature before entry to the combustion tubes, while T_e is the final gas temperature at the burner exit. T_{ba} represents the adiabatic flame temperature calculated for the corresponding value of ϕ. As is seen in the figure, there is a remarkable wall temperature distribution for the both flames. In the first flame, the temperature attains a maximum inside the tubes and decreases downstream, suggesting that the main reaction has been completed inside the tubes. In the second flame, on the other hand, the temperature continues to increase downstream, attaining a gradual maximum there, which suggests that the main reaction occurs downstream of the tube exit. These suggestions were confirmed by the observation that in the second flame the gas temperatures downstream of the exit increased downstream, while in the first flame they were lower than the maximum wall temperature in the tubes.

Figures 7 and 8 show the effects of the flow rate on the temperature distributions for the first flame and the second flame, respectively. The wall temperatures, as well

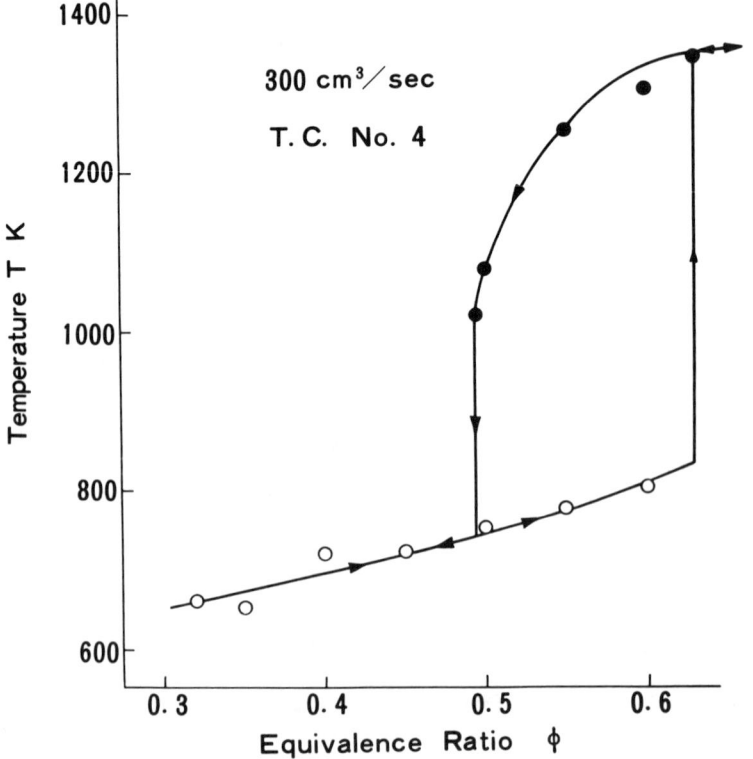

Fig. 5 Hysteresis behavior of wall temperature.

as Te, increased with the flow rate. This is presumably due to a decrease of the heat loss relative to the total heat release. This effect is more remarkable in the first flame, and the position of the maximum temperature moved downstream, approaching the tube exit. In the second flame, on the other hand, the distribution remained similar, although there is a slight indication that the gradual maximum moved farther downstream. It should be noticed that in the first flame a further increase in the flow rate will produce the flame transition (see Fig. 4). Figures 9 and 10 show the effects of the equivalence ratio on the temperature distribution. The temperatures increased with ϕ. In the first flame, the position of the maximum temperature evidently moved upstream. Although it is not obvious from the figure, this was also the case for the second flame. Here again we note that in the first flame, as ϕ is decreased to approach the critical value for the flame transition, the position of the maximum temperature approaches the tube exit. Similarly, when ϕ is increased in the second flame to approach the critical value for the transition, the position of the maximum temperature also approaches the tube exit from downstream. These results suggest that whenever the position of the maximum wall temperature approaches the combustion tube exit, flame transition will occur.

Emission Characteristics of NO_x and CO

Figure 11 shows the NO_x concentration in the burned gas plotted against ϕ for a constant flow rate. The NO_x emission naturally increased with ϕ and it also exhibited a pronounced hysteresis behavior. The solid circles represent data for the first flame while the open circles represent data of the second flame. As is seen in the figure, the emission is reduced remarkably when the flame is stablized inside the tubes. Figure 12 shows the isoconcentration curves in the flow rate/equivalence ratio plane. The solid curves represent the emission for the first flame, while the dotted curves those for the second flame. Here again we see that the first flame is accompanied by less emission. For both the flames, the emission depends mainly on the equivalence ratio, being independent of the flow rate. Figure 13 shows the hysteresis behavior of the CO emission, which is rather complicated. However, it can be seen that the CO emission also is reduced remarkably in the first flame. For the second flame, the emission decreased with a decrease of ϕ, except for an unusual increase at the leanest end. The latter may be attributed to a break of the flame stability

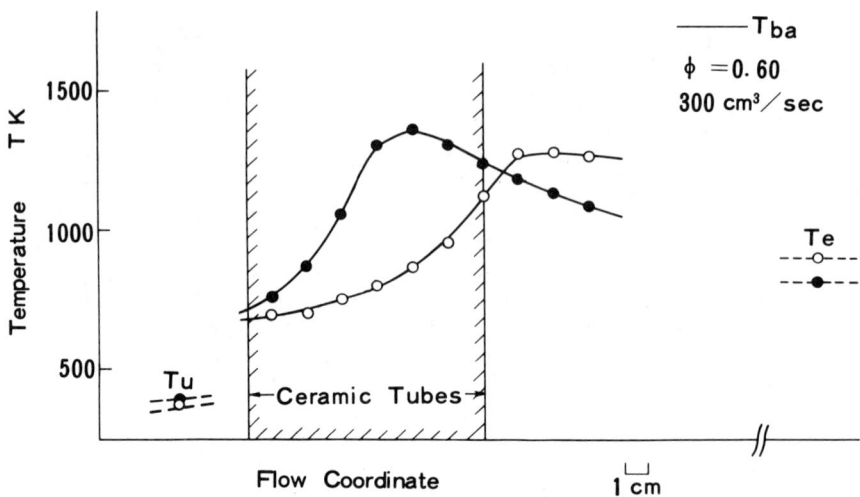

Fig. 6 Comparison of wall temperature distributions for two flames.

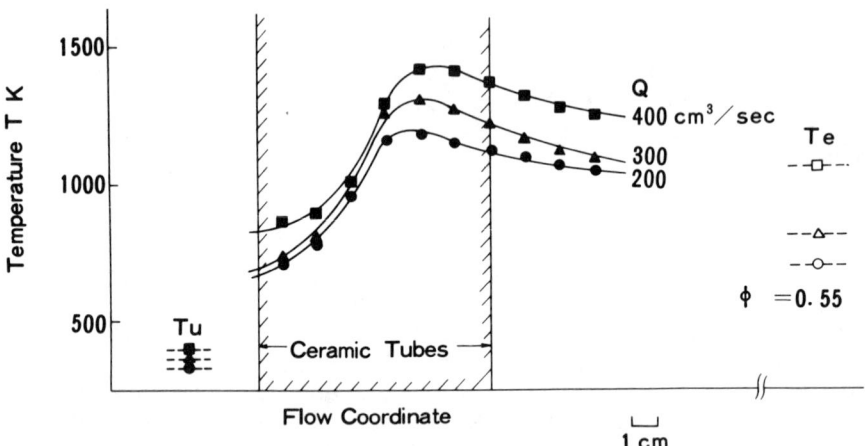

Fig. 7 Effects of flow rate on wall temperature distribution for flame stabilized inside tubes.

and the consequent intermittent discharge of the unburned gas into the hot combustion gas stream, since the gas temperatures downstream of the tube exit were observed to become unstable at this condition. In the first flame, on the other hand, the CO emission curiously enough attains a minimum at a certain value of ϕ. This feature is reflected in the complicated isoconcentration curves shown in Fig. 14. However, in general it may be said that the CO emission also is governed mainly by the equivalence ratio.

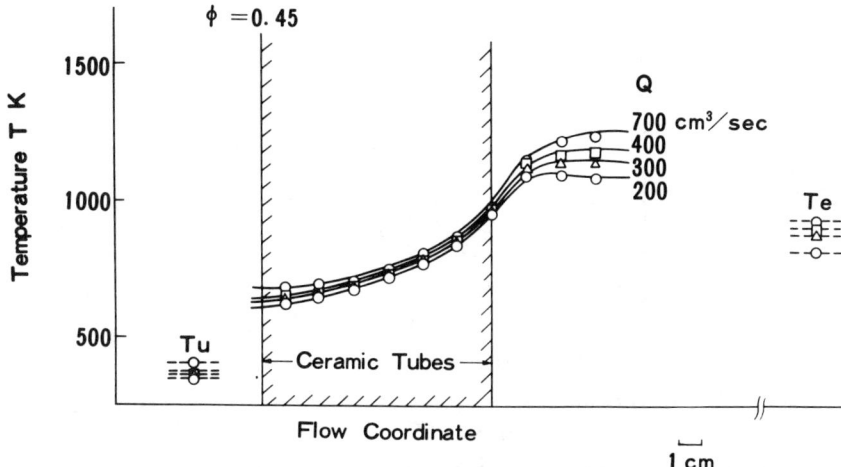

Fig. 8 Effects of flow rate on wall temperature distribution for flame stabilized downstream of tube exit.

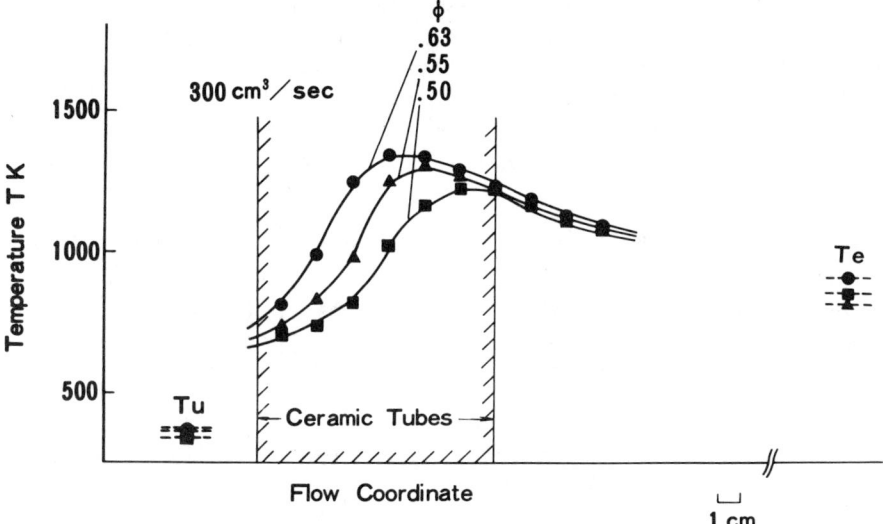

Fig. 9 Effects of equivalence ratio on wall temperature distribution for flame stabilized inside tubes.

Heat Loss Rate

The overall heat loss from the burner is an important parameter which may govern the stability of the system. The composition analysis of the burned gas indicated that the reactions were almost completed at the burner exit, and

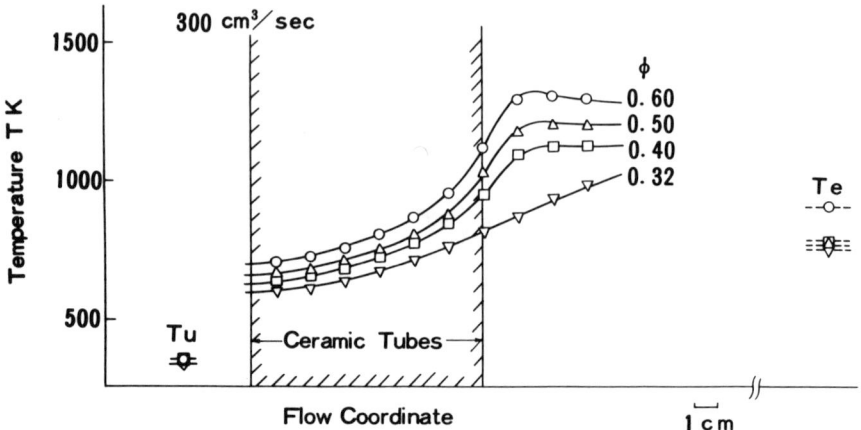

Fig. 10 Effects of equivalence ratio on wall temperature distribution for flame stabilized downstream of tube exit.

we can make use of the observed final gas temperature T_e to evaluate the overall heat loss rate in terms of the one-dimensional flame theory (Takeno, Sato, and Hase 1981; Takeno and Hase 1981). Let us introduce a reduced heat loss rate q_e (negative value) by

$$q_e = Q_e/m_s c_p (T_{ba} - T_o) \tag{1}$$

where Q_e is the heat input from the outer surroundings into the burner system; and m_s, c_p, T_{ba}, and T_o are the mass burning velocity, specific heat, adiabatic flame temperature, and initial temperature, respectively, of the mixture. Then T_e is related to q_e by

$$\frac{T_e - T_e}{T_{ba} - T_o} = 1 + \frac{q_e}{r} \tag{2}$$

where

$$r = m/m_s = Q/AS_u \tag{3}$$

is the reduced flow rate, or more exactly, the mass flux made nondimensional by the mass burning velocity; and Q, A, and S_u are the volume flow rate, the total flow area, and the burning velocity, respectively. Therefore

$$q_e = - \left(\frac{Q}{AS_u}\right)\left(\frac{T_{ba} - T_e}{T_{ba} - T_o}\right) \tag{4}$$

In the equation, Q, A, T_o, and T_e were measured in the experiments, while T_{ba} was calculated from the respective value of ϕ and the experimental burning velocity S_u (Andrews and Bradley 1973) as a function of ϕ was used.

Figure 15 shows the heat loss rate obtained, plotted against the flow rate for the various values of ϕ. The solid marks represent the data for the first flame, while the open marks represent those of the second flame. It can be seen that when r is small, the data for the second flame can be represented by a single line, which increases linearly with slope of 1/2. When r is increased further, the heat loss rate becomes saturated. The data for the first flame are rather scattered, but it can be said that

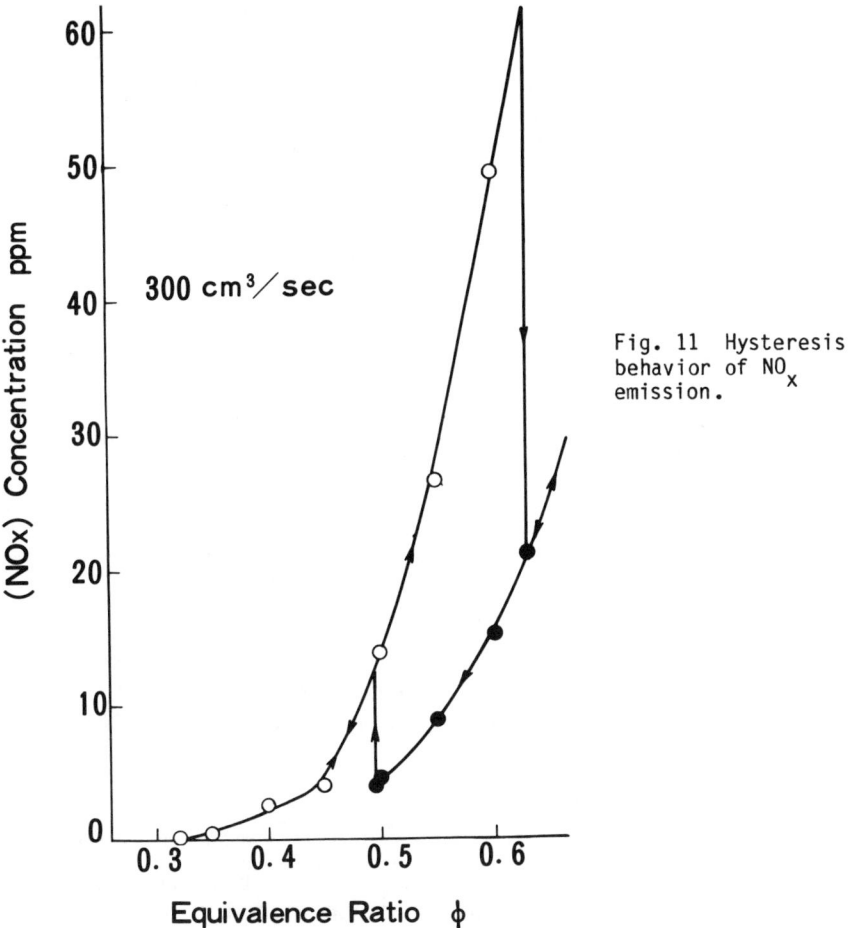

Fig. 11 Hysteresis behavior of NO_x emission.

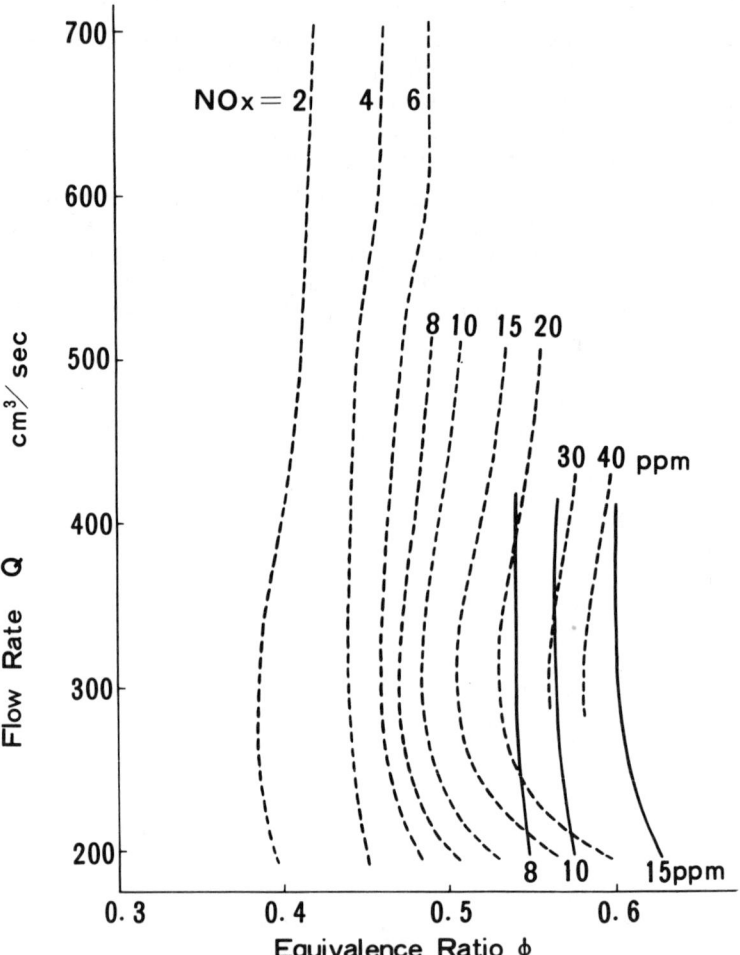

Fig. 12 Isoconcentration curves of NO_x in flow rate/equivalence ratio plane.

the heat loss is larger. This is presumably due to the higher solid temperature with the consequent increase of radiative heat transfer. The heat loss relative to the total heat release is given by

$$Q_e/m\, c_p\, (T_{ba} - T_e) = q_e/r \qquad (5)$$

In the small flow rate region, $-q_e/r = 1/2$, but about one-half of the total heat release escaped from the burner as heat loss. This value decreased with the flow rate, and it became around one-third at the maximum flow rate in this experiment. Thus the new burner design is subject to considerable heat loss.

Discussion

Although some indication of this was obtained in the previous experiment with the metal block (Sato, Hase, and Takeno 1981), the present study has made clear the existence of two distinct stable flames. Since the main observation in the present study was concerned with the wall temperature distribution rather than with the distribution of the gas temperature itself, it might be thought impossible to deduce the detailed nature of the flames observed on the basis of these experimental findings. Nonetheless, some implications will now be discussed.

Figure 16 shows the reduced flow rate as a function of ϕ for a constant flow rate. r increases with a decrease of ϕ

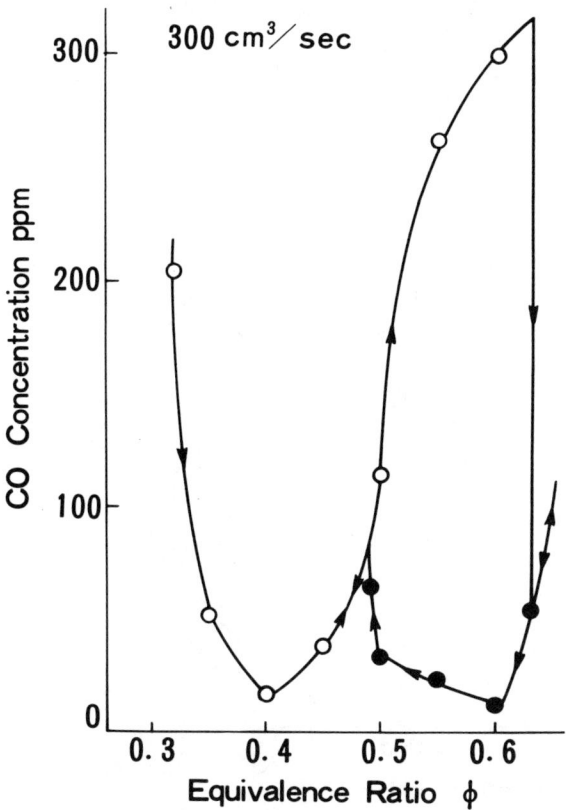

Fig. 13 Hysteresis behavior of CO emission.

because of the decrease in S_u, until at a critical value the blowoff occurs. For the first flame, the critical values reach up to around 30, although the value for 200 cc/s is exceptionally small owing to the increased relative heat loss. It is only through the proposed mechanism of the excess enthalpy flame, accompanied by a temperatue overshoot over the adiabatic flame temperature, that the flame can be stabilized up to such a fast flow rate. In the present experiment, the existence of the critical maximum flow rate, as well as the lean limit, was observed for this flame. As

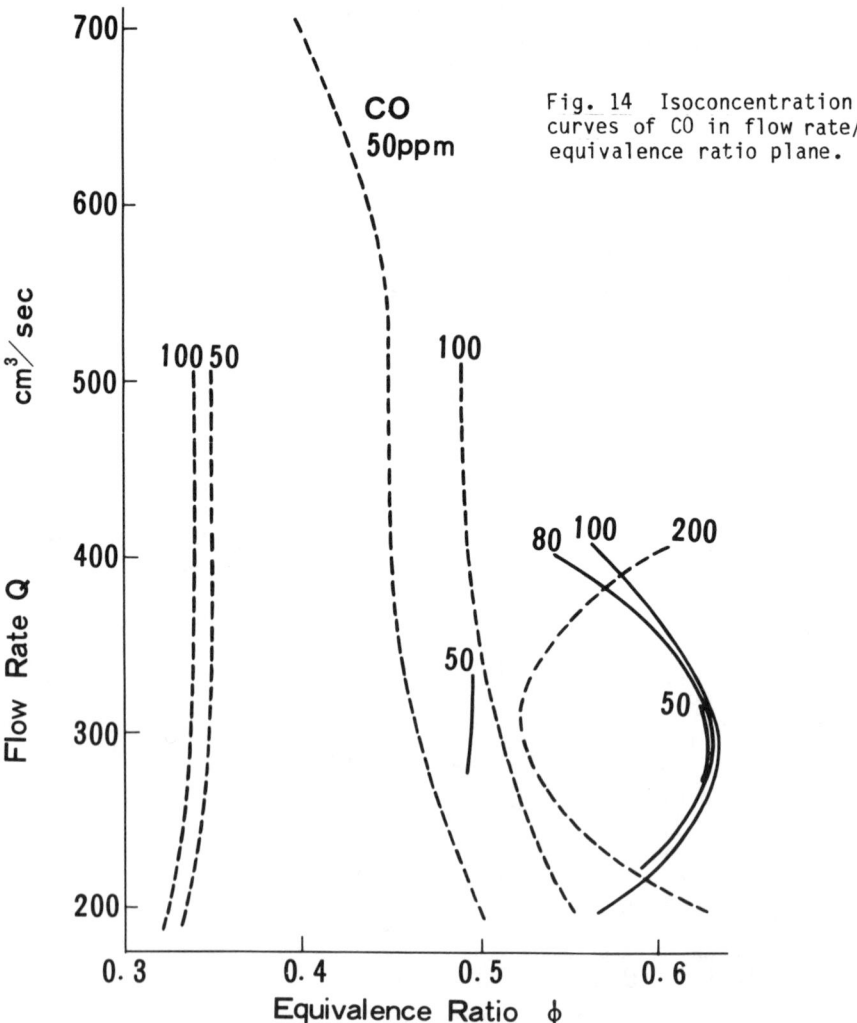

Fig. 14 Isoconcentration curves of CO in flow rate/equivalence ratio plane.

FLAME STABILIZED IN CERAMIC TUBES

the flow rate was increased for a constant value of ϕ, the main reaction zone inside the tubes was forced to move downstream until at the critical flow rate it emerged from the tubes, resulting in blowoff. This behavior is essentially in agreement with that predicted by the one-dimensional theory (Takeno, Sato, and Hase 1981). On the other hand, when ϕ was decreased for the same flow rate, the reaction temperature should decrease with consequent movement of the reaction zone downstream and blowoff at the critical value. Thus the blowoff mechanism of the first flame is understood. The problem is that the observed critical values are worse than those of flames stabilized inside the metal block. The wall temperature distributions reveal that the solid temperature at the tube inlet is much lower than that of the metal block. Then the temperature difference between the solid and the incoming unburned gas is too small to produce sufficient preheat. In other words, the long ceramic tube was insufficient to compensate for the reduced heat conduction. To make matters worse, the wall temperature

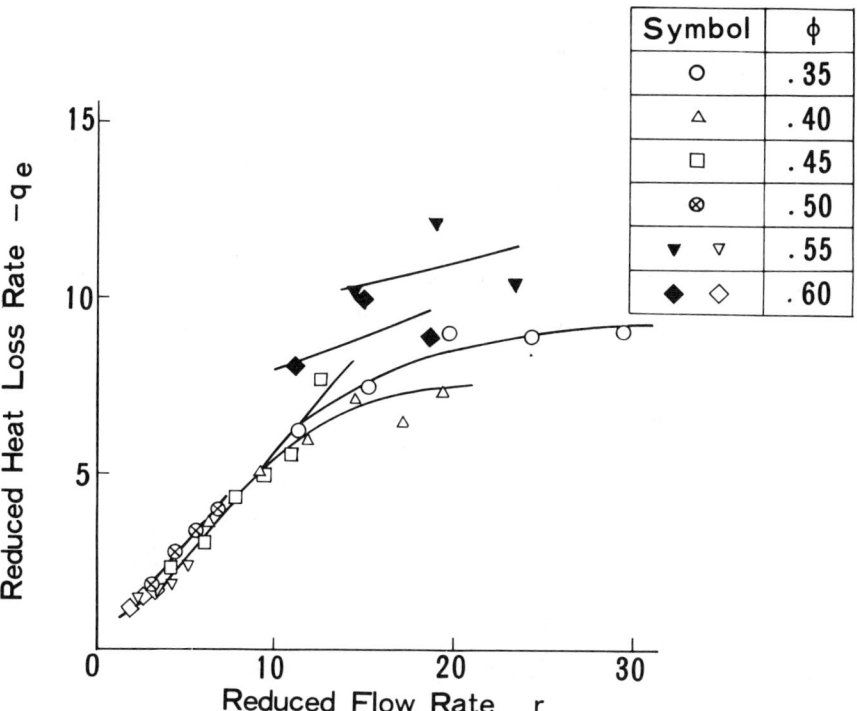

Fig. 15 Reduced heat loss rate plotted against reduced flow rate.

decreases in the downstream direction around the tube exit, presumably owing to the radiative heat loss from the heat tubes, which extracts additional heat from the reaction zone and reduces the flame stability. It should be noted that in the present burner, the perforated plate to suppress the loss was not placed downstream.

When the first flame blows off to come out of the combustion tubes, the linear velocity of the unburned gas is decreased by less than 1/6 by the increase in flow area. The second flame can then be stablized downstream of the tube exit. This is a kind of flat flame which differs from the usual one in one important aspect: There is heat exchange between the solid and the gas in the porous solid

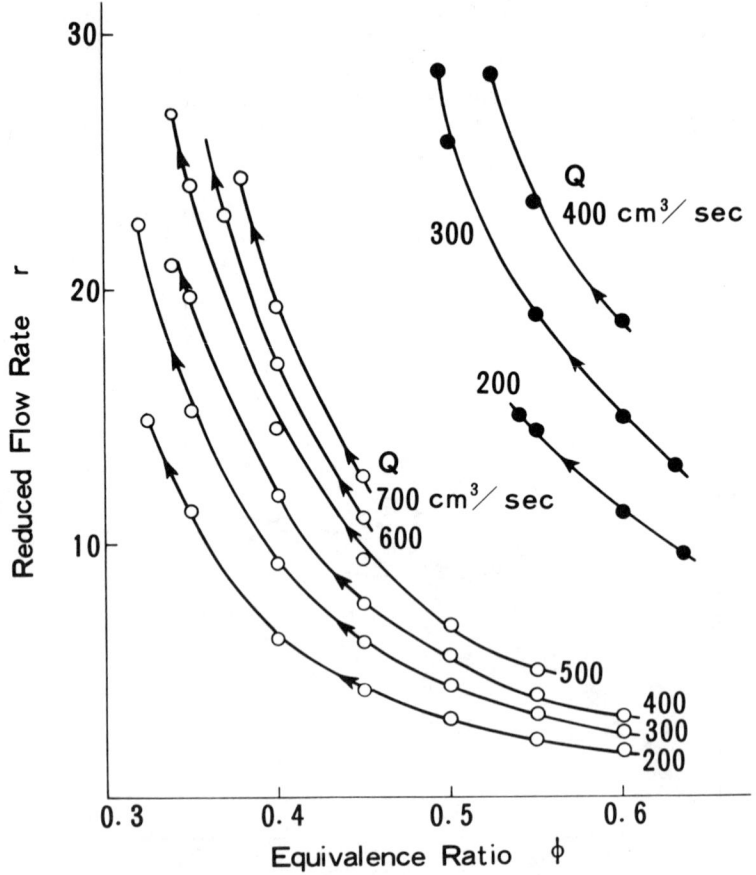

Fig. 16 Reduced flow rate plotted against equivalence ratio for constant flow rate.

upstream, and the incoming unburned gas is preheated appreciably before entering the reaction zone. This is the reason why the flame can sustain the fast velocity, up to around 25 times the burning velocity, as is seen in Fig. 16. A part of the heat released in the reaction zone is transferred to the upstream tubes and used to preheat the unburned gas. This is exactly the situation for the lower branch solution stabilized downstream of the solid, which was predicted in the one-dimensional theory. In the actual event observed in this experiment, the radiative heat transfer should play an important role in this heat feedback mechanism, although the effect was not taken into consideration in the theory.

The stability limit for the second flame may be explained in the following way. As ϕ is decreased for a constant flow rate, the reaction temperature is decreased to cause the flame to move downstream, which brings about a decrease in the feedback of heat until a critical value is reached at which the feedback mechanism cannot sustain the flame anymore. Conversely, when the flow rate is increased for a constant value of ϕ, the flame is forced to move downstream, as was predicted in the one-dimensional theory. In this process, the reaction temperature should increase slightly because of the relative decrease of the heat loss. However, the latter increase cannot compensate for the decreased feedback of heat over the critical value, resulting in the blowoff. If ϕ is increased or the flow rate is decreased, on the other hand, the flame moves upstream, approaching the tube exit. Then the feedback mechanism is increased to shift the flame farther upstream until flashback into the combustion tubes occurs at the critical value.

The present study has revealed the excellent emission characteristics of flames stablized in the tubes. As was pointed out in the first theoretical study (Takeno and Sato 1979), the peak produced in the gas temperature is so sharp that the gas becomes cooled very rapidly as it flows downstream. The short residence time in the higher-temperature region will bring about the reduced emission of NO_x. The reduction in the CO emission, on the other hand, is difficult to explain. The higher reaction temperature and the consequent accelerated reaction rate may explain this. However, more detailed experimental information will have to be obtained before the matter can be resolved.

The flame behavior described above seems to agree essentially with that predicted in the simplified one-dimensional theory, in spite of the fact that the solid temperature was assumed constant in the theory. A more sophisticated theory, in which the latter assumption is relaxed and the effects of the change in flow area at the solid exit, as well as of the radiative heat transfer, are taken into consideration, will bring a more complete agreement. In any case, the work illustrates the excellent combustion characteristics of the proposed flame system and the wide range of possibilities of the concept of excess enthalpy burning.

Acknowledgment

The authors would like to express their sincere thanks to Professor F. J. Weinberg, who read the manuscript and gave valuable advice.

References

Buckmaster, J. and Takeno, T. (1981) Blow-off and flashback of an excess enthalpy flame. Combust. Sci. Technol. 25, 153-158.

Hardesty, D. R. and Weinberg, F. J. (1974) Burners producing large excess enthalpies. Combust. Sci. Technol. 8, 201-214.

Sato, K., Hase, K., and Takeno, T. (1981) An experimental study on an excess enthalpy flame (unpublished).

Takeno, T. and Hase, K. (1981) Effects of solid length and heat loss on an excess enthalpy flame (unpublished).

Takeno, T. and Sato, K. (1979) An excess enthalpy flame theory. Combust. Sci. Technol. 20, 73-84.

Takeno, T. and Sato, K. (1980) An experimental study on an excess enthalpy flame. Bull. Inst. Space Aeronaut. Sci. Univ. Tokyo 16(1C), 573-586 (in Japanese).

Takeno, T. and Sato, K. (1981a) A further experimental study on the excess enthalpy flame. Bull. Inst. Space Aeronaut. Sci. Univ. Tokyo 17(1B), 273-285 (in Japanese).

Takeno, T. and Sato, K. (1981b) A theoretical and experimental study on an excess enthalpy flame. Combustion in Reactive Systems: AIAA Progress in Astronautics and Aeronautics (edited by Bowen, Manson, Oppenheim, and Solonkhin), Vol. 76, pp. 596-610. AIAA, New York.

Takeno, T., Sato, K., and Hase, K. (1981) A theoretical study on an excess enthalpy flame. 18th Symposium (International) on Combustion, pp. 465-472. The Combustion Institute, Pittsburgh, Pa.

Weinberg, F. J. (1971) Combustion temperature; the future? Nature 233, 239-241.

Andrews, G. E. and Bradley, D. (1973) Determination of burning velocity by double ignition in closed vessel. Combust. Flame 20, 77-89.

Chapter II. Turbulent Flames

Differential Diffusion Effects on Measurements in Turbulent Diffusion Flames by the Mie Scattering Technique

S.H. Stårner* and R.W. Bilger†
The University of Sydney, Sydney, Australia

Abstract

The Mie scattering technique has been developed to be a very useful method of measuring scalar quantities in flames. In turbulent diffusion flames, small particles are introduced with the fuel, and records of the light scattered from these particles allow, with some assumptions, the computation of time series for the fuel mixture fraction, density, and other scalars such as the fuel and oxidant mass fraction. The technique may be combined with laser-Doppler velocimetry to yield measurements of scalar fluxes and other scalar-velocity correlations. A crucial assumption of the technique is that the marker particles mix and diffuse in the same way as the fuel atoms. At flame temperatures, molecular diffusivities are as much as two orders of magnitude higher than at room temperature; and in flames of moderate Reynolds number, molecular diffusion makes a significant contribution to the total diffusion of the fuel atoms. The effective molecular diffusivity of small particles is very small and significant differential diffusion between fuel atoms and marker particles can result. A theory is presented to quantify this differential diffusion and the results used to calculate likely errors in measurements of scalar quantities and their fluxes in a hydrogen jet diffusion flame of nozzle Reynolds number 10,850. Errors in mixture fraction range to 10% on the mean and 25% on the variance and on radial flux. Errors in density are about one-half of these.

Presented at the 8th ICOGER, Minsk, USSR, Aug. 23-26, 1981. Copyright © American Institute of Aeronautics and Astronautics, Inc., 1983. All rights reserved.
  *Research Fellow, Department of Mechanical Engineering.
  †Professor, Department of Mechanical Engineering.

Introduction

The use of small particles to tag or mark one stream of fluid in turbulent mixing studies dates from the work of Rosensweig et al. (1961) and Becker et al. (1967). Light scattered from the particles (Mie scattering) yields measurements of concentration with fine spatial and temporal resolution. Becker (1977) gives a comprehensive review of the technique for which he uses the term "marker nephelometry." Application to flames has been relatively recent: Ebrahimi and Kleine (1977), Kennedy and Kent (1979), Stårner and Bilger (1981), and Stårner (1980) have used it to measure mixing and scalar quantities generally in turbulent diffusion flames, while Moss (1980) has used it to measure the degree of reaction in a turbulent premixed flame.

An understanding of the structure of turbulent flames and the development of mathematical models of turbulent combustion has been severely limited by the difficulty of making measurements in these demanding and hostile flows. Reference may be made to Libby and Williams (1980) for an introduction to current theory, while Goulard et al. (1976) give a recent review of measurement methods and goals. The laser-Doppler velocimeter (Self and Whitelaw 1976) has greatly improved our capability to measure velocity in turbulent flames, and much useful information has been gathered in recent years by this technique. However, it is scalar quantities such as reactant and product concentration, temperature, and fuel mixture fraction (mass fraction of fuel in both burnt and unburnt form) which can throw the most light on the combustion process; and, in particular, correlations of such scalars with the velocity yield direct measurements of the scalar fluxes associated with turbulent transport in the flow, a phenomenon of considerable interest and conjecture. Molecular scattering processes such as Rayleigh scattering (Dibble and Hollenbach 1981), spontaneous Raman scattering (Drake et al. 1981), and CARS (coherent anti-Stokes Raman spectroscopy) (Eckbreth 1981) show great promise of providing much of this needed scalar information. These latter two techniques are very expensive and have low data rates and are consequently not likely to cater fully for the broad scope of data collection required in this field. The Rayleigh scattering technique has a limited utility due to the usually complex relationship of Rayleigh scattering cross section to the variables of primary interest in the theory, particularly for nonpremixed combustion. The Mie scattering technique is relatively inexpensive, has a high signal strength, is conceptually quite simple, and could be readily utilized in a wide range of research laboratories.

A crucial assumption of the Mie scattering technique is that the tracer faithfully follows the mixing and diffusion of the atoms of the stream (usually the fuel) into which it is seeded. This is a different matter to the particles faithfully reproducing the velocity fluctuations of the flow; this is ensured by their small size: from 0.1 to 1 µm. At flame temperatures, molecular diffusivities are as much as two orders of magnitude higher than at room temperature; and in flames of moderate Reynolds number, molecular diffusion makes a significant contribution to the total transport of the fuel atoms. On the other hand, the effective "molecular" diffusivity of small particles, given by the Einstein equation (Green and Lane 1957) is very small. It has hitherto been argued that the effects of differences in molecular diffusivity will be confined to the high wave number, "dissipation" end of the scalar fluctuation spectrum and should have little influence on moments and correlations. While this may be so at high Reynolds number, significant effects on the mean, variance, and correlations should occur at low to moderate Reynolds number. This can be inferred from the fact that in a laminar flame the tracer particles will remain undiffused in a central core while the fuel atoms have diffused a long way into the surrounding fluid. Large "differential diffusion" in the mean has occurred, and at a somewhat higher Reynolds number where there is an onset of instability, large differences in the variance and correlations for tracer and fuel atoms will result.

In the theory developed here, the contribution of the molecular transport to the total transport is included and this gives rise to a difference in the mean concentration of tracer particles and fuel atoms. Equations for the variance of this difference and the correlation of the difference with the fuel mixture fraction are also written. The formulation follows that of Bilger (1980). Bilger and Dibble (1980) have developed a very similar analysis to predict the differential diffusion of propane and hydrogen in a jet of these gases mixing with air without reaction. These predictions are to be compared with measurements using Rayleigh scattering to detect the differential diffusion. Bilger (1981) has developed the theory for the differential diffusion of elemental species and enthalpy in a hydrogen-air diffusion flame and the results show very good agreement with the measurements of Drake et al. (1981).

The theory is applied to the case of a jet diffusion flame of hydrogen in a co-flowing stream of air. Mie scattering measurements in this flame have been reported by Kennedy and Kent (1979) who obtained conventional and density weighted means of the mixture fraction as well as higher moments and probability density functions. Stårner and Bilger (1981) report similar measurements in this flame with and without axial pressure gradients and in addition report correlations with the axial component of velocity. Stårner (1980) also reports correlations with the radial component of velocity. The theory presented here is used to predict errors in the measurement of mean mixture fraction and its variance, of mean density and its variance, and of the correlation with the radial velocity component.

Scalar Measurements by the Mie Scattering Method

The Mie scattering method, when applied to concentration measurements in two-stream flows, is attractively simple. In a typical experimental setup for a jet or flame, an unfocused laser beam intersects the jet in a plane perpendicular to its axis. The light collection equipment consists of a low-cost photomultiplier mounted in the same plane, with its axis perpendicular to the laser beam axis. A lens system is used to image a section of the laser beam on a pinhole plate, whereby the measurement volume diameter is defined by the pinhole, and its length by the laser beam diameter. The upper limit to the size of the measurement volume is set by the spatial resolution required, and the lower size limit is determined by marker shot noise arising from the random arrivals of (usually) polydisperse marker particles. A measurement volume size of around 1 mm^3 has been found useful in laboratory flows. By fitting a narrow-band interference filter to the collection optics, ambient light and flame radiation are effectively suppressed. The output signal from the photomultiplier tube is strong enough to be easily amplified for recording on FM tape, or digitizing "on-line."

The horizontal hydrogen jet flame under consideration in this study issues with a bulk velocity of 151 m/s into a coflowing airstream of 15.1 m/s. The jet nozzle diameter is 7.62 mm and the jet Reynolds number is 10,850. The nozzle plane velocity profiles are those given by Kent and Bilger (1973).

For this two-stream flow, we define here the conserved scalar ξ (Bilger 1980) as the mass fraction of hydrogen atoms, i.e., the mass fraction of atoms originating from the fuel stream. This conserved scalar is also known as the mixture fraction, and takes the limit values one at the nozzle and zero in the external stream. The mass concentration of nozzle fluid is $\rho\xi$, where ρ is the instantaneous density. Assuming that there is no differential diffusion, the instantaneous light-scatter signal J can be related to $\rho\xi$ by

$$J/J_j = (\rho\xi)/(\rho\xi)_j \equiv (\rho\xi)/\rho_j \qquad (1)$$

Subscript j denotes the nozzle values. Clearly, J can be calibrated instantaneously by

$$J(t) = (\rho\xi)(t) \times J_j/\rho_j \qquad (2)$$

provided that the nozzle values are known and constant.

In data processing of the recorded light-scatter signal, it is desirable to decouple ρ from ξ. For this purpose, a number of assumptions must be used, relating to the nature of the combustion mechanism. A good approximation to real processes in the hydrogen-air system is given by the "shifting equilibrium" model in which infinitely fast, reversible, adiabatic reactions are assumed to occur wherever both fuel and oxidant are present, so that the system is always in thermodynamic equilibrium. Shifts in concentration are instantly accompanied by chemical reactions which restore equilibrium. With the two additional assumptions of equal diffusivity of species and unity Lewis number (Williams 1965), other relevant scalar quantities (temperature, species mass fractions, and density) become functions only of ξ. Thus, by means of equilibrium data (Gordon and McBride 1971), time-traces and probability density functions of these scalars can be generated in the digital reduction of the light-scatter signal. Figure 1 shows a representative set of traces derived by this procedure for a point at 120 diameters downstream from the nozzle plane.

Some of the limitations and difficulties of the light-scattering method are common to measurements in isothermal flows and in flames. The particles must be small enough to be free of inertial effects, and the effect of agglomeration must be small, or well enough known to enable correction. The number of particles in the measurement volume must not be so great as to cause turbulence modulation by momentum drag, or to cause significant attenuation of the incident and scattered laser light. Attenuation and agglomeration can usually be rendered negligible, so that the output signal varies linearly with the number density of the marker particles in the measurement volume. All of these effects have been studied in detail by Becker et al. (1967) and by Shaughnessy and Morton (1977).

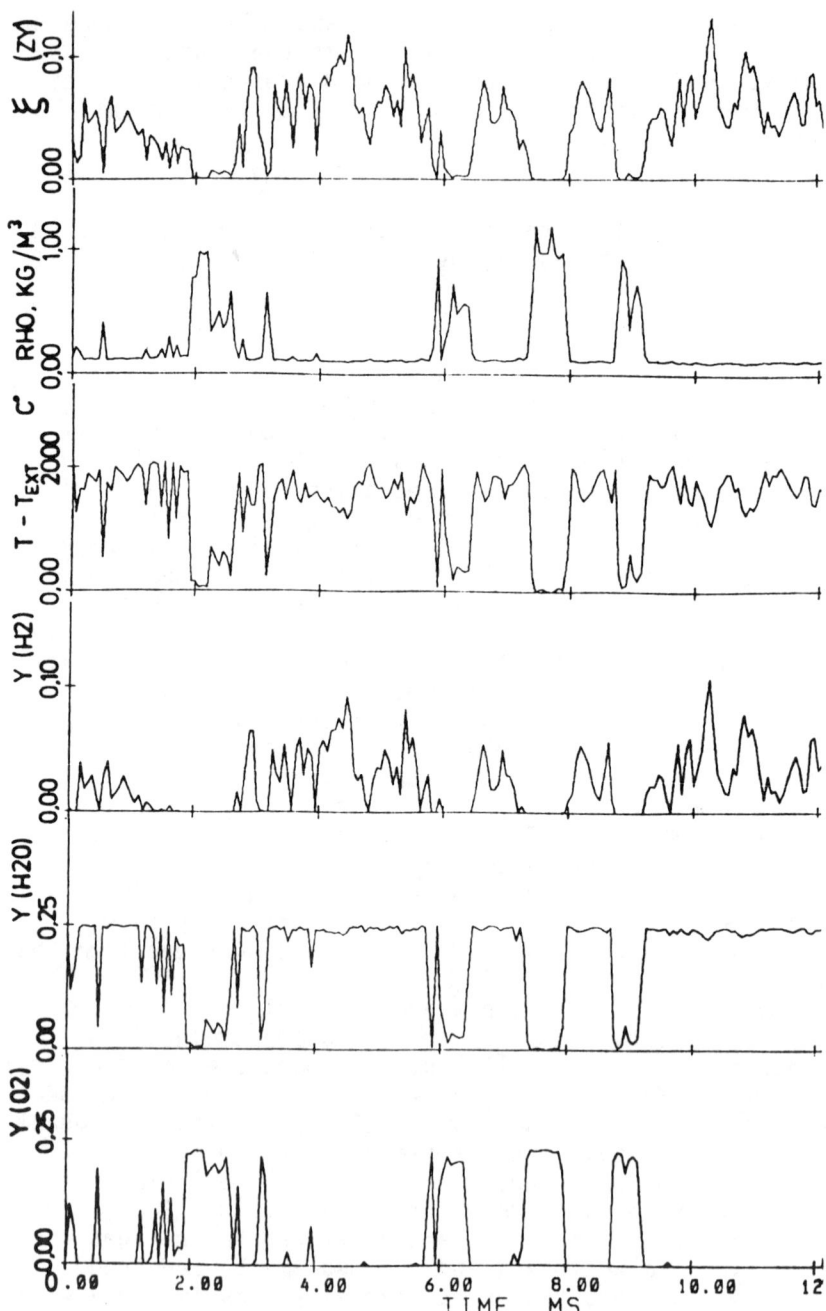

Fig. 1 Typical time histories of density, excess temperature, and mass fractions computed by means of equilibrium data.

There are several difficulties peculiar to the application of Mie scattering measurements to flames. The marker must be refractory and must have a constant scattering cross section throughout the flame. Suitable seeding materials are, for instance, alumina and magnesium oxide powders, which can be dispersed in the fuel stream by means of a cyclone seeder or some similar device. It has, however, been found by Kennedy and Kent (1979) that titanium dioxide seeding, produced by reacting titanium chloride with water vapor in the fuel stream, is not stable in the flame.

As outlined in the Introduction, in turbulent flames the molecular diffusivity of the fuel is increased drastically. In addition, the Reynolds number of turbulence[‡] decreases with increasing temperature, reducing turbulent fluxes so that the effect of molecular diffusion is further enhanced. The marker particles, however, have very low "molecular" diffusivity; the particle Schmidt number is of order 10^6, while the fluid Schmidt number is near unity. Thus, while the particles may follow the flow in a no-slip sense, they do not follow the marked fluid faithfully, because of differential diffusion.

Theory

The theoretical development here follows that of Bilger (1980), in which the differential diffusion of two conserved scalars is considered. The approach differs slightly from that of Bilger and Dibble (1980) and Bilger (1981) in that the reference conserved scalar is chosen as the mean of the two scalars of interest rather than arbitrarily choosing one of them. This leads to some simplification in the dissipation terms in the variance and correlation equations, but does not result in any essential difference in the theory or the modeling. Its choice here is mainly from historical causes. Bilger (1981) gives a full treatment of the differential diffusion of the elements and the enthalpy in the hydrogen/air diffusion flame studies here. It would be possible to repeat those calculations including equations for the tracer material. This has not been done here. Only the differential diffusion between the tracer material and the hydrogen element has been considered. The effects calculated in Bilger (1981) can be considered additive to the ones calculated here when going from the hydrogen

---

[‡]$Re_t = k^2/(\varepsilon \nu)$, where k is the turbulent kinetic energy, ε is the dissipation rate of k, and ν is the kinematic viscosity of the fluid.

element conserved scalar to other conserved scalars such as oxygen element and enthalpy (for adiabatic flow), and to reactive scalars such as density, temperature, and molecular species concentrations.

For two conserved scalars, A and B, with molecular diffusivities D_A and D_B in a two-stream flow, the mixture fractions ξ_A and ξ_B are defined in the usual way. They are assumed to have identical initial boundary conditions. We define the scalars

$$Z \equiv \tfrac{1}{2}(\xi_A + \xi_B) \qquad D \equiv \tfrac{1}{2}(D_A + D_B)$$
$$z \equiv \tfrac{1}{2}(\xi_A - \xi_B) \qquad d \equiv \tfrac{1}{2}(D_A - D_B) \tag{3}$$

Balance equations may be derived for the four Favre§ quantities $\tilde{Z}$, $\tilde{z}$, $\widetilde{z''^2}$, and $\widetilde{Z''z''} = \tfrac{1}{4}(\widetilde{\xi_A''^2} - \widetilde{\xi_B''^2})$. Here we are interested in the special case of differential diffusion of hydrogen element and tracer material. Let the hydrogen element be species A and the tracer species, B. Then, as the hydrogen diffusivity is very much larger than the tracer diffusivity ($D_A \simeq 10^6 D_B$), we neglect D_B.

For axisymmetric, stationary turbulent flow, and with Favre averaging, the equations take the form

$$\bar{\rho}\tilde{u}\frac{\partial \tilde{Z}}{\partial x} + \bar{\rho}\tilde{v}\frac{\partial \tilde{Z}}{\partial r} - \frac{1}{r}\frac{\partial}{\partial r}\left(r\bar{\rho}(D_t + \tilde{D})\frac{\partial \tilde{Z}}{\partial r}\right) = 0 \tag{4}$$

$$\bar{\rho}\tilde{u}\frac{\partial \tilde{z}}{\partial x} + \bar{\rho}\tilde{v}\frac{\partial \tilde{z}}{\partial r} - \frac{1}{r}\frac{\partial}{\partial r}\left[r\bar{\rho}(D_t + \tilde{D})\right]\frac{\partial \tilde{z}}{\partial r} - \frac{1}{r}\frac{\partial}{\partial r}\left(r\bar{\rho}\tilde{d}\frac{\partial \tilde{Z}}{\partial r}\right) = 0 \tag{5}$$

where D_t is the turbulent diffusivity. Gradient transport of the conserved scalars is assumed, and $\tilde{z}$ is assumed small compared with $\tilde{Z}$. Correlations between molecular diffusivities and gradients in $\tilde{Z}$ are neglected.

For the covariance $\widetilde{Z''z''}$, we have

$$\underbrace{\bar{\rho}\tilde{u}\frac{\partial \widetilde{Z''z''}}{\partial x}}_{(1)} + \underbrace{\bar{\rho}\tilde{v}\frac{\partial \widetilde{Z''z''}}{\partial r}}_{(2)} + \underbrace{\bar{\rho}\widetilde{v''z''}\frac{\partial \tilde{z}}{\partial r}}_{(3)} + \underbrace{\bar{\rho}\widetilde{v''z''}\frac{\partial \tilde{Z}}{\partial r}}_{(4)}$$

---

§In Favre averaging, all quantities except pressure are weighted by the instantaneous density before averaging, e.g., $\tilde{u} \equiv \overline{\rho u}/\bar{\rho}$. The resulting equations become considerably simplified, and the main density effects become implicit.

$$+ \frac{1}{r}\frac{\partial}{\partial r}(r\bar{\rho}v''\widetilde{Z''z''}) + \chi_{Zz} = 0 \qquad (6)$$

where

$$\chi_{Zz} \approx \bar{\rho}\,\varepsilon/k\, C_\chi\, \widetilde{Z''z''}$$

and terms 3-5 are gradient modeled. Bilger and Dibble (1980) discuss the problems in modeling the dissipation terms in this equation. For the variance of z, the equation is

$$\bar{\rho}\tilde{u}\frac{\partial \widetilde{z''^2}}{\partial x} + \bar{\rho}\tilde{v}\frac{\partial \widetilde{z''^2}}{\partial r} + 2\bar{\rho}\widetilde{v''z''}\frac{\partial \tilde{z}}{\partial r} + \frac{1}{r}\frac{\partial}{\partial r}\left(r\bar{\rho}\widetilde{v''z''^2}\right) + \chi_{zz} = 0 \qquad (7)$$

where

$$\chi_{zz} \approx \bar{\rho}C_\chi\,\varepsilon/k\,\widetilde{z''^2}$$

The boundary conditions for $\tilde{z}$, $\widetilde{z''^2}$, and $\widetilde{Z''z''}$ are zero values at the nozzle plane and in the external stream with zero derivatives at the axis of symmetry. Equations (4-7) are solved in conjunction with the usual equations for the velocity, turbulence kinetic energy and dissipation, and the hydrogen element mixture fraction $\tilde{\xi}_A$ and its variance $\widetilde{\xi''^2_A}$. The formulation and computer code of Kent and Bilger (1977) has been used. The gradient modeling used in Eqs. (4-7) is the same as that used in the remainder of the model, with a turbulent Prandtl number of 0.7 used in the turbulent flux terms and a somewhat higher value used in the production terms, so that

$$\widetilde{v''Z''}\frac{\partial \tilde{z}}{\partial r} = -C_g\frac{\mu_t}{\bar{\rho}}\frac{\partial \tilde{Z}}{\partial r}\frac{\partial \tilde{z}}{\partial r}$$

$$\widetilde{v''z''}\frac{\partial \tilde{z}}{\partial r} = -C_g\frac{\mu_t}{\bar{\rho}}\left(\frac{\partial \tilde{z}}{\partial r}\right)^2 \qquad (8)$$

with $C_g = 1.35$. In the dissipation terms, $C_\chi = 1.79$.

The effective molecular diffusivity for hydrogen element is that used by Bilger (1981)

$$D_A = \sum_i \mu_{H,i}\, D_i\, \frac{\partial Y_i^0}{\partial \xi_A} \qquad (9)$$

where $\mu_{H,i}$ is the mass fraction of hydrogen in the molecular species i which has a molecular diffusivity D_i in the diluent nitrogen at the temperature T. The temperature T and composition Y_i^0 are those for unity Lewis numbers and fast chemistry for the mixture fraction ξ_A. The molecular diffusivities as functions of temperature are computed from simple formulas given by Mitchell (1979). The diffusivity D_A is tabulated as a function of ξ_A, and the Favre average $\tilde{D}$ is computed by weighting this function by the Favre probability density function of ξ_A in the normal manner (Kent and Bilger 1977).

From Eq. (5) it can be seen that since the boundary conditions on z are zero, the only thing that gives rise to non-zero values of z is the term

$$-\frac{1}{r}\frac{\partial}{\partial r}\left(r\bar{\rho}\tilde{d}\frac{\partial \tilde{Z}}{\partial r}\right)$$

which is variously a source or sink term with an integral of zero across the flow. Estimates of the magnitude $\tilde{z}_0$ of the maximum of $|\tilde{z}|$ can be made by balancing this term against the turbulent diffusion term, yielding

$$\tilde{z}_0 = A_1 \frac{\tilde{d}}{D_t + \tilde{D}} \tilde{Z}_0 \qquad (10)$$

where $\tilde{Z}_0$ is the magnitude of $\tilde{Z}$, taken as its value on the centerline and A_1 is a "constant" of proportionality. Now

$$D_t = C_\mu k^2/(\varepsilon \sigma_t) \qquad (11)$$

where C_μ is the modeling constant for the turbulent diffusivity and σ_t is the turbulent Prandtl number. Thus at moderate Reynolds number we have

$$\tilde{z}_0 = A_2 \frac{\tilde{d}}{\tilde{\nu}} Re_t^{-1} \qquad (12)$$

where $\tilde{\nu}$ is the Favre averaged kinematic viscosity and Re_t is the turbulence Reynolds number $[k^2/(\tilde{\nu}\varepsilon)]$ defined earlier. Bilger and Dibble (1980) also note that $(\widetilde{z''^2})^{\frac{1}{2}}$ and $\widetilde{z''Z''}$ will show a similar Reynolds number dependence.

Results

Predictions of errors arising from differential diffusion of marker and fuel have been computed for three sets of boundary conditions, in which the streamwise pressure gradient, $d\bar{p}/dx$, is varied while the nozzle plane conditions are unaltered. Variation of $d\bar{p}/dx$ is of interest, as it alters turbulence levels, turbulence length scales, and the ratio of advection of production of turbulent kinetic energy. Consequently, errors arising from differential diffusion can be expected to vary with the pressure gradient. The values of $d\bar{p}/dx$ used for the present study, 23, -18, and -102 Pa/m, are those which apply to the experiments by Stårner and Bilger (1980, 1981).

The predicted errors are presented as fractional differences between the true quantities and those which would be derived from Mie scattering measurements. Figures 2 and 3 show centerline values for the mean mixture fraction and its rms of fluctuation, as

$$\Delta_\xi \equiv (\tilde{\xi}_B - \tilde{\xi}_A)/\tilde{\xi}_A = -2\tilde{z}/\tilde{\xi}_A \tag{13}$$

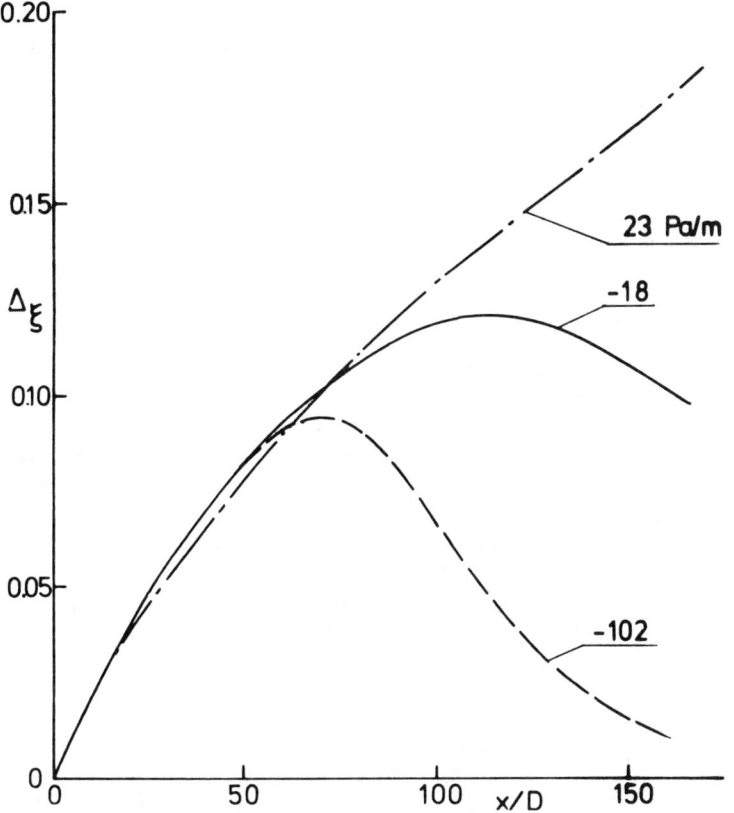

Fig. 2 Centerline variation of error in mean mixture fraction, Δ_ξ; $d\bar{p}/dx$ = 23, -18, and -102 Pa/m, Re_j = 10,850, u_j/u_e = 10.

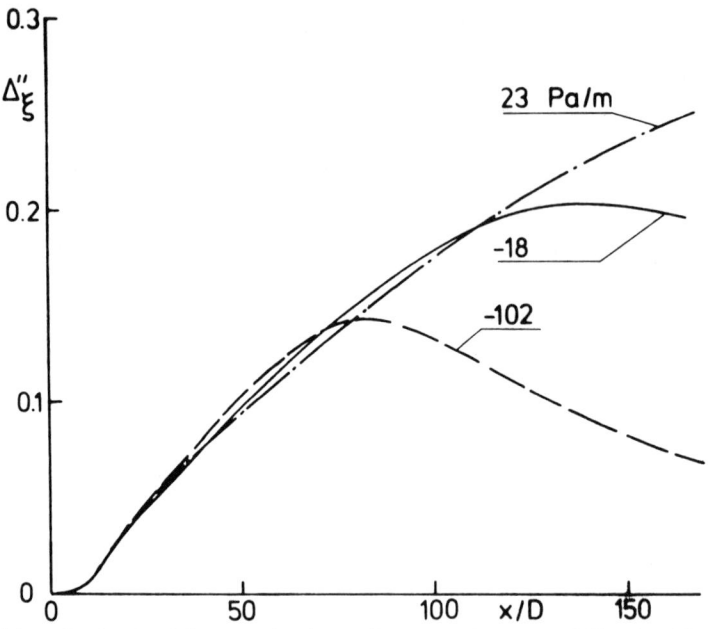

Fig. 3 Centerline variation of error in rms of fluctuation of mixture fraction, Δ_ξ''; conditions as in Fig. 2.

and

$$\Delta_\xi'' \equiv \left[(\widetilde{\xi_B''^2})^{\frac{1}{2}} - (\widetilde{\xi_A''^2})^{\frac{1}{2}}\right]/(\widetilde{\xi_A''^2})^{\frac{1}{2}}$$
$$= \left[(\widetilde{\xi_A''^2} - 4z''\widetilde{Z''})/\widetilde{\xi_A''^2}\right]^{\frac{1}{2}} - 1 \quad (14)$$

The subscripts A and B can be thought of here as denoting true and measured values, respectively. Both figures show the same trend; the errors rise monotonically with the distance from the nozzle to around $x/D = 60$ (D here is the nozzle diameter), from which point onwards the effect of the axial pressure gradient becomes increasingly obvious. For the accelerated flow, the errors peak in midrange, where the temperature is high and the Reynolds number of turbulence, Re_t, has a minimum of around 500. For the case of near-zero pressure gradient, -18 Pa/m, the errors have peaks further downstream, since this is a longer and more slender flame. In the retarded flow, 23 Pa/m, the turbulence levels decrease with increasing x/D and the flow eventually stagnates on the centerline (Stårner and Bilger 1980). Also, Re_t falls monotonically from 1350 at $x/D = 10$ to 200 at x/D

= 160. As a result, the differential diffusion errors continue to rise with x/D over the whole range for the retarded flow.

Figure 4 shows radial profiles for the error in mean mixure fraction, Δ_ξ. The radial profiles are plotted against the normalized radius $\eta = r/L$, where L is the radius at which $\tilde{\xi}$ is half of its centerline value. Here, the errors are normalized by the centerline mean mixture fraction, $\tilde{\xi}_{A_o}$. Maxima occur on the flame axis, and negative errors are found in the outer regions. This is consistent with the requirement that the axial flux of $\tilde{z}$ must integrate to zero across the flow, since both ξ_A and ξ_B are conserved scalars, and their difference ($\equiv 2z$) is zero at the nozzle plane. Physically, Fig. 4 means that the measured mean mixture fraction will be distorted to yield narrowed

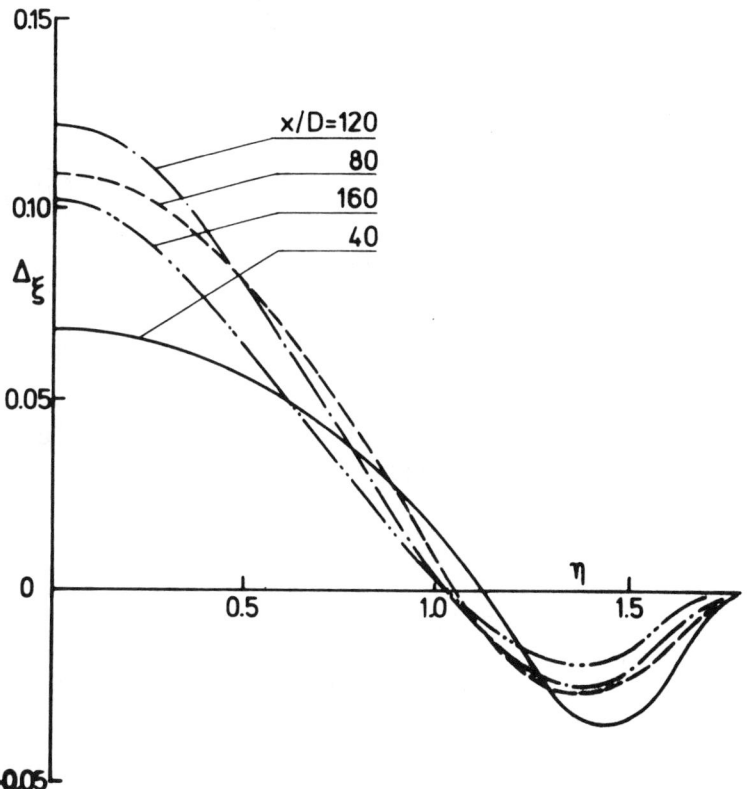

Fig. 4 Radial profiles of error in mean mixture fraction; $d\bar{p}/dx = -18$ Pa/m; other conditions as in Fig. 2.

profiles with too high values on the axis. This is obviously the effect one may expect from low total marker diffusivity.

The error in the rms of fluctuation, Δ_ξ'', in Fig. 5 is normalized by the centerline value $(\tilde{\xi}_A^{''2})_0^{\frac{1}{2}}$. The peaks occur off-axis and values are positive over most of the radial profiles. This is also to be expected; lower small-scale mixing rate must result in larger fluctuations.

Computed values of

$$\tilde{z}^{''2} = \tilde{Z}^{''2} - \tilde{\xi_A'' \xi_B''} \approx \tilde{Z}^{''2}(1-R_{AB}) \approx \tilde{\xi}_A^{''2}(1-R_{AB}) \quad (15)$$

where R_{AB} is the correlation coefficient for A and B, may be used to obtain the phase difference between A and B. R_{AB} is everywhere found to be > 0.99, indicating that the phase difference is very small; this is of significance in assessing the effects of differential diffusivity on velocity-scalar correlations.

The effect on the radial flux $\widetilde{v''\xi''}$ has also been computed, using gradient modeling assumptions:

$$\Delta_{v\xi} \equiv (\widetilde{v''\xi_B''} - \widetilde{v''\xi_A''})/\widetilde{v''\xi_{A(max)}''} = -2\widetilde{v''z''}/\widetilde{v''\xi_{A(max)}''} \quad (16)$$

with

$$\widetilde{v''z''} = -D_t\, \partial \tilde{z}/\partial r$$

and $\widetilde{v''\xi_{A(max)}''}$ taken as the value of $\eta = 0.7$.

The results in Fig. 6 for -18 Pa/m indicate a fractional error of 0.11 in the measured flux at $x/D = 40$, rising to 0.25 at $x/D = 120$. Since the phase difference between A and B was found to be small, i.e., $R_{AB} \approx 1.0$, it is easy to show that, approximately,

$$\Delta_\xi'' \approx \Delta_{v\xi}\, (\widetilde{v''\xi_{A(max)}''}/\widetilde{v''\xi_A''})\, (\tilde{\xi}_A^{''2}/\tilde{\xi}_{Ao}^{''2})^{\frac{1}{2}} \quad (17)$$

so that the behavior of $\Delta_{v\xi}$ should be similar to Δ_ξ'' in the middle of the shear layer; this is also seen to be the case.

It is of interest in this context to assess the effects on density, since this quantity may also be derived from Mie scattering experiments, as outlined in the scalar measurements section. This evaluation has been done here, making use of the experimental data by Stårner and Bilger (1981), by the following method: For a particular point in

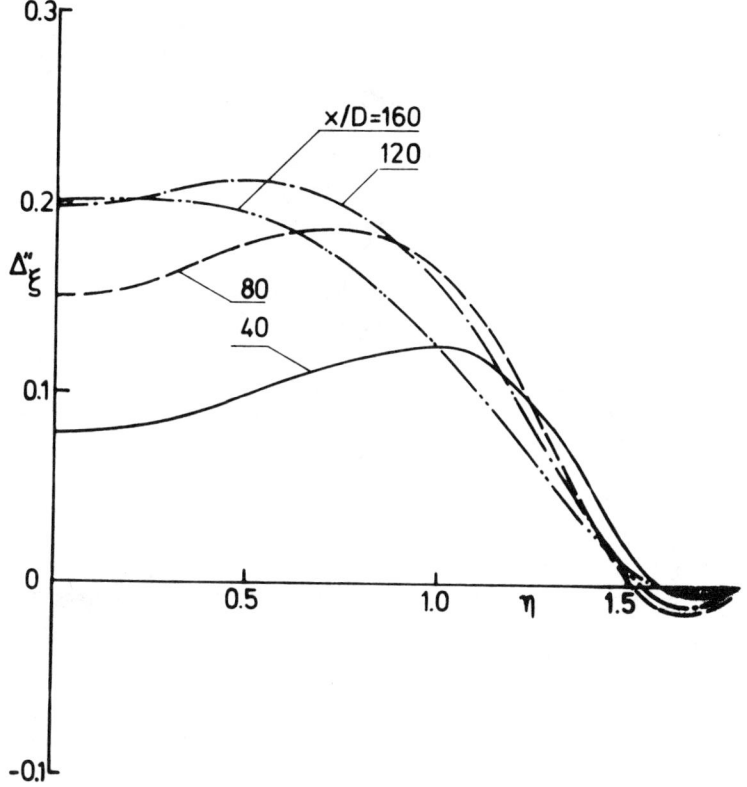

Fig. 5 Radial profiles of error in rms of fluctuation of mixture fraction; conditions as in Fig. 4.

the flame, values of $\tilde{\xi}$ and $\widetilde{\xi''^2}$ are corrected, using the results shown in Figs. 4 and 5. New values of $\bar{\rho}$ and $\overline{(\rho'^2)}^{\frac{1}{2}}$ are computed by using a measured probability density function from a suitable nearby point in the flame which has the same intensity $(\widetilde{\xi''^2})^{\frac{1}{2}}/\tilde{\xi}$, scaled to give the corrected mean $\tilde{\xi}$. The results are shown in Fig. 7 for -18 Pa/m, and x/D = 120. Here the definitions are

$$\Delta_\rho = (\bar{\rho}_m - \bar{\rho}_c)/\bar{\rho}_c \qquad \Delta_\rho' = \left[(\overline{\rho_m'^2})^{\frac{1}{2}} - (\overline{\rho_c'^2})^{\frac{1}{2}}\right] / (\overline{\rho_c'^2})^{\frac{1}{2}}$$

where m and c denote measured an corrected values. It is seen that both the mean density and its rms of fluctuation are increased by the effect of low marker diffusivity. The indicated errors are approximately one-half of the magnitude of the corresponding quantities obtained for the mixture fraction.

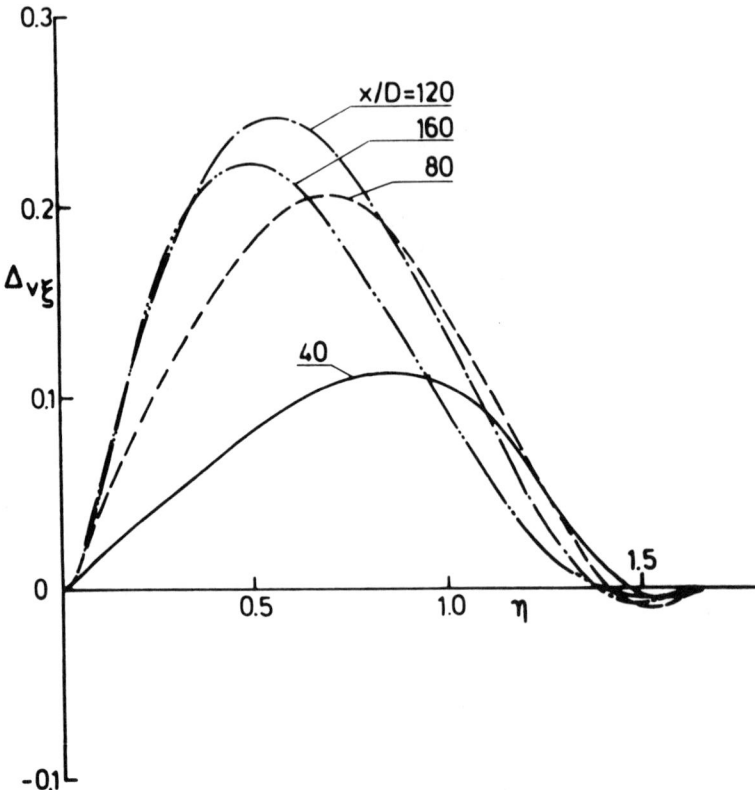

Fig. 6 Radial profiles of error in turbulent radial flux, $\Delta_{v\xi}$; conditions as in Fig. 4.

The effects on density found here are likely to be specific for the hydrogen-air system, for which the correlation of density and the light-scatter signal is highly nonlinear around stoichiometric composition (Kennedy and Kent 1979).

Discussion

In Eq. (10), for estimating differential diffusion errors in the mean mixture fraction, it is of interest to determine the "constant" of proportionality A_1. If A_1 can be shown not to vary greatly from flow to flow, or if its variation is predictable, then Eq. (10) may be a useful tool for obtaining order-of-magnitude assessments of differential diffusion errors in other Mie scattering experiments, past and planned. The present results indicated that A_1 is nearly constant, with a weak jet Reynolds number dependence. At x/D = 60, A is 1.03, 0.84, and 0.80 for Re_j = 10,850, 21,700, and 43,400, respectively (with $d\bar{p}/dx \doteq$ -18 Pa/m). This compares well with the corresponding value 0.92 computed by Bilger (1981) for the differential diffusion of oxygen element and nitrogen in a "lazy" (Re_j = 1500) hydrogen diffusion flame studied by Drake et al. (1981). However, an earlier calculation of A_1 by Bilger and Dibble (1980) for an isothermal hydrogen-propane jet in a coflowing airstream yields values around 0.5; whether this low value is an anomaly or a characteristic of isothermal flows is not clear at present.

For the correlation $(Z''\tilde{z}'')_o$, Bilger and Dibble (1980) suggest a formula for error estimation analogous to Eq. (10), which in the present context takes the form

$$(Z''\tilde{z}'')_o = A_3 \frac{\tilde{d}}{D_t + \tilde{D}} \tilde{Z}_o^2 \qquad (18)$$

They find values of A_3 around 0.080 for the flow of Drake et al. (1981). The present results yield good agreement; A_3

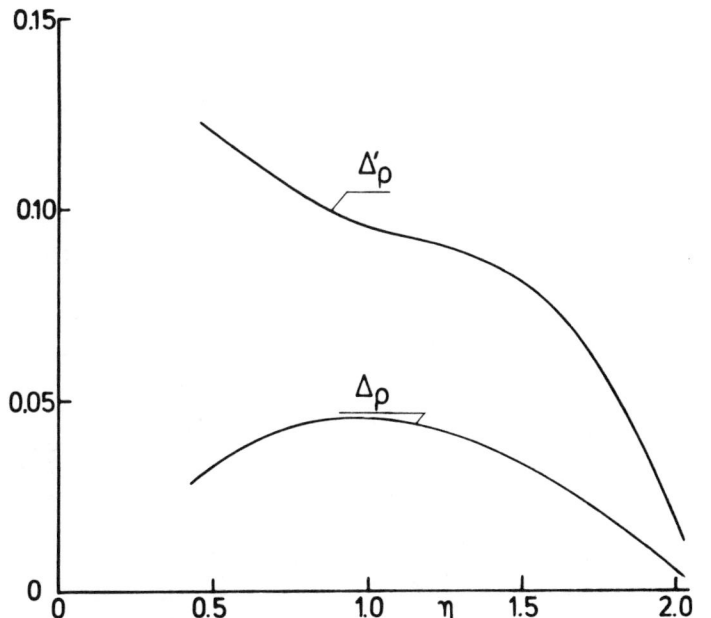

Fig. 7 Radial profiles of errors in mean density, Δ_ρ, and its rms of fluctuation, Δ'_ρ, at x/D = 120; conditions as in Fig. 4.

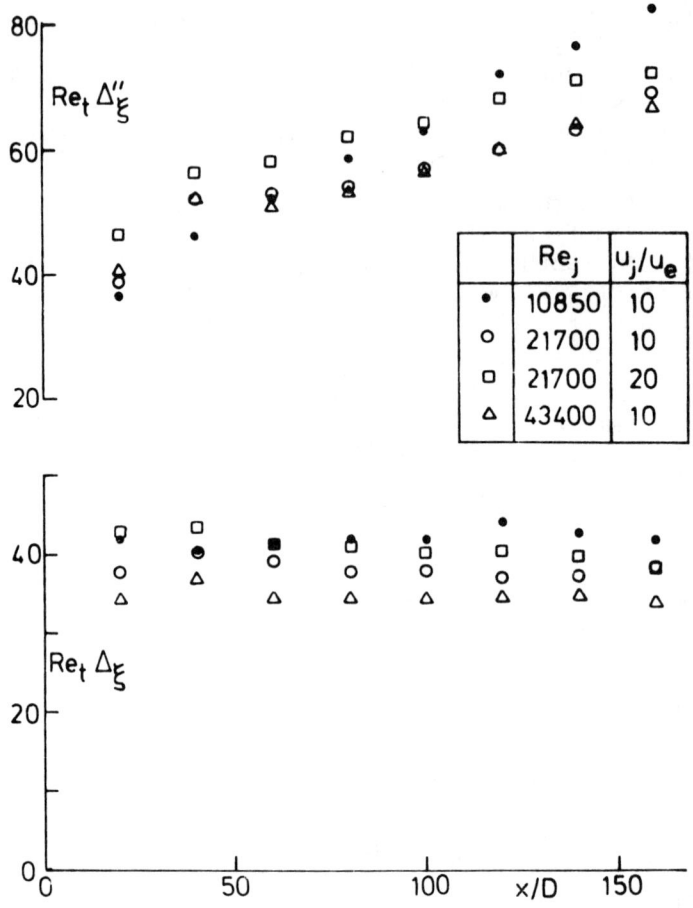

Fig. 8 Centerline variation of the products $Re_t \Delta_\xi$ and $Re_t \Delta_\xi''$; $d\bar{p}/dx = -18$ Pa/m.

here is 0.088, 0.074, and 0.074 for Re_j = 10,850, 21,700, and 43,400, respectively, at x/D = 60.

As indicated in the theory section, Eq. (10) implies that the error in mean mixture fraction should vary as the inverse of the Reynolds number of turbulence Re_t. That this is so is clearly shown in Fig. 8, in which centerline values of the product $Re_t \Delta_\xi$ are seen to be close to constant for x/D > 20 for a wide range of jet Reynolds numbers. The constancy of $Re_t \Delta_\xi$ also implies [see Eq. (12)] that the molecular Schmidt number for the hydrogen element does not vary much over the range investigated. The initial velocity ratio $\bar{u}_j/\bar{u}_e$ does not appear to be a significant parameter,

DIFFERENTIAL DIFFUSION IN TURBULENT DIFFUSION FLAMES 99

so the results are likely to have at least first-order validity also for buoyant flames into still air.

The product of Re_t with the error in rms of fluctuation of mixture fraction, Δ_ξ'', (Fig. 8) also shows good scaling, although there is a rising trend which is close to linear with distance from the nozzle.

For completeness, centerline profiles of the products of Δ_ξ and Δ_ξ'' with the jet Reynolds number Re_j are shown in Fig. 9. Although these plots scale reasonably, it should be stressed that this scaling pertains only to the hydrogen flame, and other fuels would certainly give much lower values of the products $Re_j \Delta_\xi$ and $Re_j \Delta_\xi''$, as would also be the case for isothermal jets. A similar caution also applies to the plots of Fig. 8, though to a lesser extent, since the local value of Re_t is a much closer measure of diffusivity ratios than Re_j, particularly in flames.

However, the variation in molecular Schmidt number with different fuels is not accounted for in the plots of Fig. 8, which must therefore be seen as specific to the hydrogen flame. The only general expressions for differential diffusion errors are Eqs. (10) and (18).

The foregoing discussion of error estimates enables a brief review to be made of previously published results of Mie scattering experiments. Becker et al. (1967) and Shaughnessy and Morton (1977) have dealt with high Reynolds number (>30,000) isothermal air jets. In these flows, the ratio of molecular to turbulent diffusivities is of order $<10^{-3}$, so differential diffusion errors would be quite negligible. Ebrahimi and Kleine (1977) report on natural gas flames of Re_j from 16,900 to 37,500. For these flames we have used data by Mitchell (1980) to estimate the effective fuel element diffusivity, obtaining a value of 2.6 cm^2/s at $x/D = 60$, on the centerline (Favre averaging has been neglected). To estimate D_t, use is made of equilibrium data for the CH_4-air system, yielding a density of 0.23 kg/m^3 and a temperature of 1250 K. Kinematic viscosity is calculated using standard data for the major species and a temperature dependence $\mu \propto T^{0.6756}$ (Mitchell 1979). Setting the dissipation length scale $\overline{L_\varepsilon \approx 2L_c}$, where L_c is the concentration half-radius, $(u'^2)^{\frac{1}{2}}/\bar{u}_0 \approx 0.16$ and $\bar{u}_0/\bar{u}_j = 0.3$, the turbulence Reynolds number becomes $Re_t = 473$ for the

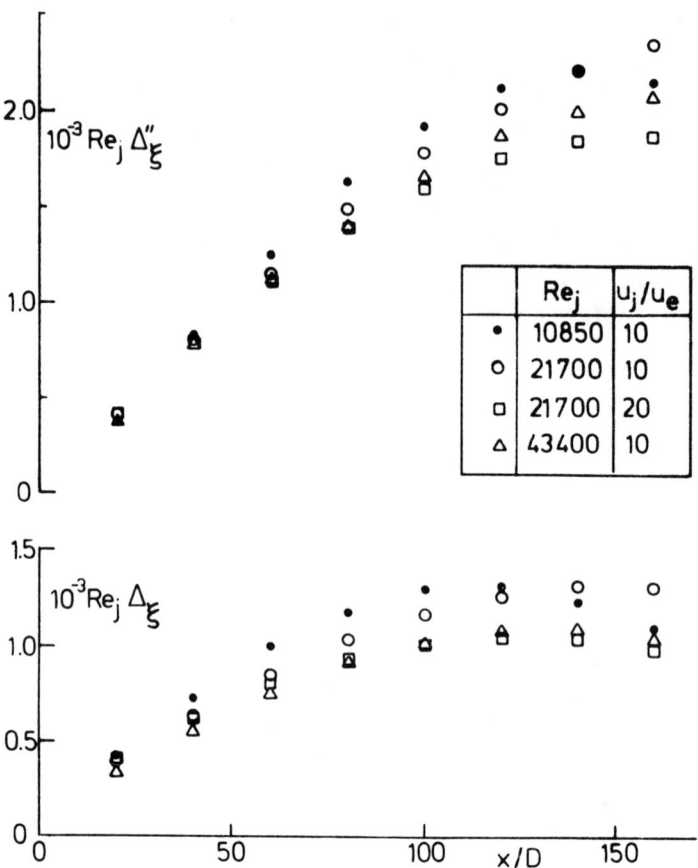

Fig. 9 Centerline variation of the products $Re_j \Delta_\xi$ and $Re_j \Delta_\xi''$; $d\bar{p}/dx = -18$ Pa/m.

lowest jet Reynolds number tested, 16,900. Using

$$D_t = (C_\mu/\sigma_t)\nu\, Re_t^{-1}$$

with $C_\mu = 0.07$, and the turbulent Prandtl number $\sigma_t = 0.7$, we get $D_t = 92$ cm^2/s, and the ratio of molecular to turbulent diffusivity becomes 0.028. Equation (10) indicates that the error in mean concentration is close to this value for A_1 near unity. Using Eqs. (14) and (18) to assess the error in rms of fluctuation of concentration, we obtain, with A_3 set to 0.08, a value $\Delta_\xi'' = 0.030$ for the same location (x/D = 60, on the axis). While these errors are quite small, somewhat larger values may be expected further downstream in this flame.

The experiments by Haberda et al. (1978) and Günther and Wittmer (1981) are similar to those of Ebrahimi and Kleine (1977) discussed above, but with a lowest jet Reynolds nuber of 24,000, so errors can be expected to be only of order 2%, or less.

The buoyant free hydrogen diffusion flame studied by Kennedy and Kent (1979) is an extreme case, with Re_j as low as 4100 and 6600. The present results should yield reasonably accurate predictions for this kindred flow. Errors in ξ of 19 and 12% are indicated at $x/D = 60$, and for $(\widetilde{\xi"^2})^{\frac{1}{2}}$, the predicted errors are 27 and 17%, respectively.

The later experiments by Kennedy and Kent (1980) were conducted on the same flame as discussed in the present work ($Re_j = 10,850$), for which detailed error predictions have been presented earlier in this paper.

Conclusions

A method has been developed for predicting errors arising from differential diffusion effects of marker particles and fluid in Mie scattering experiments in two-stream flows, where only one stream is seeded. This method is an adaptation of a general theory of differential diffusion of species for mixing and reacting flows. The predictions show that in isothermal flows the effects are negligible for jet Reynolds numbers above 10^4, while in diffusion flames the increased molecular diffusivity results in errors of several percent at $Re_j = 10^4$. In particular, hydrogen diffusion flames are susceptible to large errors at $Re_j < 10^4$, with errors for mean mixture fraction rising to above 10%, and for the rms of fluctuation to above 20%. It appears that the formulation of the results in this paper is sufficiently general to enable correction of previously published experiments, as well as forecasting the likely errors in planed studies.

Acknowledgment

This work is supported by a grant from the Australian Research Grants Committee.

References

Becker, H. A. (1977) Mixing, concentration fluctuations, and marker nephelometry. Studies in Convection (edited by B. E. Launder, Vol. 2, pp. 45-139. Academic Press, New York.

Becker, H. A., Hottel, H. C., and Williams, G. C. (1967) The nozzle-fluid concentration field of the round turbulent, free jet. J. Fluid Mech. 30, 285-303.

Bilger, R. W. (1980) On diffusion and reaction in turbulent shear flows. Paper WSS80-1 presented at the Spring Meeting, Western States Section, The Combustion Insitute, University of California, Irvine, Calif.

Bilger, R. W. (1981) Molecular transport effects in turbulent diffusion flames at moderate Reynolds number. AIAA paper 81-0104 presented at the AIAA 19th Aerospace Sciences Meeting, St. Louis, Mo.

Bilger, R. W. and Dibble, R. W. (1980) Differential molecular diffusion effects in turbulent mixing and combustion: I - nonreacting flow. SAND 80-8809, Sandia National Laboratories, Livermore, Calif.

Dibble, R. W. and Hollenbach, R. E. (1981) Laser Rayleigh thermometry in turbulent flames. 18th Symposium (International) on Combustion, pp. 1489-1499. The Combustion Institute, Pittsburgh, Pa.

Drake, M., Lapp, M., Penney, C. M., Warshaw, S., and Gerhold, B. (1981) Measurements of temperature and concentration fluctuations in turbulent diffusion flames using pulsed Raman spectroscopy. 18th Symposium (International) on Combustion, pp. 1521-1531. The Combustion Institute, Pittsburgh, Pa.

Ebrahimi, I. and Kleine, R. (1977) The nozzle fluid concentration fluctuation field in round turbulent free jets and jet diffusion flames. 16th Symposium (International) on Combustion, pp. 1711-1723. The Combustion Institute, Pittsburgh, Pa.

Eckbreth, A. C. (1981) Recent advances in laser diagnostics for temperature and species concentration in combustion. 18th Symposium (International) on Combustion, pp. 1471-1481. The Combustion Institute, Pittsburgh, Pa.

Gordon, S. and McBride, B. J. (1971) Computer program for calculation of complex chemical equilibrium compositions, rocket performance, incident and reflected shocks, and Chapman-Jouguet detonations. NASA SP-273.

Goulard, R., Mellor, A. M., and Bilger, R. W. (1976) Combustion measurements in air breathing propulsion engines. Survey and research needs. Combust. Sci. Technol. 14, 195-220.

Green, H. L. and Lane, W. R. (1957) Particulate Clouds: Dusts, Smokes and Mists, p. 68, Spon, London.

Günther, R. and Wittmer, V. (1981) The turbulent reaction field in a concentric diffusion flame. 18th Symposium (International) on Combustion, pp. 961-967. The Combustion Institute, Pittsburgh, Pa.

Haberda, F., Günther, R., and Ebrahimi, I. (1978) Probability density functions of nozzle fluid concentration in free diffusion flames. High Temp. High Pressures 10, 571-580.

Kennedy, I. M. and Kent, J. H. (1979) Measurements of a conserved scalar in turbulent jet diffusion flames. 17th Symposium (International) on Combustion, pp. 279-287. The Combustion Institute, Pittsburgh, Pa.

Kennedy, I. M. and Kent, J. H. (1980) Laser scattering measurements in turbulent diffusion flames. AIAA paper 80-0206 presented at the AIAA 18th Aerospace Sciences Meeting, Pasadena, Calif.

Kent, J. H. and Bilger, R. W. (1973) Turbulent diffusion flames. 14th Symposium (International) on Combustion, pp. 615-625. The Combustion Institute, Pittsburgh, Pa.

Kent, J. H. and Bilger, R. W. (1977) The prediction of turbulent diffusion flame fields and nitric oxide formation. 16th Symposium (International) on Combustion, pp. 1643-1656. The Combustion Institute, Pittsburgh, Pa.

Libby, P. A. and Williams, F. A. (Eds.) (1980) Fundamental Aspects. Turbulent Reacting Flows, pp. 1-43. Springer-Verlag, Heidelberg.

Mitchell, R. E. (1979) A theoretical model of chemically reacting recirculating flows. SAND 79-8236, Sandia National Laboratories, Livermore, Calif.

Mitchell, R. E. (1980) Chemical element diffusion factors for use in the conserved scalar approach to diffusion flame modeling. SAND 80-8751, Sandia National Laboratories, Livermore, Calif.

Moss, J. B. (1980) Simultaneous measurements of concentration and velocity in an open premixed turbulent flame. Combust. Sci. Technol. 22, 119-129.

Rosensweig, R. W., Hottle, H. C., and Williams, G. C. (1961) Smoke-scattered light measurement of turbulent concentration fluctuations. Chem. Eng. Sci. 15, 111-129.

Self, S. A. and Whitelaw, J. H. (1976) Laser anemometry for combustion research. Combust. Sci. Technol. 13, 171-198.

Shaughnessy, E. J. and Morton, J. B. (1977) Laser light-scattering measurements of particle concentration in a turbulent jet. J. Fluid Mech. 80, 129-148.

Stårner, S. H. (1980) Investigations in turbulent diffusion flames. Ph.D. Thesis. The University of Sydney, Sydney, Australia.

Stårner, S. H. and Bilger, R. W. (1980) LDA-measurements in a turbulent diffusion flame with axial pressure gradients. Combust. Sci. Technol. 21, 259-276.

Stårner, S. H. and Bilger, R. W. (1981) Measurement of scalar-velocity correlations in a turbulent diffusion flame. 18th Symposium (International) on Combustion, pp. 921-930. The Combustion Institute, Pittsburgh, Pa.

Williams, F. A. (1965) Combustion Theory. pp. 9-13. Addison-Wesley, Reading, Mass.

Concentration and Velocity Measurements in a Turbulent Reacting Mixing Layer

J.L. Bousgarbies* and J. Nérault†
Laboratoire d'Études Aérodynamiques et Thermiques, Poitiers, France

Abstract

An experimental technique is carried out to study hydrodynamic and concentration fields inside a turbulent mixing layer between two plane reactive flows. Simultaneous measurements of the instantaneous velocity components and concentration are performed by combining the conductivity technique and laser velocimetry. They allow us to determine directly the local values of the turbulent diffusion flux of matter in the studied liquid flow. Results for mean and fluctuation conductivity and velocity profiles are also given.

Introduction

While the velocity field, and particularly the coherent structures, have been studied for some time, it is only recently that mixing and the chemical reaction have been considered. The mixing of passive species has been studied in gas flows (Brown and Roshko 1974) and between two water streams having different salinities (Koop and Browand 1979). A chemically reacting turbulent shear layer has been also investigated; using optical densitometry techniques, Breidental (1979) measured the amount of reaction product as a function of the Reynolds number. With these reacting flows, an injection of a pH indicator in one of the two

---

Presented at the 8th ICOGER, Minsk, USSR, Aug. 23-26, 1981. Copyright © American Institute of Aeronautics and Astronautics, Inc., 1982. All rights reserved.
*Charge de Recherche C.N.R.S.
†Assistant.

liquid streams allows the flow to be seen (Breidenthal 1981; Dimotakis and Brown 1976).

The present paper concerns the experimental study of mass transfer within a plane turbulent mixing layer between two liquid reactive streams. We aimed mainly to determine directly the local values of the turbulent diffusion fluxes of matter in a liquid flow. These values must be known if a comparison is desired with the corresponding values in the appropriate mixing models used for mass balance equation closure.

Few investigations have been carried out in this field. The most important work is that of Alber and Batt (1976) and Batt (1977), who have studied the turbulent mixing of passive and chemically reacting species in a gaseous shear layer. The experiments presented in this paper are part of a similar study (Gatard 1980), also conducted in a shear layer, but in the case of liquid flows. Another characteristic of our experiments is the high turbulence level of the two freestreams which mix together.

Experimental Conditions

The determination of the turbulent diffusion fluxes required the simultaneous measurement of the instantaneous velocity components and the concentration values at the same point of the flow. This objective was reached by combining the electrical conductivity technique with the laser Doppler velocimetry by means of a data acquisition system and a computer.

The plane turbulent mixing layer chosen for this study was formed by two uniform parallel streams (thickness e = 1 cm) flowing at different velocities into a rectangular duct (Fig. 1). The streams were constituted by aqueous solutions of ammonium hydroxide and of acetic acid, both of low concentration (0.05 N). The experimental arrangement and the measuring techniques have already been described (Gatard 1980) in more detail.

The flow was characterized by the ratio α between the velocities of the streams at the entry into the rectangular duct. In all the experiments, the upper stream, the aqueous solution of ammonium hydroxide, had the lower velocity. The results presented below concern two ratio velocity values respectively equal to 0.5 and 0.75. In both cases, the upper velocity U_2 is 6 cm/s.

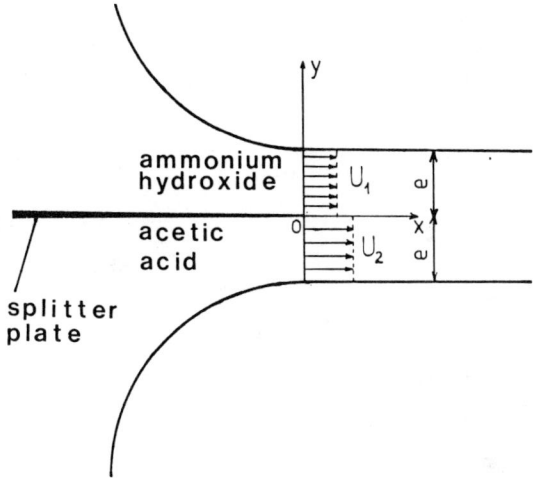

Fig. 1 Schematic diagram of the reacting mixing layer.

Results

Velocity Field

Figure 2 shows profiles of the mean velocity longitudinal component at different points along the mixing zone. For this example the velocity ratio equals 0.5. The dimensionless velocity u^* is deduced from the corresponding time average value by

$$u^* = (\overline{u} - U_1)/(U_2 - U_1)$$

U_1 and U_2 are respectively the low and the high freestream velocities. From these profiles it can be observed that the confinement effect caused by the horizontal walls of the duct starts disturbing the flow in the mixing zone as soon as x reaches 25 mm (x is the coordinate along the mean flow direction). Consequently, the results which follow concern only the upstream part of the mixing zone, located between the trailing edge of the splitter plate and the point of abscissa x = 25 mm.

In the region thus defined, the well-known self-similarity property of the velocity field can be observed (Fig. 3). The value of the dimensionless longitudinal velocity u^* is plotted against the $\overline{y}$ reduced ordinate, defined by the relation

$$\overline{y} = (y - y_{0.1})/(y_{0.9} - y_{0.1})$$

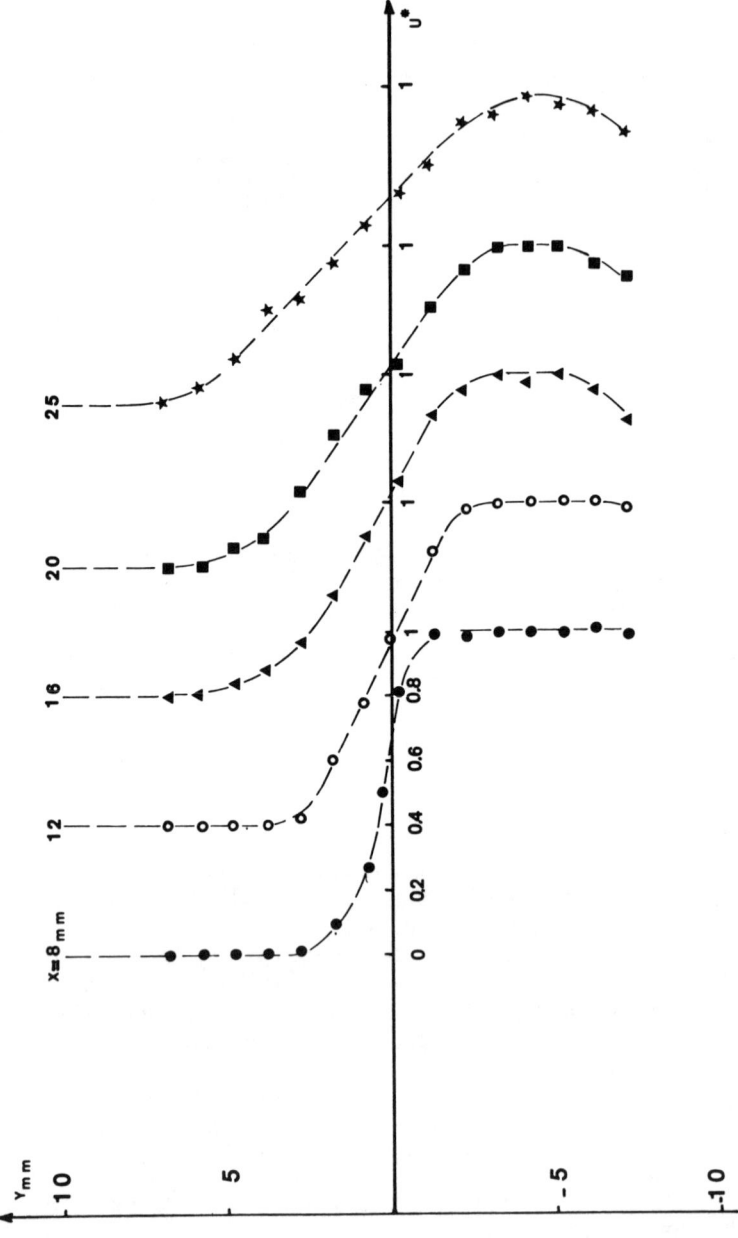

Fig. 2 Mean velocity profiles at different x, α = 0.5.

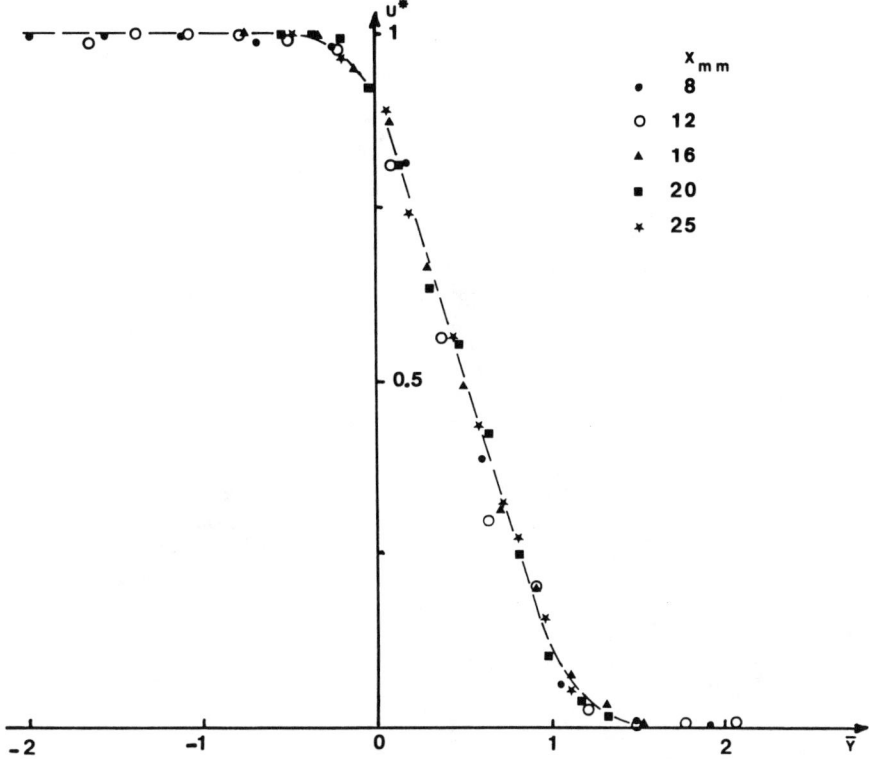

Fig. 3 Dimensionless velocity distribution, $\alpha = 0.5$.

$y_{0.1}$ and $y_{0.9}$ being the points where the reduced velocity u^* equals respectively 0.1 and 0.9.

A characteristic of our experimental apparatus was the high turbulence level of the two initial flows which generated the mixing zone. The velocity fluctuations in the entrance section of the duct exceed 10% as can be seen on Fig. 4. A screen with regularly spaced holes, placed in each flow upstream from the splitter plate trailing edge, caused these fluctuations. Like the mean velocity profiles, the intensity velocity fluctuation distribution is also more or less self-similar.

Concentration Field

As the conductivity of the chemical reaction product is much greater than the conductivity of reactants (Gatard 1980), the measured conductivity is approximately proportional to the local product concentration.

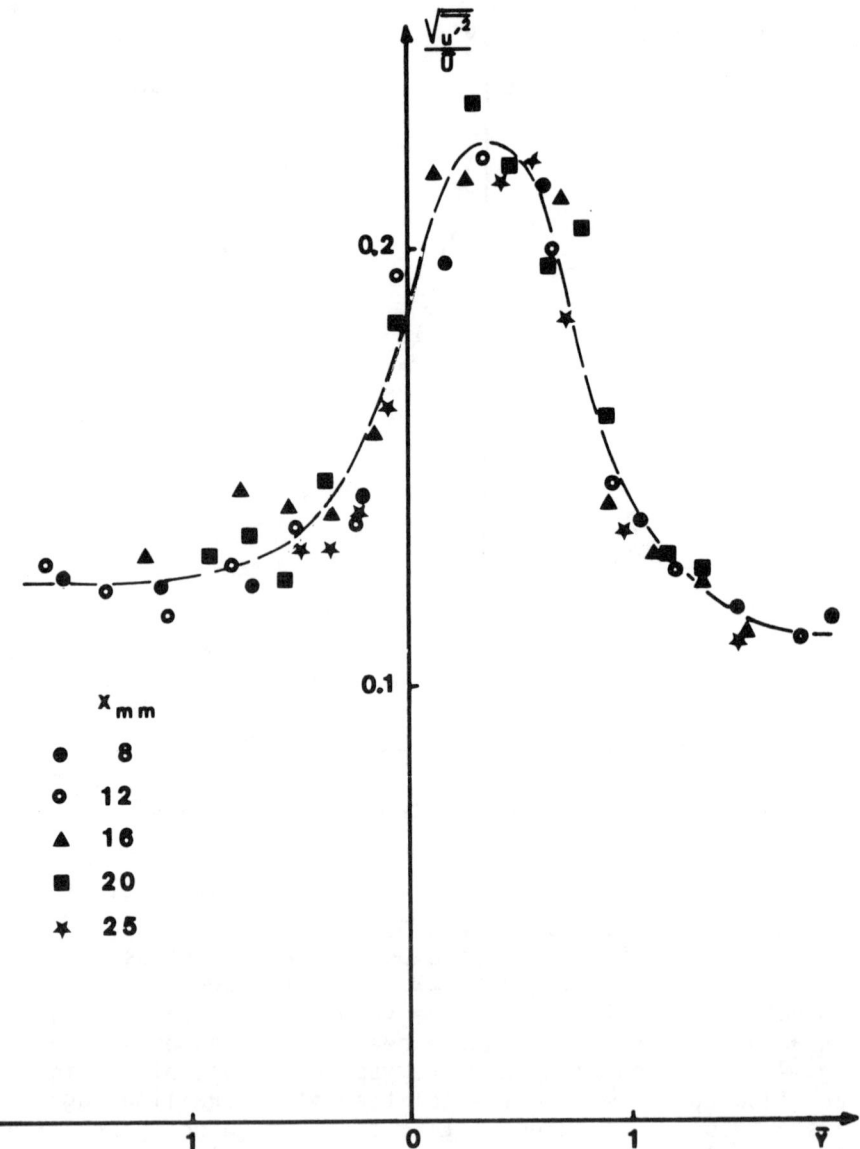

Fig. 4 Root mean square velocity distribution, $\alpha = 0.5$.

From instantaneous conductivity measurements performed at different stations in the upstream part of the mixing zone, mean and fluctuating values were calculated. The dimensionless conductivity $\tilde{C}$ is deduced from the time

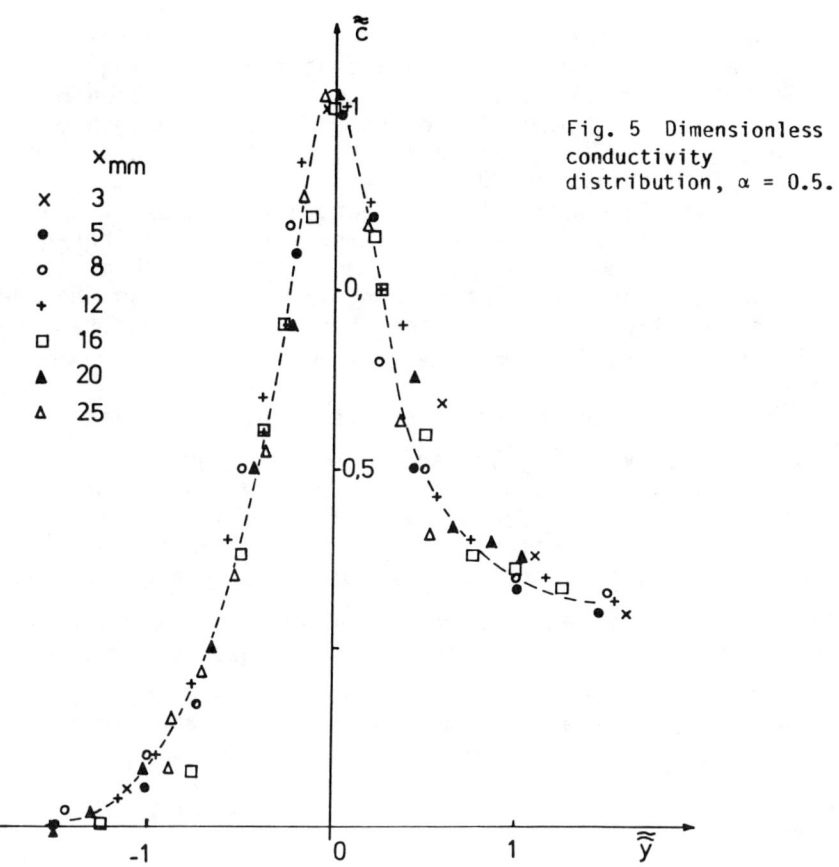

Fig. 5 Dimensionless conductivity distribution, $\alpha = 0.5$.

averaged conductivity by the expression

$$\tilde{c} = (\overline{c} - C_2)/(C_M - C_2)$$

C_2 indicates the lower initial conductivity (acetic acid solution) and C_M the maximum conductivity (reaction product).

The results presented in Fig. 5 show that, as was the case for the velocity field, the time averaged conductivity profiles are similar. The dimensionless ordinate $\tilde{y}$ is calculated from the ordinary location y with the reaction zone thickness as reference length. The thickness of the reaction zone, noted δ_r, is defined by $\delta_r = y_+ - y_-$, y_+ and y_- being the two points where the dimensionless conductivity equals 0.5.

Experimental data show that the growth of the reaction layer is approximately linear, and that the spreading velocity of the reaction zone is a decreasing function of the α parameter. It can also be noted that the origin of the reaction layer is situated at the trailing edge of the splitter plate (i.e., x = 0).

A comparison between the respective thickness of the reaction zone and the turbulent mixing layer, at different points along the mean flow direction, is given on Fig. 6. The growth rate of the reaction zone is greater than the one of the mixing layer. On the other hand, the similarity origins of the velocity and the concentration fields coincide.

The fluctuation intensity of the conductivity is plotted against $\tilde{y}$ in Figs. 7 and 8, where α equals respectively 0.5 and 0.75. In spite of the scatter of experimental data, the conductivity fluctuation intensity distribution is approximately self-similar. For each α parameter value, the corresponding profile shows a minimum at a point where $\tilde{y}$ equals zero and presents a maximum on either side of this value. The peaks correspond to maximum values of $\partial \tilde{C}/\partial \tilde{y}$. Those of higher intensity are located on the ammonium hydroxide stream side, i.e., for the highest conductivity gradient. These results are similar to those

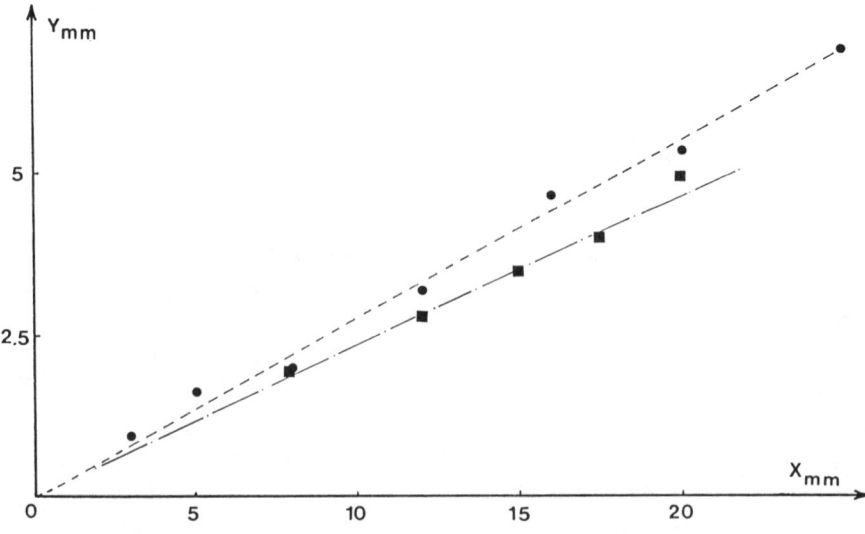

Fig. 6 Comparison between the growth rate of the mixing layer δ and the reaction zone δ_r (α = 0.5); δ - ■, δ_r - ●.

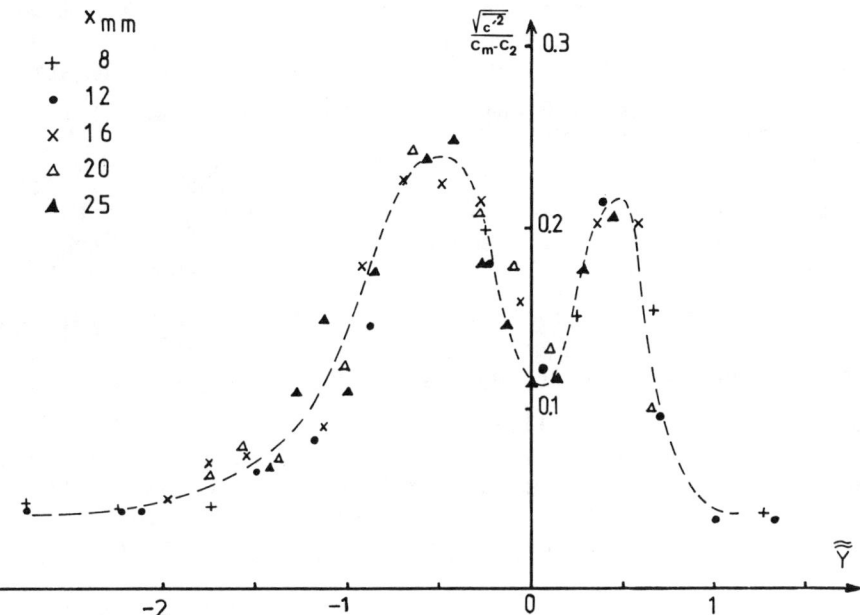

Fig. 7 Root mean square conductivity distribution, $\alpha = 0.5$.

obtained in a turbulent mixing layer with an asymmetrical temperature distribution (Beguier et al. 1978) and in a turbulent mixing of chemically reacting gas flows (Alber and Batt 1976).

Figures 7 and 8 show that the intensity of the peak decreases when the α parameter increases, whereas the low peak intensity is an increasing function of the velocity ratio. This phenomenon can be explained by the fact that the velocity and the conductivity gradients vary in the same direction for the ammonium hydroxide but in opposite directions for the acetic acid flow.

From the instantaneous values of the concentration product and of the velocity components, longitudinal or transverse turbulent diffusion fluxes can be calculated. The corresponding correlation coefficients are deduced by means of the relations

$$R_{u'c'} = \overline{u'c'}/\overline{u}\,\overline{c} \quad R_{v'c'} = \overline{v'c'}/\overline{v}\,\overline{c}$$

The distributions of longitudinal and transverse correlation coefficients are respectively given in Figs. 9 and 10.

In each figure the dotted line corresponds to the mean profile deduced from the whole of experimental data. Both $R_{u'c'}$ and $R_{v'c'}$ curves have similar forms; they are negative on the acetic acid side (high velocity stream) and positive on the low velocity side flow.

Nevertheless, from Fig. 10, it can be noted that $R_{v'c'}$ conserves a negative value when $\tilde{y}$ is between zero and 0.1. On the other hand, Fig. 5 shows that the sign of the mean concentration profile slope, which is positive for negative values of $\tilde{y}$ and negative when $\tilde{y}$ is positive, becomes zero on the ordinate axis. Consequently, a displacement occurs between the points where $R_{v'c'}$ and the derivative of the mean concentration profile are zero.

Thus there exists a zone where the transverse turbulent diffusion flux cannot be assumed to be proportional to the mean concentration gradient. This hypothesis, often admitted in mass transport calculations, would imply a

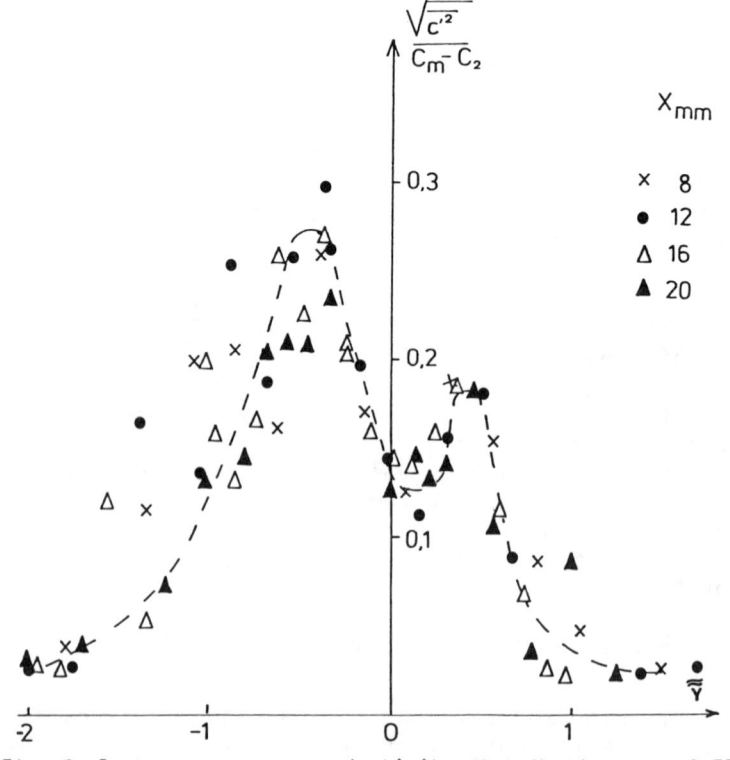

Fig. 8 Root mean square conductivity distribution, $\alpha = 0.75$.

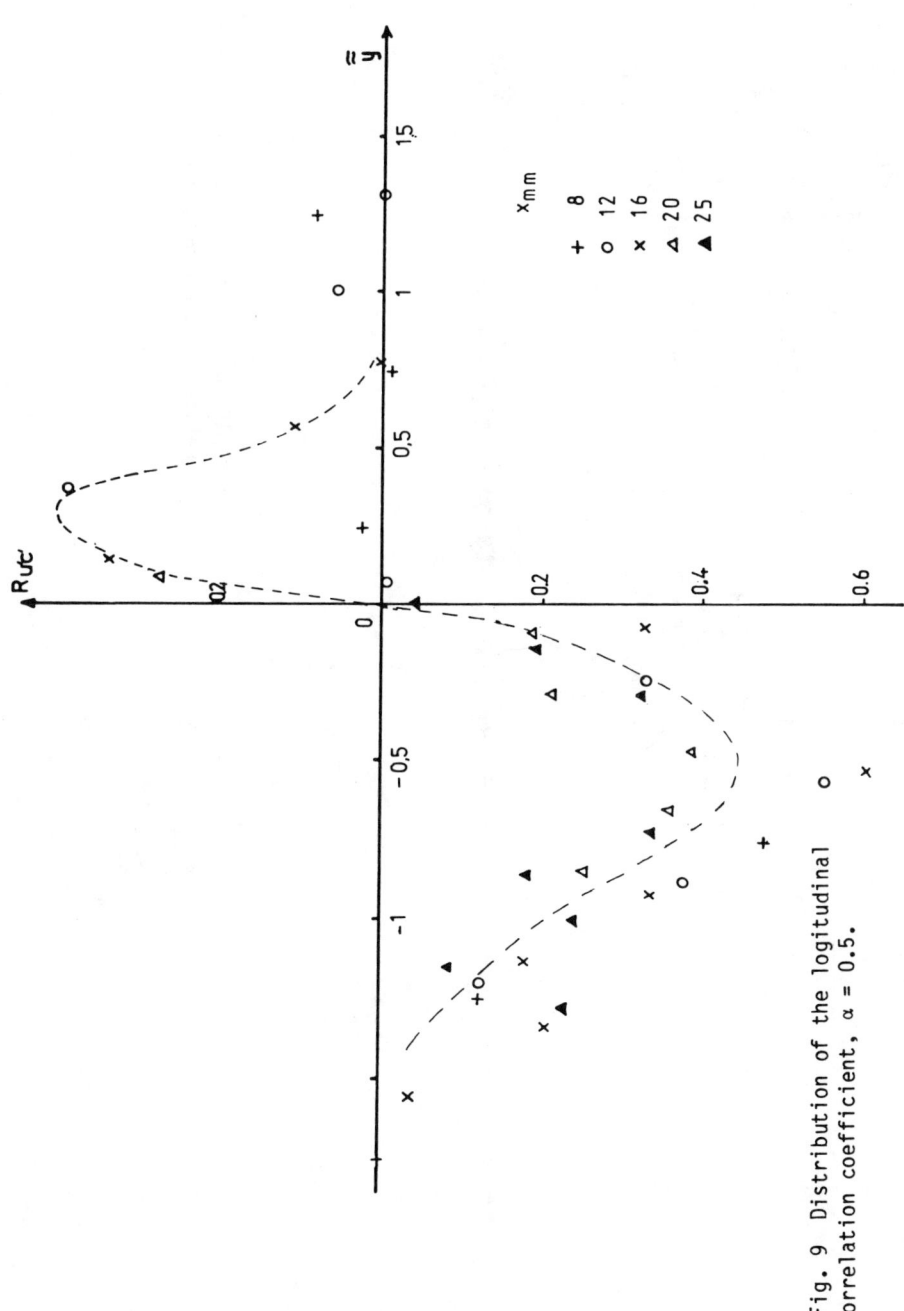

Fig. 9 Distribution of the logitudinal correlation coefficient, $\alpha = 0.5$.

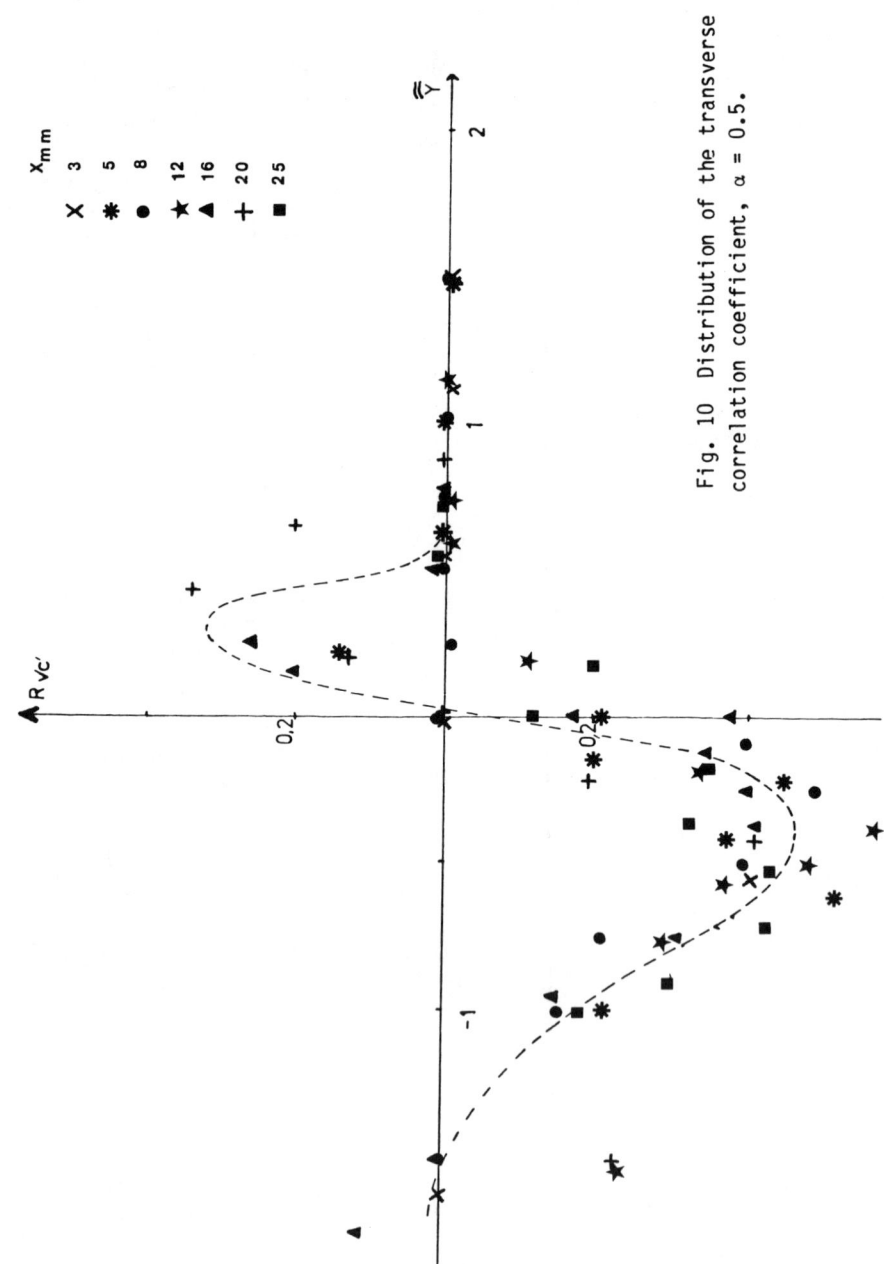

Fig. 10 Distribution of the transverse correlation coefficient, $\alpha = 0.5$.

negative coefficient of eddy mass transfer. A similar situation has been encountered in the case of a turbulent mixing layer with an asymmetrical distribution of temperature (Beguier et al. 1978).

Conclusions

Additional results on the turbulent mixing of chemically reacting species in a shear layer are given. The experimental method employed, combining the laser velocimetry and the conductimetry technique, allows us to obtain the simultaneous values of the instantaneous velocity and conductivity. From these two fluctuating values its is possible to calculate directly the turbulent diffusion fluxes, whose values must be known if we are to test the mathematical models used in mass transport studies. The results obtained in this work show that there is a zone within the turbulent mixing layer where the turbulent transport term cannot be proportional to the mean concentration gradient as is commonly assumed.

The experiments allowed us to observe at the same time that the time averaged and the fluctuating velocity and concentration fields are self-similar in the upstream part of the reacting mixing layer. They have the same similarity origin located at the trailing edge of the splitter plate. It must also be emphasized that the present method can be used to study the diffusion of passive species in various turbulent flows.

References

Alber, J. E. and Batt, R. G. (1976) Diffusion limited chemical reaction in a turbulent shear layer. AIAA J. 14, 70-76.

Batt, R. G. (1977) Turbulent mixing of passive and chemically reacting species in a low-speed shear layer. J. Fluid Mech. 82, 53-93.

Beguier, C., Fulachier, L., and Keffer, J. (1978) The turbulent mixing layer with an asymmetrical distribution of temperature. J. Fluid Mech. 89, 561-587.

Breidenthal, R. (1979) Chemically reacting turbulent shear layer. AIAA J. 17, 310-311.

Breidenthal, R. (1981) Structure in turbulent mixing layers and wakes using a chemical reaction. J. Fluid Mech. 109, 1-24.

Brown, G. L. and Roshko, A. (1974) On density effects and large structure in turbulent mixing layers. J. Fluid Mech. 74, 775-816.

Dimotakis, P. and Brown, G. L. (1976) The mixing layer at high Reynolds number: Large structure dynamics and entrainment. J. Fluid Mech. 78, 535-560.

Gatard, J. M. (1980) Contribution à l'étude expérimentale des champs de vitesse et de concentration au sein de la zone de mélange de deux écoulements plans, parallèles, réactifs. Thèse de 3ème cycle, Poitiers, France.

Koop, C. G. and Browand, F. K. (1979) Instability and turbulence in a stratified fluid with shear. J. Fluid Mech. 93, 135-159.

Turbulent Combustion Zone in a Tubular Reactor

Y. Chauveau,* P. Cambray,† E. Gengembre,* M. Champion,† and J.C. Bellet‡
Université de Poitiers, Poitiers, France

Abstract

A turbulent combustion zone is stabilized in a tubular reactor by fast mixing of a burnt gas flux with a propane-air mixture. To know whether the combustion is kinetically controlled or diffusion controlled, i.e., whether it takes place in an almost homogenous medium or whether it occurs at the interface of unburnt and fully burnt gas packets, the temperature fluctuations were measured in the flow; indeed their amplitude is expected to be small in the first case [quasi-Gaussian probability density function (PDF)] and large in the second case (bimodal PDF). Two measurements methods were used: the infrared pyrometry and the compensation of a thin thermocouple time constant. The statistical analysis of the signals shows that the temperature fluctuation amplitude is small, and therefore that the combustion is kinetically controlled. This property is used to compare the measured mean temperature streamwise variation with the results of calculations based on different simplified kinetic models published in the literature.

Introduction

According to the physical mechanism which controls the interaction between combustion and a turbulent flowfield,

---

Paper presented at the 8th ICOGER, Minsk, USSR, Aug. 23-26, 1981. Copyright © American Institute of Aeronautics and Astronautics, Inc., 1982. All rights reserved.

*Boursier Docteur-Ingénieur au C.N.R.S., Laboratoire d'Energétique et de Détonique.

†Chargé de Recherche au C.N.R.S., Laboratoire d'Energétique et de Detonique.

‡Maître de Recherche au C.N.R.S., Laboratoire d'Energétique et de Détonique.

and whose nature is a function of the different mechanical and chemical properties of the reactive flow itself, turbulent combustion can be divided into different types. Following the pioneering work of Damkohler (1947) and the classification introduced by Schtetnikov (1965), by Barrère and Borghi (1973), and by Barrère (1974), two limiting regimes of turbulent combustion can be identified in which the characteristic length scale of turbulence is either very large or very small in comparison with the thickness of the combustion zone.

The "wrinkled flame" corresponds to a turbulence whose length scale is large compared to the thickness of the reaction zone and whose intensity is small. In this case combustion is controlled by the transport processes, and the effect of turbulence on the reaction rates can be simply expressed through a progress variable. When the turbulence intensity becomes high enough, the thin wrinkled reaction zone is broken into parcels of unburnt and fully burnt gases, combustion occurring at the interfaces.

The "distributed combustion zone" (Libby and Williams 1976) corresponds to the other limiting regime when the combustion is kinetically controlled, i.e., the chemical length scales are large. In this case the chemical structure of the combustion zone can be deeply modified by the turbulence, and thus the use of a detailed kinetic scheme to describe such a zone keeps all its importance.

In the first case the probability density function (PDF) of temperature is found to be bimodal with well-defined peaks at the temperatures of unburnt (T_u) and fully burnt (T_b) mixture (Fig. 1). On the other hand, in a distributed combustion zone the PDF is expected to be quasi-Gaussian (Bray and Moss 1977; Borghi and Moreau 1977).

In the present work the goal was to define such a distributed combustion zone by an investigation of the main statistical properties of the flowfield as well as the structure of the reaction zone itself. The experimental setup which will be described below was designed to model the combustion in the primary zone of a turbojet combustor, which is stabilized by hot gas recirculation, and has the following main characteristics: 1) the ignition of hydrocarbon-air mixture is obtained by very rapid mixing with already burnt gas; 2) the mixing time is short when compared to the reaction time; and 3) the flow can be considered as quasi-one-dimensional (plug flow reactor).

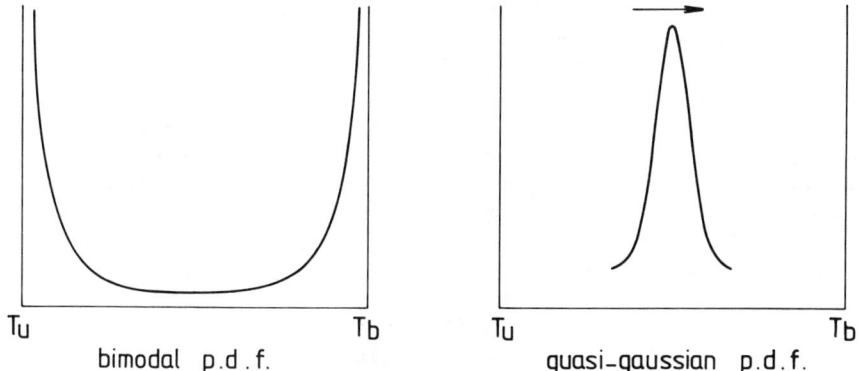

Fig. 1 Typical shapes of bimodal and quasi-Gaussian temperature probability density functions.

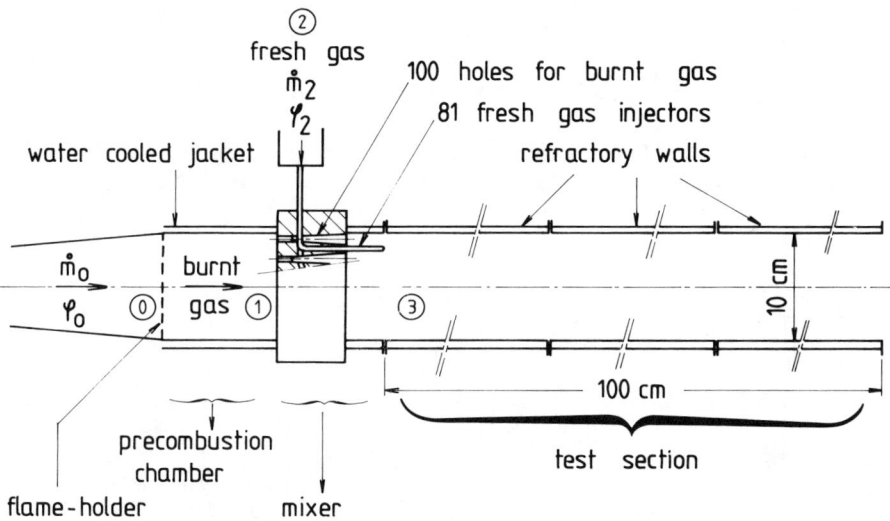

Fig. 2 Schematic diagram of the experimental setup.

The Tubular Combustor

The experimental system consists of three main parts (Fig. 2):
1) A precombustion chamber which provides the burnt gas from a propane-air mixture of equivalence ratio ϕ_o;
2) A mixing section into which hot combustion products from the precombustion chamber are injected through 100 holes positioned on a 1-cm square mesh and mixed with a fresh mixture whose equivalence ratio is ϕ_2 and which has

been injected through 81 holes located on a 1-cm square grid, as shown in Fig. 3; and

3) A 10 cm square test channel, 1 m in length. The walls of the precombustion chamber and the mixer are water-cooled. To improve the initial mixing, the end of each injector has four holes, the axis of each making a 45-deg angle with the axis of the mean flow. The test channel itself is made of three identical elements whose walls are refractory bricks linked together by water-cooled flanges. The lateral bricks are provided with holes which allow sampling and windows for optical measurements.

For a given total mass flow rate $\dot{m}$, it is possible to stabilize a combustion zone at a given distance downstream of the mixer by adjusting the mixing temperature T_3 through the equivalence ratio ϕ_0 and the burnt-unburnt gas flow rate ratio $R = \dot{m}_0/\dot{m}_2$ (Bellet et al. 1980; Chauveau et al. 1981). The position of the combustion zone is strongly dependent on the values of these two parameters.

Characterization of the Mixing of Burnt and Unburnt Gas

To know whether this combustion zone is a "distributed combustion zone" or a "packet combustion zone," the

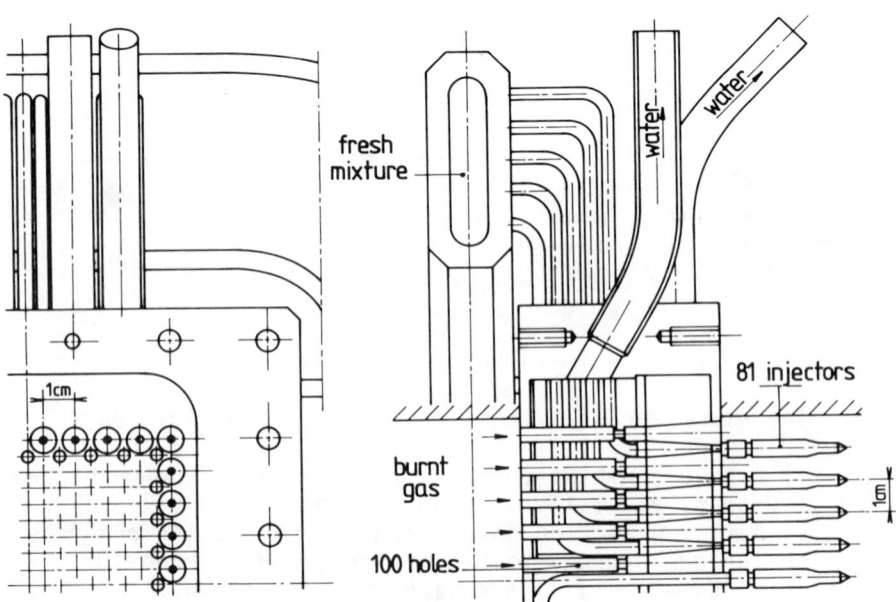

Fig. 3 Sketch of the mixer.

temperature fluctuations were measured and the type of PDF (bimodal or quasi-Gaussian) was determined. Two complementary measurements methods were used, the digital compensation of a thin thermocouple time constant and i.r. pyrometry.

Thermocouple with Time Constant Compensation

This method allows the determination of the temperature PDF at a point of the flow, but its response is limited by the attenuation of the signal at high frequency, and the compensation is valid only when the signal-to-noise ratio is higher than unity (Ballantine and Moss 1977; Yoshida and Tsuji 1979; Tanaka and Shimamoto 1981; Lockwood and Moneib 1980). To check the capability of this method to detect a bimodal PDF, simulations (Chauveau et al. 1981) were made with 1) a square function generator representing a succession of unburnt and fully burnt gas packets, and 2) a low-pass filter representing the thermal inertia of the thermocouple.

Since the flow velocity is close to 100 m/s and the injection mesh is 1 cm, simulations were made with a signal frequency of 10 kHz. The following results were obtained:

1) The two peaks are separated if the sampling rate is higher than 100 kHz for a time constant τ equal to 5 ms (Fig. 4a) and if the difference ΔT between these two peaks is at least 70 K (Fig. 4b), i.e., the critical value is much lower than the difference $T_b - T_u$.

2) The electronic noise causes an increase of the measured variance σ_T which depends on ΔT (Fig. 4b).

Furthermore, the lowest temperature jump ΔT detectable with spectral analysis has been calculated. In a "packet combustion zone," a thermocouple is exposed by a succession of hot and cold packets, which produces a series of peaks in the power density spectrum. The first of them (fundamental) can be detected when, in spite of its attenuation by thermal inertia, it culminates at about the same power density as that of the noise. If almost all the signal energy is accumulated in the fundamental peak (a hypothesis that is well verified for a quasi-square signal), the following result may be deduced from Parseval's theorem:

$$H \cdot \Delta f \sim \Delta T^2/4$$

where H is the strength of the peak and Δf its width.

If f is the central frequency of the peak and f_c the cutoff frequency of the thermocouple, the attenuation is

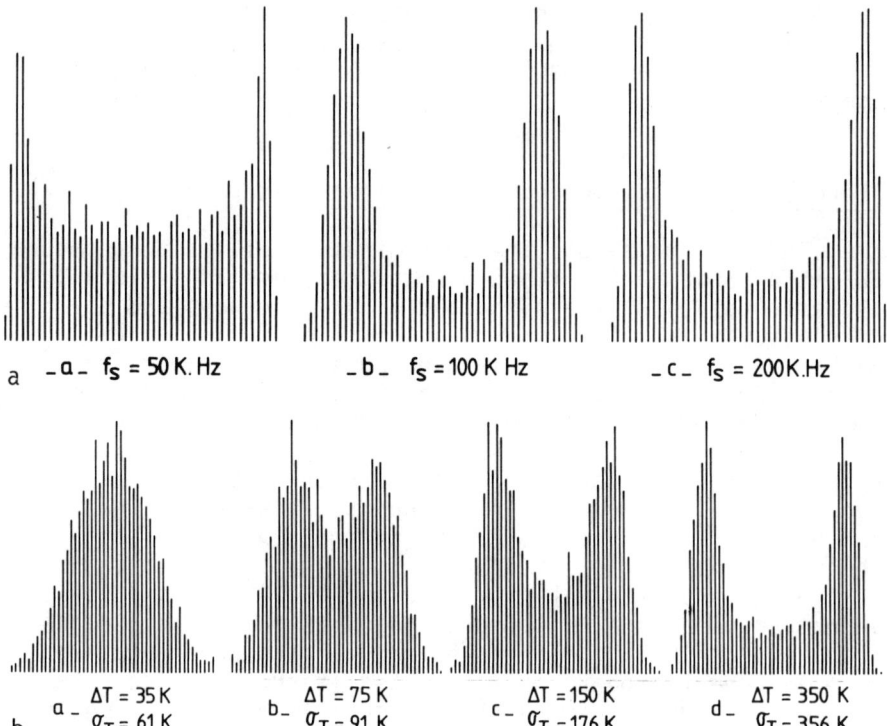

Fig. 4a) Simulation of a bimodal PDF: influence of the sampling rate f_s for $\Delta T = 350$ K. b) Simulation of a bimodal PDF: influence of the fluctuation amplitude (sample rate 200 kHz).

$(f_c/f)^2$, so that the minimum strength H for a detectable peak is

$$H \sim B \cdot (f/f_c)^2$$

where B stands for the noise power density. Accordingly, the following relations hold

$$\Delta T^2 \sim [(4B \cdot \Delta f)/f_c^2] \, f^2$$

or

$$\Delta T \sim (2/f_c) \sqrt{B \, (\Delta f/f)} \, f\sqrt{f}$$

Figure 5 shows the variations of ΔT vs f for various values of $\Delta f/f$ (approximate packets size relative to dispersion). The two temperature jump scales have been calculated for $f_c = 10$ and 26 Hz, these values being deduced from measurements with 50- and 25-µm-diam thermocouples. As can be seen, a "packet combustion zone" can be easily detected with a 25-µm thermocouple even when the frequency

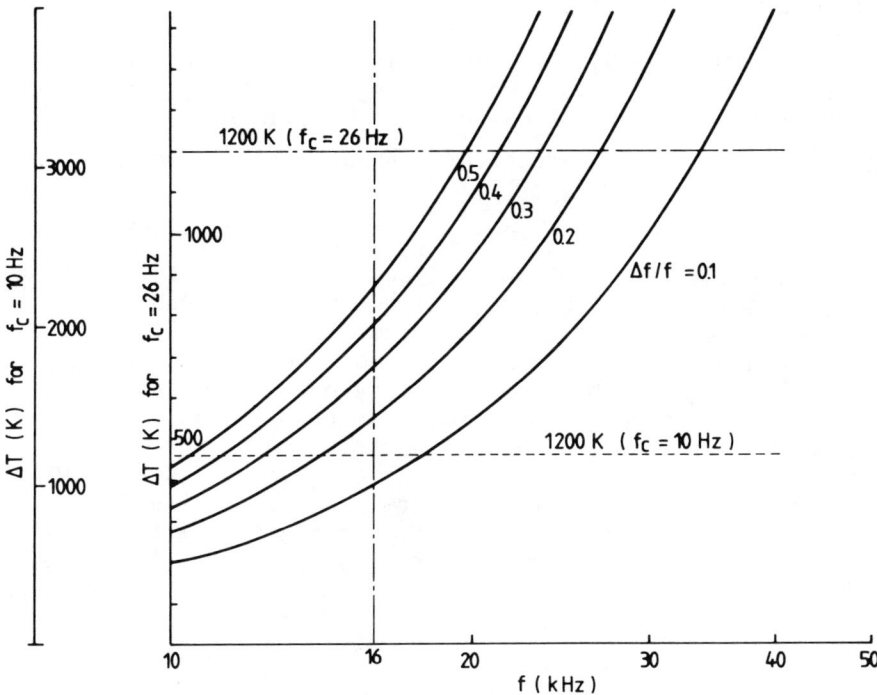

Fig. 5 Minimal temperature jump detectable with thermocouples.

dispersion is as large as 50%, but cannot be detected with a 50-μm thermocouple if the relative dispersion of the packets size is greater than 15%.

Measurements were carried out with chromel-alumel thermocouples, made with 50-μm diameter wires. The values of the time constant measured by electrical overheating (Ballantine and Moss 1977) are in the range 10-15 ms. Figure 6 shows the PDF obtained at a distance x = 15 cm downstream of the injectors, where ṁ = 300 g/s, R = 1.57, ϕ_0 = 1, and ϕ_2 = 0. The corresponding rms value of the signal fluctuations is 70 K and gives, after noise correction, a rms value of temperature fluctuations of 45 K. Spectral analysis (Fig. 7) showed that the signal-to-noise ratio tends to unity between 2 and 3 kHz, and that the energy spectrum decreases from a frequency of about 500 Hz.[§]

[§]Recent measurements using 25-μm thermocouples confirm the lack of any peak at about 16 kHz (corresponding to the actual local velocity 130 m/s and the integral turbulence scale 8 mm) which is strong enough to be the consequence of the existence of burnt and fresh lumps exhibiting a 1200 K ($T_b - T_u$) temperature difference (Chauveau et al. 1983).

Infrared Pyrometry

Charpenel (1979) has shown that if a gas (of low emissivity for the chosen wavelength) is statistically homogenous along the optical path, the measured signal variance $\overline{\Omega'^2}$ and the temperature variance $\overline{T'^2}$ are related by

$$\overline{\Omega'^2}/\overline{T'^2} = 2\Lambda/\ell$$

where ℓ is the length of the optical path and Λ the integral scale of turbulence, which can be measured by crossed beams correlation. Measurements are made for the same experimental conditions that, for the thermocouple measurements, lead to an integral scale $\Lambda = 8$ mm and to a rms value of temperature fluctuations of 38 K, which agrees within 20% with the results of thermocouple measurements. Spectral analysis of the i.r. emission (Fig. 7) is also in agreement with thermocouple spectral analysis: the energy spectrum decreases from a frequency near 500 Hz with a slope $-(5/3 + 1)$ due to the spatial integration effect (Charpenel 1979).

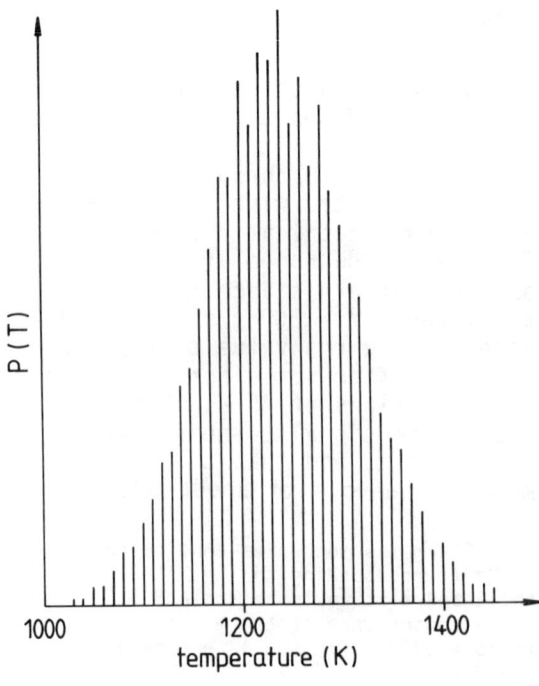

Fig. 6 Temperature PDF measure at x = 15 cm (sample rate 200 kHz).

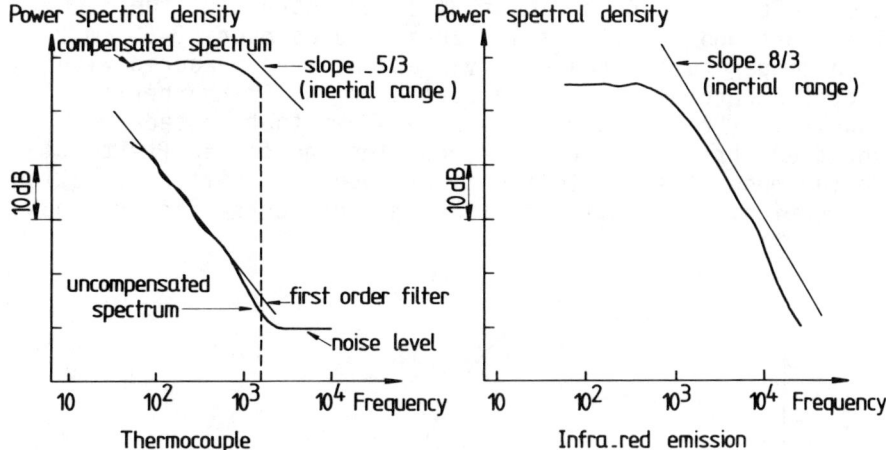

Fig. 7 Power spectra of the thermocouple and i.r. emission signals (mixture of burnt and fresh gas at x = 45 cm).

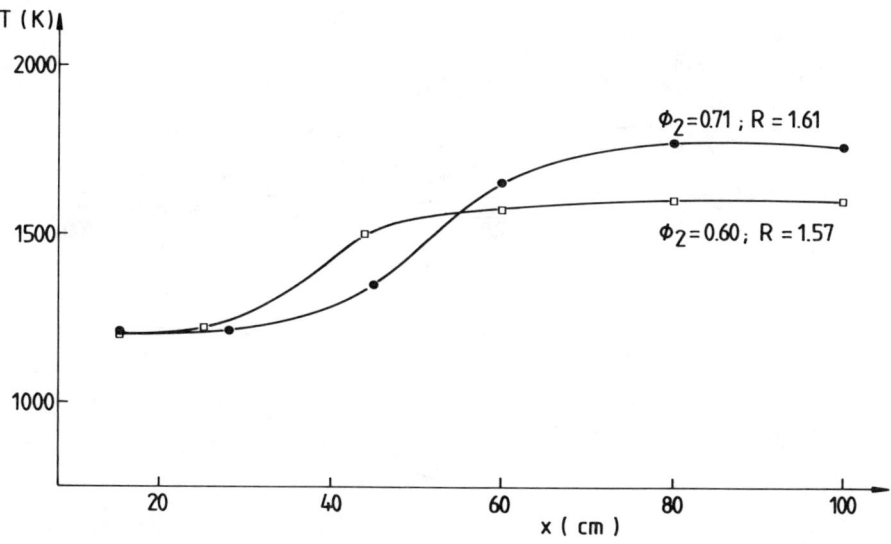

Fig. 8 Mean temperature streamwise variations for ϕ_2 = 0.6 and 0.7.

Combustion Experiments

Combustion experiments were carried out for ϕ_2 = 0.6 and 0.7 with industrial propane (81.2% C_3H_8; 7.6% C_3H_6; 5.6% C_2H_2; 4.7% C_2H_6; 0.9% C_4H_{10}). For $\dot{m}$ = 300 g/s and R ≃ 1.6, the ignition occurs at about 30 cm downstream of the injectors and the combustion zone spreads over 50-60 cm. Mean temperature streamwise variations were measured with uncoated chromel-alumel and coated Pt-Pt 10% Rh thermocouples. (Preliminary tests have shown that coated and uncoated chromel-alumel thermocouples and coated Pt-Pt 10% Rh thermocouples led to identical results.) After radiation correction, the results of these measurements (Fig. 8) show

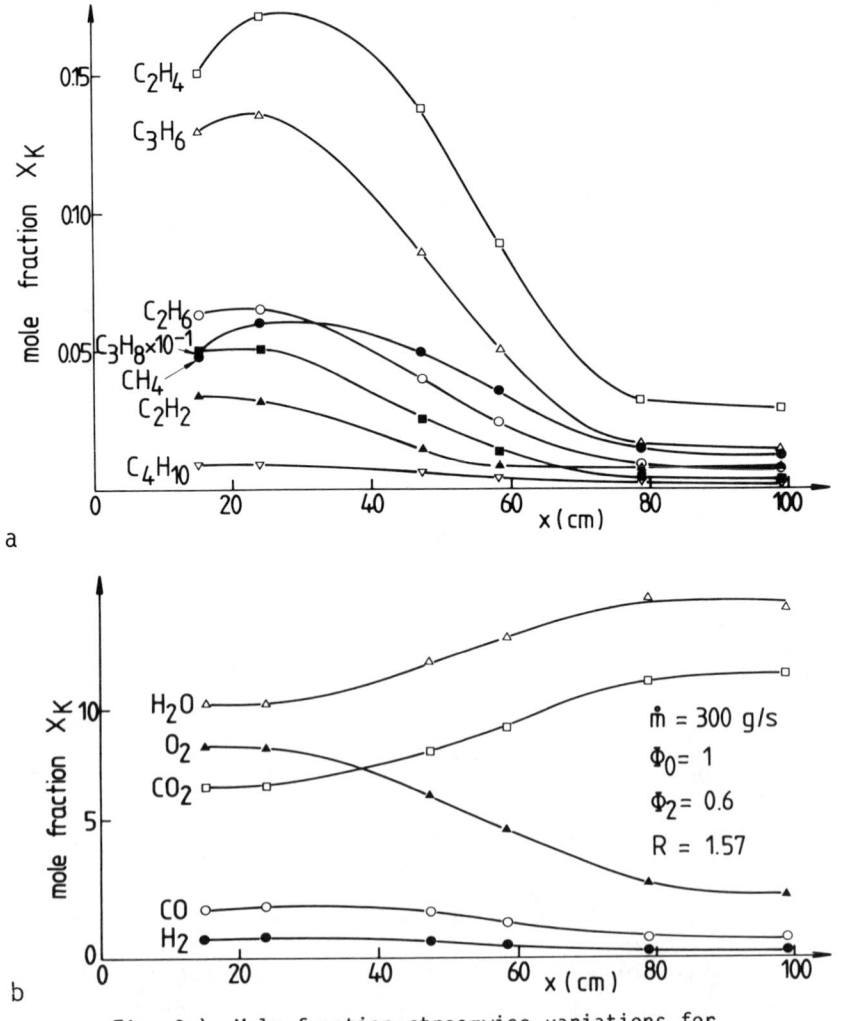

Fig. 9a) Mole fraction streamwise variations for $\phi_2 = 0.6$. b) Mole fraction streamwise variations for $\phi_2 = 0.6$.

that the ignition length increases when the equivalence ratio ϕ_2 increases, its value being in agreement with the results of measurements made in the shock-tube (Myers and Bartel 1969; Burcat et al. 1971).

Mole fractions streamwise variations of the main species were also measured by water-cooled probe sampling and chromatographic analysis. The results show that the consumption of fuel takes place over the whole combustion zone and that intermediate species (CO, H_2, C_2H_4) are present in relatively significant concentrations (Fig. 9).

Discussion and Conclusion

From these results it appears that the reactive flow in the tubular reactor is characterized by 1) a good homogeneity of mixing of burnt and unburnt gas, with small amplitude to temperature fluctuations; 2) an ignition length which is very sensitive to the mixing temperature and increases as the equivalence ratio increases; and 3) relatively significant concentrations of intermediate species.

All these properties lead to the conclusion that the combustion is effectively kinetically controlled. Then the theoretical model developed by Champion et al. (1978), Champion and Bellet (1978), and Champion (1980) shows that for initial temperature fluctuations which are as low as those observed in this work, the effect of turbulence on

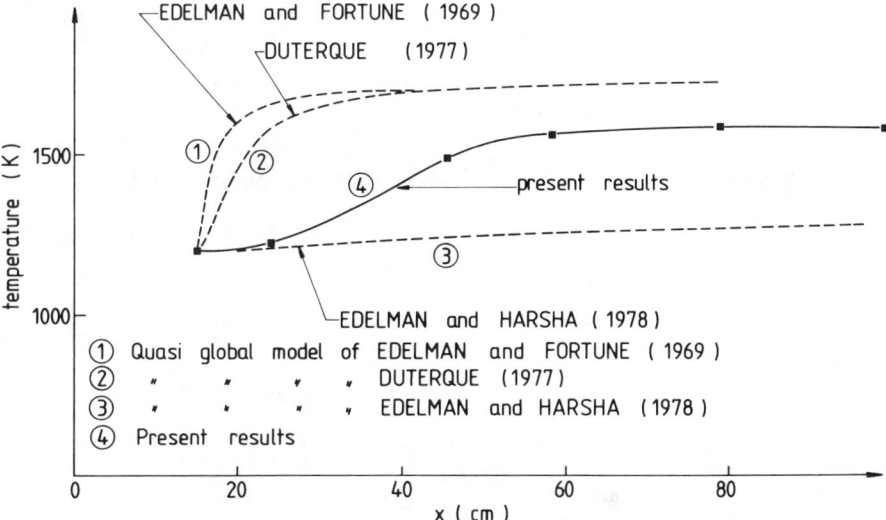

Fig. 10 Comparison of the measured mean temperature variation with a plug-flow calculation using three different quasiglobal kinetic models.

chemical production rates can be neglected as a first approximation.

A comparison of the experimental results with plug-flow calculations shows that no "quasiglobal" model of propane-air combustion (Edelman and Fortune 1969; Duterque 1977; Edelman and Harsha 1978) leads to a correct numerical simulation of the chemical evolution along this combustion zone, even if chemical reactions are assumed to begin after a mixing zone whose length is about 15 cm, and if the nonadiabaticity of the combustor is taken into account (Fig. 10). In these "quasiglobal" models the ignition phase is represented only by one global reaction:

$$C_3H_8 + 3/2\ O_2 \rightarrow 3CO + 4H_2$$

associated with a certain number of chain branching-propagating and recombination reactions for the oxidation of CO and H_2. To obtain a more general applicability, it seems necessary to use at least two global reactions to represent the formation of CO and H_2 (Hautman et al. 1981) or to use a detailed kinetic model (Cathonnet et al. 1981).

Acknowledgments

This work has been sponsored by DRET under Contracts 79/065 and 80/022, and by CNRS under Contract ATP 3257. The authors would like to express their thanks to SNECMA Combustion Department for its contribution to the development of the tubular reactor.

References

Ballantine, A. and Moss, J. B. (1977) Fine wire thermocouple measurements of fluctuating temperature. Combust. Sci. Technol. 17, 63.

Barrére, M. and Borghi, R. (1973) Taux de production chimique en régime turbulent. Entropie 52, 7.

Barrere, M. (1974) Modèles de combustion turbulente. Rev. Gen. Therm. 148, 295.

Bellet, J. C., Merlin, J., Bauer, P. and Kageyama, T. (1977) Experiments on the turbulent combustion of propane in a tubular reactor. Paper presented at the 6th International Colloquium on Gasdynamics of Explosion and Reactive Systems, Stockholm, Sweden.

Borghi, R. and Moreau, P. (1977) Turbulent combustion in a premixed flow. Acta Astron. 4, 321.

Bray, K. N. C. and Moss, J. B. (1977) A unified statistical model of premixed turbulent flame. Acta Astron. 4, 291.

Burcat, A., Lifshitz, A., Scheller, K., and Skinner, G. B. (1971) Shock-tube investigation of ignition in propane-oxygen-argon mixtures. 13th Symposium (International) on Combustion, p. 745. The Combustion Institute, Pittsburgh, Pa.

Cathonnet, M., Boettner, J. C., and James, H. (1981) Experimental study and numerical modeling of high-temperature oxidation of propane and n-butane. 18th Symposium (International) on Combustion, p. 903. The Combustion Institute, Pittsburgh, Pa.

Champion, M., Bray, K. N. C., and Moss, J. B. (1978) The turbulent combustion of a propane-air mixture. Acta Astron. 5, 1063.

Champion, M. and Bellet, J. C. (1978) Influence des fluctuations de température initiales sur le développement des réactions dans zone de combustion distribuee. C. R. Acad. Sci. Paris 287, 159.

Champion, M. (1980) Premixed turbulent combustion controlled by complex chemical kinetics. Combust. Sci. Technol., 24, 23.

Charpenel, M. (1979) Measures instantanees par pyrometrie infrarouge de temperature de gas de combustion, application a la turbulence termique. Rev. Phys. Appl. 14, 491.

Chauveau, Y., Merlin, J., Gengembre, E., Champion, M., and Bellet J. C. (1981) Inflammation d'un melange hydrocarbure-air-gaz brules en ecoulement turbulent. Colloque Berthelot-Mallard Vieille et Le Chatelier, p. 413. Bordeaux, France.

Chauveau, Y., Cambray, P., Champion, M., and Bellet, J. C. (1983) Experimental study and modelling of the reactive flow in a constant section reactor. Phys. Chem. Hydrodynamics (in press).

Damkohler, G. Z. (1947) The effect of turbulence on the flame velocity in gas mixture. NACA TN 1112.

Duterque, J. (1977) Cinetique de la combustion des hydrocarbures. Rapport technique 11/1847 EY., ONERA, Chatillon, France.

Edelman, R. B. and Fortune, O. F. (1969) A quasi-global chemical kinetic model for the finite rate combustion of hydrocarbon fuels with application to turbulent burning and mixing in hypersonic engines and nozzles. AIAA Paper 69-86. New York, N.Y.

Edelman, R. B. and Harsha, P. T. (1978) Some observations on turbulent mixing with chemical reactions. Turbulent Combustion: Progress in Astronautics and Aeronautics (edited by L. Al Kennedy), Vol. 58, p. 55. AIAA, New York.

Hautman, D. J., Dryer, F. L., Schug, K. P., and Glassman, I. (1981) A multi-step overall kinetic mechanisms for the oxidation of hydrocarbons. Combust. Sci. Technol. 25, 219.

Libby, P. A. and Williams, F. A. (1976) Turbulent flows involving chemical reactions. Ann. Rev. Fluid Mech. 6, 351.

Lockwood, F. C. and Moneib, H. A. (1980) Fluctuation temperature measurements in a heated round free jet. Combust. Sci. Technol. 22, 63.

Myers, B. F. and Bartle, E. R. (1969) Reaction and ignition delay time in the oxidation of propane. AIAA J. 7, 1862.

Schtetnikov, E. S. (1965) Physics of Combustion, Izd, Nauka, Moscow, USSR.

Tanaka, Y. and Shimamoto, Y. (1981) Simultaneous measurements of multifluctuating components in an open-jet flame. Combustion in Reactive Systems: Progress in Astronautics and Aeronautics (edited by J. R. Bowen, N. Manson, A. K. Oppenheim and R. I. Soloukhin), Vol. 76, pp. 314-333. AIAA, New York.

Yoshida, A. and Tsuji, H. (1979) Measurements of fluctuating temperature and velocity in a turbulent premixed flame. 17th Symposium (International) on Combustion, p. 945. The Combustion Institute, Pittsburgh, Pa.

Premixed Turbulent Flames—Interplay of Hydrodynamic and Chemical Phenomena

A.M. Klimov*
Academy of Sciences, Moscow, USSR

Abstract

The structure and propagation mechanism of turbulent flame fronts are considered. The most general features of the phenomenon are demonstrated on the basis of a simplified model of turbulence, characterized by a single velocity scale and a single length scale, with the additional assumption of constant density of the fluid. The results obtained are used to analyze the effects of real turbulence and the role played by gas thermal expansion. It is shown that the structure of the local reaction zones is intimately connected with the internal intermittency of turbulence and with the action of temperature increase on the small-scale eddies, while the propagation velocity of the flame as a whole is augmented by interaction of gas density inhomogeneities with large-scale eddies. The emerging picture is hardly compatible with the conventional formalisms of the statistical theory of turbulence.

Introduction

The theory of turbulent flame fronts in a premixed gas is central of the problem of describing combustion in turbulent media. Most important, the turbulent flame front is probably the most simple situation theoretically, and therefore its approximate realization in experiments can be readily achieved. Most frequently the basic phenomena of turbulent combustion are studied in the framework of a stationary plane front, propagating through a combustible

---

Presented at the 8th ICOGER, Minsk, USSR, Aug. 23-26, 1981. Copyright © 1983 by A. M. Klimov. Published by the American Institute of Aeronautics and Astronautics, Inc. with permission.
*Chief Turbulent Reactive Flows Group.

mixture which is in a state of homogeneous turbulent motion. Another typical case is the flame front in a channel at high flow velocity. Here turbulence is inhomogeneous and its generation in the shear layer between fresh mixture and combustion products is significant. In the limiting case of a weakly turbulent oncoming flow the flame front is interconnected with the turbulence front by a single interface. Though these two typical situations have much in common, the present work is oriented toward the first one.

Of immediate interest is combustion in a field of large-scale turbulence, that is, at integral length scales of turbulence much larger than the laminar flame thickness. All the available evidence indicates that the processes taking place at the fine "molecular" level (chemical reaction, molecular transfer) and the processes connected with large- and small-scale convective motions are intimately coupled together. The wide range of the relevant length scales results in considerable diversity of possible conclusions.

Since the formation of laminar flame fronts is a fundamental feature of premixed combustion, the initial theories, beginning with Damköhler (1940) and Schelkin (1943), used the so-called wrinkled laminar flame model. Qualitatively this brought some understanding of the phenomenon and helped to solve the practical problems of the time. Analytically this model led to the problem of turbulent diffusion of a self-propagating surface, which remained unsolved. Later the models of "distributed reaction zone" due to Summerfield et al. (1955) and Schetinkov (1956) deny the possibility of the chemical reaction field degenerating into thin reaction zones.

In connection with this contradiction, Kovasznay (1956) on dimensional grounds proposed a criterion of relative importance of these two limiting ("surface" and "volume") mechanisms of turbulent combustion:

$$\Gamma \sim (u' \delta_n / u_n \lambda) \tag{1}$$

where u_n, δ_n are the propagation speed and the thickness of the so-called normal (plane stationary laminar) flame, and u', λ are the root-mean-square velocity and Taylor microscale of turbulence. The physical meaning of this criterion became clear after the development of the theory of laminar flame in a nonuniform velocity field (Klimov 1963). However, owing to the existence of a whole spectrum of length and velocity scales, the subject of the interrelation of surface and volume elements cannot be considered fully solved. Further analysis of this problem is one of the aims of this work.

The problem of turbulent flame speed is trivial for purely volume burning mechanisms, but in the case of dominating surface elements it is related to the problem of turbulent diffusion of self-propagating isothermal surfaces. For $u' \gg u_n$, the approximate closure of this problem is possible along the quasi-Lagrangian approach (Klimov 1975). In this case the mean position of the leading isothermal surfaces is advanced by the envelopments of unburnt mixture, which are not a purely hydrodynamic phenomena, but depend also on the laminar flame properties. Burnout proceeds by stretching the layers of fresh mixture between self-propagating reaction zones. A conceptually rather similar approach has been pursued by Spalding (1978), who directed his attention to the conditions of shear flow (ducted flames) and stirred reactors. Later in this paper, a more rigourous consideration of the subject is presented and the interaction of turbulent eddies with the inhomogeneities of gas density is analyzed.

In view of the large number of studies applying the methods of statistical theory of turbulence to premixed flames, the problem of the compatibility of these methods with the nature of the phenomena is also discussed.

The "Basic" Model

The essential physics of the turbulent flame will be first considered in a simplified situation which permits a rather rigorous treatment. This allows one to investigate the basic concepts in uncomplicated circumstances and facilitates the analysis of a more realistic situation.

Suppose the homogeneous turbulence is characterized by a single velocity scale u' and a single length scale ℓ. Suppose also that the fluid density is constant.

When $\ell \gg \delta_n$, we may expect that under certain conditions the turbulent flame consists of wrinkled laminar flame fronts. Two properties of turbulent motion should exert the greatest influence on combustion: the ability to cause dispersion of the flame surface and the ability to change the structure of a laminar flame by the action of strain rates.

The second factor will be discussed first. In turbulent media the area of material surfaces is changing. The internal structure of laminar flames in these conditions can be exactly described, as for $\ell \gg \delta_n$ the flame front is mostly locally flat. With the logarithmic rate of change of the material surface area,

$$-\phi = \frac{1}{A}\frac{dA}{dt}$$

the equations describing the structure of the flame front are (Klimov 1963):

$$\frac{\partial \theta}{\partial \sigma} + [m(\sigma) + \gamma n]\frac{\partial \theta}{\partial n} = \frac{\partial^2 \theta}{\partial n^2} + f(\theta, \alpha_i)$$

$$\frac{\partial \alpha_i}{\partial \sigma} + [m(\sigma) + \gamma n]\frac{\partial \alpha_i}{\partial n} = \frac{D_i}{a}\frac{\partial^2 \alpha_i}{\partial n^2} + \psi(\theta, \alpha_i) \quad (2)$$

where θ and α_i are respectively the nondimensional temperature and species concentration: n and σ are respectively the coordinate normal to the flame front and the time made nondimensional with u_n and δ_n; f and ψ are respectively the source terms for θ and α_j and D_i and a are respectively the diffusivities for the i-th species and heat. The equations are written in the reference frame of the flame front; $m(\sigma)$ is the eigenfunction of the problem (an analog of the flame speed); and $\gamma = \phi\delta_n/u_n$. In comparison with a normal flame, Eqs. (2) contain new convective terms corresponding to the straining motion (rotational components of the velocity field do not reveal themselves, since isoscalar surfaces rotate with the fluid). For the case of variable fluid density, the equations are qualitatively the same as Eqs. (2). In a turbulent medium the stretching of the surfaces is dominating. Evidently, the mean value $\langle\gamma\rangle \sim u'\delta_n/u_n \ell$. Structural similarity between Γ and γ is obvious.

A detailed theory of such a flame front (Klimov 1963) shows that for $\langle\gamma\rangle \ll 1$ the flame front does not appreciably differ from the normal flame, i.e., it represents a propagating wave. For $\langle\gamma\rangle \sim 1$ or for $\langle\gamma\rangle \gg 1$, the flame resembles a mixing layer and does not depart from material surfaces. Specific heat release (per unit flame area) is not generally constant and depends on γ and on the detailed chemical kinetics. For stretching the flat layer of combustion products between two flame fronts, Eqs. (2) do not have a stationary solution if the absolute magnitude $|\gamma|$ exceeds some critical value (combustion is quenched). Such quenching events, occurring locally inside turbulent flames, explain how transition from the "surface" mode of combustion to the "volume" mode proceeds.

In a system of interwoven laminar flames (alternating layers of fresh mixture and combustion products), such as those which emerge in the turbulent combustion zone, the combined action of stretching and molecular transfer causes sufficiently large $\langle\gamma\rangle$ to complete mixing before noticeable burning occurs; then a thermal explosion takes place. For small $\langle\gamma\rangle$, combustion proceeds by means of propagating flames. Numerical analysis (Klimov and Lebedev 1980) shows that if the reaction rate depends explicitly on the temperature alone, the transition from surface to volume mode evolves gradually when $\langle\gamma\rangle > 2$. Specific heat release per unit flame area does not change appreciably in this case.

The subject of local reaction rate needs, however, specific consideration. Usually reaction proceeds through the active intermediate particles (mostly free radicals), and its rate may generally depend on their concentration. The straining motion changes the distance between isoscalar surfaces and hence can change the concentration of radicals in the reaction zone as compared to normal flame (for the same isothermal surfaces). The extent of this effect depends on

$$\Omega = \delta_n/u_n t_a$$

where t_a is the active particle lifetime.

When $\Omega \gg 1$, the generation and decay of radicals occur so quickly that they are in local equilibrium with the main species and diffusion has little influence on their concentration [in normal flame it corresponds to $\Omega \approx 100$ (Williams 1965)]. Therefore, if Ω is sufficiently large for all the radicals involved, the chemical reaction rate in turbulent fluid is related to local conditions exactly in the same way as in normal flame.

If local equilibrium is absent, the stretching of the flame bring isoscalar surfaces closer and increases the loss of radicals from the reaction zone. Interaction of adjacent flame fronts can diminish this effect and can even increase the concentratioon of radicals and reaction rate in comparison with normal flame (Klimov and Lebedev 1980). These effects are especially pronounced when the diffusivity of an active particle is greater than that of the main species. Specific production of weakly interacting stretched flame fronts is greatly decreased when $\langle\gamma\rangle > 1$. Transition from the surface to volume mode occurs sharply in the range $\langle\gamma\rangle = 1.5$-$2.0$.

Next the mechanism of the turbulent flame front propagation as a whole is considered, beginning with $\langle\gamma\rangle$ 1. Since $\ell \gg \delta_n$, the process can be treated as the movement in turbulent media of the mathematical surface propagating relative to fluid with normal velocity u_n. (Under these conditions the propagation velocity of this surface is not affected by strain or curvature.)

For the case $u' \gg u_n$ (the most interesting condition in practice) since on average all the material surfaces are stretching. In the reference frame attached to any of them there exists straining motion with a characteristic rate of strain u'/ℓ, and the flame front cannot depart farther than the distance $\Delta\chi \sim u_n\ell/u' \ll \ell$ from any material surface with which it once coincided. In turbulent dispersion it can be treated as a passive surface, and Taylor's theory of diffusion by continuous movements can be used.

Dispersion of the initially flat flame front (and also the volume of fluid drawn into combustion region) grows with time as $\sqrt{\tau}$ for large $\tau = tu'/\ell$, while the surface area of the front is increasing as $\exp \tau$ for large τ. The exponential growth of its area would quickly lead to complete burnout in the dispersion zone. Evidently there exists a stationary thickness of the combustion region for which both these trends are balanced and the region moves ahead with the turbulent flame speed u_T, corresponding to the total area of the flame front. For small u_n/u' and $\Delta\chi/\ell$, the dispersion of the flame surface is considerable, so that turbulent flame consists of interwoven layers of unburnt mixture and combustion products. The burnout can be viewed locally as the disappearance of a fresh mixture layer between two opposite flame fronts resulting from their propagation and hydrodynamic stretching of the layer itself. The time for burnout of such a layer with the initial width y is

$$t_y \simeq \ell/u' \ln [1 + (y/\ell)(u'/u_n)] \qquad (3)$$

Equation (3) with $y \sim \ell$, gives a very rough estimate of characteristic combustion time in the turbulent flame front.

In principle, it is possible to calculate u_T from the balance between burnout and the growth of dispersion region, but Klimov (1975) proposed a preferable method, since it emphasizes a very important feature of the process.

Turbulent dispersion leads to irregular disposition of volumes of fresh mixture and combustion products divided by the flame front. Because of the chaotic nature of the process, some parts of the flame front are closer together than others and meet at an early stage. The result is the formation of isolated flame front sections which surround enveloped volumes of unburned mixture and which fall behind

the leading front. The mean position of the latter is
advanced by such outflankings at the speed u_T. The process
described by Eq. (3) can serve as a qualitative model of the
closing phase of the outflankings (when the nearest
protruding parts of the flame front meet). The main
contribution to the event comes from the stretching of the
interstice between the flame fronts (hydrodynamic factor),
but motion of the fronts relative to fluid (physicochemical
factor) is indispensable: $t_y \to \infty$ when $u_n \to 0$ and the parts
of a material surface in continuous media cannot touch each
other. This is the way by which u_T depends on u_n when $u' \gg u_n$.

The physical picture described leads to the following
procedure for estimating u_T. Taylor dispersion Y of the
initally flat leading flame front increases with time.
After time t_* outflanking occurs and the cycle is repeated
again. This gives

$$u_T \approx Y(t_*)/t_*$$

The average distance between adjacent parts of the material
surface at which the leading flame front was at $t = 0$ is

$$\delta_o \approx 2Y(\tau)/[S(\tau) - 1]$$

where $S(\tau)$ is the area of the material surface related to
the area of the initial plane surface. The flame front
advance relative to fluid decreases the width of the fresh
mixture layer by twice the ratio of the fluid volume taken
up by the leading flame during the time τ to the flame area
at the moment of time τ:

$$\delta_1 = 2\ell(u_n/u')[1/S(\tau)] \int_0^\tau S(\tau) \, d\tau$$

The condition for outflanking is

$$\alpha \, \delta_o = \delta_1$$

where α is a constant appreciably smaller than unity. It
accounts for the nonuniformity of the actual width of the
interstices and is determined by the properties of
turbulence alone.

Thus

$$\frac{\alpha Y(\tau_*)}{S(\tau_*) - 1} = \ell \frac{u_n}{u'} \frac{1}{S(\tau_*)} \int_0^{\tau_*} S(\tau) \, d\tau \qquad (4)$$

The approximate solution of this equation (Klimov 1975) gives

$$u_T/u_n \approx (1/\alpha)^{0.3} (u'/u_n)^{0.7} \qquad (5)$$

Burnout of enclosed volumes proceeds approximately in accordance with Eq. (3). Yet in real flames it can be slack because of the turbulent energy decay due to dissipation and dilatational effects of thermal expansion. This underlines the advantage of singling out the leading front which operates in the region of unweakened turbulence.

So far, the most comprehensive experiments of Kozachenko (1960) and Talantov (1964) give for $u' \gg u_n$,

$$u_T/u_n \approx 3.5 \, (u'/u_n)^m \qquad (6)$$

with $m = 0.67$ (Kozachenko 1960) and $m = 0.75$ (Talantov 1964).

One of the reasons for the large factor which appears in Eq. (6) could be the circumstance not accounted for in Eq. (4). The irreversible nature of outflankings should give more weight to faster than average turbulent pulsations which make the effective hydrodynamic velocity greater than the rms u'. Another cause will be discussed in the next section.

As was proved earlier in this section, the flame front cannot "depart" from any material surface with which it was associated in the past. This apparently contradicts the fact that the mean position of the flame front advances as τ, while the dispersion of the material surface increases as $\sqrt{\tau}$. Actually there is no contradiction, since the parts of the material surface "attached" to the flame front constitute an exponentially decreasing share of the whole material surface area (the flame front area is constant, but the material surface area is growing exponentially) and behave as the collisionless elements in the finite velocity diffusion, that is, they move with a constant speed.

This description of turbulent flame propagation is valid also when $\langle \gamma \rangle \sim 1$, with the only difference that u_n no longer has significance here and should be replaced by the specific productivity of the laminar flame front.

For $\langle \gamma \rangle \gg 1$ (volume mode), the Reynolds number $u'\ell/u_n\delta_n$ acts as the main parameter, while in the surface mode of combustion it does not play an independent role.

When $u' \sim u_n$, the quasi-Lagrangian approach is not valid, since the flame front cannot be treated as passive during the dispersion process. The necessity of accounting for both Eulerian and Lagrangian aspects of the phenomena leaves little hope for the construction of a relevant formalism. Qualitatively it may be expected here that $u_T \sim u'$. Additional considerations on this subject will be presented in connection with the role of thermal expansion.

Equation (5) was obtained for large Reynolds numbers ($u' \gg u_n$, $\ell \gg \delta_n$) and Damköhler numbers ($u'\delta_n/u_n\ell \ll 1$). In the majority of current theoretical approaches they are regarded as the conditions for the limiting case where the time-averaged heat release is controlled by turbulent mixing [see Libby et al. (1979) for an example]. The physical picture described above and in Eq. (5) show that this limiting case does not exist in premixed combustion. Care should also be taken with the notion of "fast chemistry," otherwise it could be misleading. Premixed flames are always reaction rate limited, and this notion means only that the surface mode of combustion is realized.

The Role of Thermal Expansion

Thus far the analysis has ignored both the existence of the fine structure of turbulence and the interaction of the velocity field with the gas density field. The former is always present, if the turbulence Reynolds number $Re_\lambda = u'\lambda/\nu$ is large. The latter may be absent, since the combustion with small relative temperature change is possible. But in practice the density change is usually appreciable. It is preferable to first discuss its influence on combustion at the level of large eddies.

Concepts of additional turbulence caused by thermal expansion have a long history, but convincing theoretical or experimental evidence is still absent. The obvious exception is combustion in a channel with pressure gradient, though in this case thermal expansion acts indirectly by changing the general structure of the flow.

Probably the flame-generated velocity pulsations (being mostly potential) are quickly leveled off. Anyway, for $u' \gg u_n$, they are masked by the high level of the cold-flow turbulence.

More significant is the anomalously fast propagation of flame along a vortex core first discovered by McCormack et al. (1972) and explored also by Margolin and Karpov (1974), who indicated its possible importance in turbulent combustion. The hydrodynamic estimate of Chomiak (1977) shows that combustion products are pulled along the vortex axis with the velocity $u'\sqrt{\theta}$, where θ is the thermal expansion ratio. Chomiak suggested that flame propagation inside the Kolmogorov eddies is a dominant feature of turbulent combustion.

In the next section it will be shown that the role of Kolmogorov eddies is vanishingly small in combustion. The phenomenon of pulling the light combustion products inside vortex cores undoubtedly is important for turbulent flame propagation, but it acts on the level of energy-containing eddies; small-scale events are precluded by quenching.

The estimate obtained by Chomiak seems justified. It can be reformulated in the following way: the pressure fluctuation level Δp is determined by cold-flow turbulence, so that

$$\Delta p \sim \rho u'^2 \sim \tilde{\rho}\tilde{u}'^2 \qquad \tilde{u}' \sim u'\sqrt{\theta}$$

where ρ is the density of a fresh mixture, $\tilde{\rho}$ is the density of combustion products, and $\tilde{u}'$ is the pulsation velocity of combustion products induced by pressure fluctuations Δp.

The "sucked" flame takes on the cylindrical shape and can be easily quenched by concomitant straining motion; the critical strain rate is several times lower here than in the case of a flat layer. Large eddies have the double advantage of less quenching and greater $\tilde{u}'$.

Vortex tubes are either closed or very entangled, so the advance of flames along their cores does not lead to straight-through marching of ignition sites across the fresh mixture; the direction of advance is frequently changed. This means that the mechanism of flame propagation is the same as that described above, but with larger effective hydrodynamic velocity. In particular, when $u' \gg u_n$, the flame advance along the vortices cannot be viewed as "ignition," since transfer of flame between vortices by propagation with speed u_n is too slow to compete with the mechanism at which flame propagation over fluid is coupled with hydrodynamic stretching of interstices.

This density related effect and the action of faster than average eddies taken together provide a quite realistic explanation of the large factor in Eq. (6), although for exact calculation more rigorous formalism is needed.

When $u' \sim u_n$, the flame transfer between vortex tubes is fast enough, and the process indeed resembles "ignition" by advancement of flame along the fastest vortices. This can readily result in the experimentally observed $u_T/u' \simeq 4\text{-}5$ for a typical θ. It should be noted that a hypothesis of ignition by fast moving points has been suggested by Zel'dovich (1947).

Small-Scale Turbulence

Developed turbulence is characterized by a wide spectrum of the length and velocity scales of fluid motion, and $\langle\gamma\rangle$ is not a number but a spectral function. Characteristic strain rates vary by orders of magnitude along the spectrum.

Formally made estimates of $\langle\gamma\rangle$ at different wave number ranges suggest that in immediately practical conditions of air-fuel mixture combustion (i.e., for $Re_\lambda > 100$) laminar flames cannot exist, since at the level of length scales of the order of δ_n, both $\langle\gamma\rangle$ and the spectral Reynolds number (i.e., the turbulent diffusivity) are large. Instead, locally there should exist reaction zones dominated by turbulent transport. They are dispersed and strained by larger eddies. In view of the very slow decrease of characteristic spectral velocity with increasing wave number, a hierarchy of outflanking activity is not ruled out.

This picture, however, contradicts the important experimental evidence. First, molecular transport phenomena (preferential diffusion, thermal-diffusive effects) are important even in strong turbulence (Karpov and Severin 1980). Second, the involvement of the whole turbulence spectrum in outflanking dynamics of isothermal surfaces cannot be reconciled with the following considerations (Klimov 1975). When $\Omega \gg 1$ reaction zones inside turbulent flames occupy a relative volume, $\xi = u_T \delta_n / u_n \delta_T$, where δ_T is the turbulent flame thickness; since $u_T \lesssim u'$ and $\delta_T \gg \ell$ (empirical evidence), it follows that $\xi \ll u'\delta_n/u_n\ell = \gamma_o$ and the chemical reaction up to $\gamma_o \sim 1$ is concentrated in thin layers. The latter advance with normal velocity $\bar{u}(\ell)$, which is the normal velocity of isothermal surfaces averaged over volume ℓ^3. To satisfy $\delta_T \gg \ell$, it is necessary to have $u(\ell) \ll u'$, while a cascade of outflankings would give $\bar{u}(\ell) \sim u'$.

All these facts are undoubtedly related to internal intermittency of turbulence. Turbulent eddies are not space-filling (except for the large ones), and the velocity

field is not a continuous hierarchy of vortices, but rather
a superposition of structures of rather disparate scales.

A review of experimental data on intermittency (Antonia
et al. 1976) shows that at $Re_\lambda > 100$, high-frequency
activity spots occupy only about half of the space (the
spacings between them scale with λ). This means that
isothermal surfaces must consume about half of the fluid in
a "laminar" way, and this becomes a limiting step in the
combustion process.

Kolmogorov eddies could hardly influence combustion at
all, since they are completely lost in the preheat zone of
laminar flames. Their turbulent diffusivity is much smaller
than molecular diffusivity in a high-temperature reaction
zone:

$$\nu \ll u_n \delta_n$$

They are also weakened by dilatation, and of course are
highly intermittent. It can therefore be seen that the use
of the strain rate u'/λ in Eq. (1) is not a relevant choice.

Larger small-scale eddies are also weakened by
dilatation, and their influence on isothermal surfaces is
diminished by the stretching of the latter caused by large
eddies and the repulsive action of thermal expansion [this
last effect was not accounted for in Eq. (4), but it should
lead to faster decrease of δ_0].

All this explains the success of the simplified model
of turbulent flame. Yet some more subtle phenomena cannot
be understood without small-scale effects. For instance,
preferential diffusion takes place at fine corrugations of
laminar (or quasilaminar) flames. High-frequency spots
introduce elements of "volume" combustion with the
dependence of u_T on gas pressure and so on. At very large
Reynolds numbers, the eddies of intermediate scale may
become important (and even lead to outflankings on their
level). The role of the turbulence spectrum remains a major
problem in turbulent combustion, and the main hope there
lies in modern gas diagnostics.

Concluding Remarks

The approaches based on the direct use of modern
statistical theories of turbulence have been under
development for the last decade [Libby and Williams (1981)
and O'Brien (1982) provide the latest assessment of the
current status]. They are undoubtedly appropriate for the
analysis of some fluid dynamical aspects of turbulent
premixed combustion. At the same time they ignore the fact

that in premixed flames the molecular terms in averaged
conservation equations are in principle nontrivial, because
combustion introduces its own length and time scales, which
results in the emergence of a kinematic element -
propagation of isoscalar surfaces relative to fluid with a
speed depending on the physicochemical properties of the gas
mixture. So, even at large Reynolds and Damköhler numbers,
these terms cannot be neglected. The physical reasons for
this are clear from the described model of turbulent flame
propagation, whose speed and thickness always depend on
chemical and molecular transport phenomena. Moreover, the
whole process seems to be too ordered at certain levels
(outflankings, pulling in vortex cores, kinematic element,
in general) to be compatible with the existing formalism of
statistical theories. In that sense probably there is an
analogy with the problem of coherent structures in turbulent
shear flows.

References

Antonia, R. A., Danh, H. Q., and Prabhu, A. (1976) Bursts in turbulent shear flows. Phys. Fluids 19, 1680-1686.

Chomiak, J. (1977) Dissipation fluctuations in the structure and propagation of turbulent flames in premixed gases at high Reynolds numbers. 16th International Symposium on Combustion, pp. 1665-1675. The Combustion Institute, Pittsburgh, Pa.

Damköhler, G. (1940) Der einfluss der turbulenz auf die flammengeschwindigkeit in gasgemischen. Z. Elektrochem. 46, 601.

Karpo, V. P. and Severin, E. S. (1980) Influence of molecularr transport properties on turbulen burning rate. Phys. Combust. Explos., 45-51.

Klimov, A. M. (1963) Laminar flame in turbulent flow. Zh. Prikl. Mekh. Tech. Fiz., 49-58.

Klimov, A. M. (1975) Flame propagation under conditions of strong turbulence. Dokl. Akad. Nauk SSSR 221, 56-59.

Klimov, A. M. and Lebedev, V. N. (1980) On the local reaction rate in combustion in turbulent media. Chemical Physics of Combustion and Explosion Processes. Combustion of Gases and Naturla Fuels, pp. 25-29. Chernogolovka.

Kovasznay, L. S. (1956) A comment on turbulent combustion. Jet Propol. 26, 485.

Kozachenko, L. S. (1960) Combustion of gasoline-air mixtures in turbulent flow. The Third All-Union Conference on Combustion Theory, Vol. 1, pp. 127-137. Moscow, USSR.

Libby, P. A., Bray, K. N. C., and Moss, J. B. (1979) Effects of finite reaction rate and molecular transport in premixed turbulent combustion. Combust. Flame 34, 285-301.

Libby, P. A. and Williams, F. A. (1981) Some implications of recent theoretical studies in turbulent combustion. AIAA J. 19, 261-274.

Margolin, A. D. and Karpov, V. P. (1974) Combustion of rotating gas. Dokl. Akad. Nauk SSSR 216, 346-349; see also SAE Paper 741165.

McCormack, P. D., Scheller, K., Mueller, G., and Tisher, R. (1972) Flame propagation in a vortex core. Combust. Flame 19, 297-303.

O'Brien, E. E. (1981) Statistical methods in reacting turbulent flows. AIAA J. 19, 366-371.

Schelkin, K. I. (1943) On combustion in turbulent flow. Zh. Tekh. Fiz. 13, 520-530.

Schetinkov, E. S. (1956) Theoretical study of premixed combustion in turbulent flow. O Turbulentnom Gorenii Gomogennoi Smesi, p. 50. Oborongiz.

Spalding, D. B. (1978) A general theory of turbulent combustion. J. Energy 2, 16-23.

Summerfield, M., Reiter, S. H., Kebely, V., and Mascolo, R. W. (1955) The physical structure of turbulent flames. Jet Propul. 25, 377.

Talantov, A. V. (1964) Izd. Mashinost. Theor. i raschet pryamotochnikh kamer sgoraniya (edited by S. M. Iljashenko and A. V. Talantov), Chap. 6, pp. 148-207.

Williams, F. A. (1965) Combustion Theory. Addison-Wesley, Reading, Mass.

Zel'dovich, Y. B. (1947) On the theory of initiation of detonation in gases. Zh. Tekh. Fiz. 17, 3-26.

Modification of Turbulent Flowfield by an Oblique Premixed Hydrogen-Air Flame

D. Escudie,* M. Trinite,† and P. Paranthoen‡
Laboratoire de Thermodynamique, Mont Saint Aignan, France

Abstract

In this paper, results concerning a flowfield modified by a flame front are presented, including the mean temperature and concentration field, the mean values, and the fluctuations and correlations of velocity components [lased Doppler velocimetry (LDV) measurements]. The combustion apparatus allows us to distinguish several different reasons for velocity change such as the wake of the flame stabilizer; turbulence generation associated with the shear or the mean pressure gradient; and direct turbulence combustion interaction. Analysis of the results shows a typical loss of turbulent isotropy in the flame on the part of the longitudinal velocity component. A strong increase in the longitudinal component of velocity fluctuation is related to the mean pressure gradient. The deflection of the mean streamlines is demonstrated and interesting results concerning the gradient transport assumption are also obtained.

Introduction

The influence of turbulence on flame is usually examined in terms of the correlation between the apparent

---

Presented at the 8th ICOGER, Minsk, USSR, Aug. 23-26, 1981. Copyright © American Institute of Aeronautics and Astronautics, Inc., 1982. All rights reserved.
*Boursier D.G.R.S.T. Faculté des sciences et Techniques de Rouen.
†Maitre de Recherche au C.N.R.S. Faculté des Sciences et Techniques de Rouen.
‡Chargé de Recherche au C.N.R.S. Faculté des Sciences et Techniques de Rouen.

burning velocity and the turbulence intensity. Dispersion of common relations is probably due to the difficulty of defining exactly the turbulence level in the reaction zone and of taking into account the role of the flame-generated turbulence.

This turbulence generation must be associated with the shear and with the pressure gradients which are generated by the flame itself and depend on combustor geometry and boundary conditions. Thus for turbulence modeling and numerical computation of the combustion flowfield, details of the flow conditions must be properly taken into account.

With the advent of LDV techniques, important progress is possible in turbulent flame measurements. In the case of a diffusion turbulent flame, interesting results have been published by Starner and Bilger (1980). The results of Moss (1980) concerning a premixed flame must be also pointed out. A theoretical approach concerning turbulence balance in this premixed case was recently presented by Bray et al. (1981a).

The measurements presented here are an extension of the works of Beaudet (1979) and Trinite and Beaudet (1981). The present results concerning mean velocity and turbulent parameters were made using a laser Doppler velocimeter, whose basic principle of operation is now well known.

Experimental Conditions

The basic experimental apparatus was presented earlier at the Göttingen Symposium (Trinite and Beaudet 1981) and is summarized in Fig. 1; following a settling chamber, the combustion chamber has a 8 x 8-cm cross section. The planar oblique flame is stabilized without recirculation effect on a 0.4-mm-diam catalytic wire. Control over the turbulence level was achieved by means of grids located at the entry of the chamber.

The value of the equivalent ratio (0.25) of hydrogen was such that the maximum mean temperature never exceeded $900°C$.

The mean temperature was measured with a coated chromel-alumel thermocouple and the mean hydrogen concentration by means of an isokinetic sampling probe. For mean velocity or turbulence measurements a classical forward scatter dual beam LDV technique was used. Two remarks pertaining to this method follow, one concerning particles seeding and the other concerning bias effect.

Particles Seeding

Because of the low velocity in the settling chamber, seeding of the whole flow was not possible and a local

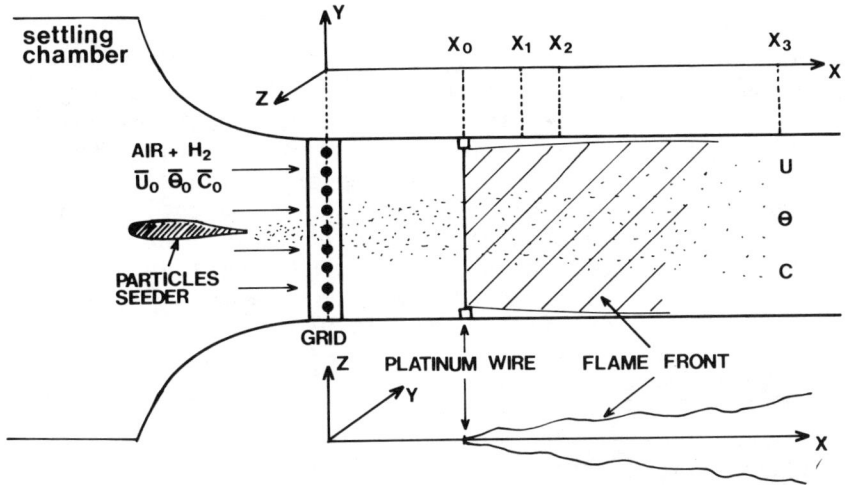

Fig. 1 Experimental conditions.

seeder was used. The flow was locally injected with alumina particles, perpendicular to the mean flame plane owing to an air foil seeder; downstream, the wake of the seeder was compensated by adjustment of the injection rate.

Bias Effect

The presence of an all-burned and unburned mixture introduces large fluctuations in the particles concentration which are probably correlated to the velocity fluctuations. Consequently a bias can be introduced with conventional arithmetic averaging.

Correct values may be obtained with the so-called residence time weighted averaging in which the residence time of each sample is a weighting factor for each individual data point (Buchhave and George 1978).

The main parameters of this experiment are summarized below:
1) Upstream flow: Mean velocity $\overline{U}$ = 5 m/s. Equivalent ratio of hydrogen = 0.25 (C_0 = 7.65% in volume).
2) Grid: Round bar, square mesh, M_G = 5 mm.
3) Typical distances: Stabilization of the flame, X = 12 M_G; measurements between X/M_G = 20 and X/M_G = 35.
4) Characteristics of turbulence: Turbulence intensity I from 3 up to 5%; Taylor microscale λ_f from 2 up to 4 mm; integral macroscale L_f from 3 up to 8 mm; and turbulent Reynolds number Re = $\lambda_f u'/\nu$ from 20 up to 40.

Experimental Results

Measurements were made with and without combustion and can be compared by means of our experimental device. This allows us to distinguish the following different reasons for velocity field change: 1) the wake of the wire used to stabilize the flame; 2) turbulence generation associated with the shear or with the mean pressure gradient; and 3) direct turbulence combustion interaction.

Two typical sections are considered: $x_1 = 4.5$ cm, downstream of the wire, where the wake is important; and $x_2 = 11$ cm, where the wake is negligible.

Mean Measurements

Mean Velocity. The mean velocity $\overline{U}$ vs the transversal coordinate z is presented in Figs. 2 and 3 for sections x_1 and x_2 respectively, with and without combustion with the same upstream conditions. (In the figures presented in this paper, the solid lines and the dashed lines correspond to the combustion case and to the noncombustion case, respectively.)

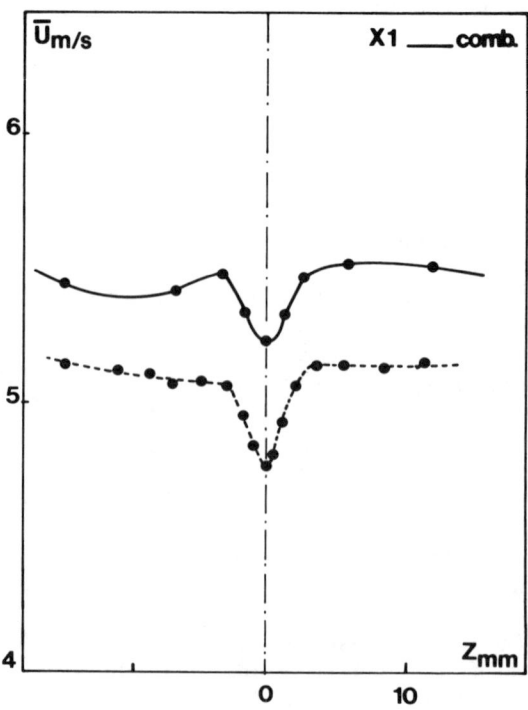

Fig. 2 Longitudinal mean velocity profiles with and without combustion in the x_1 section.

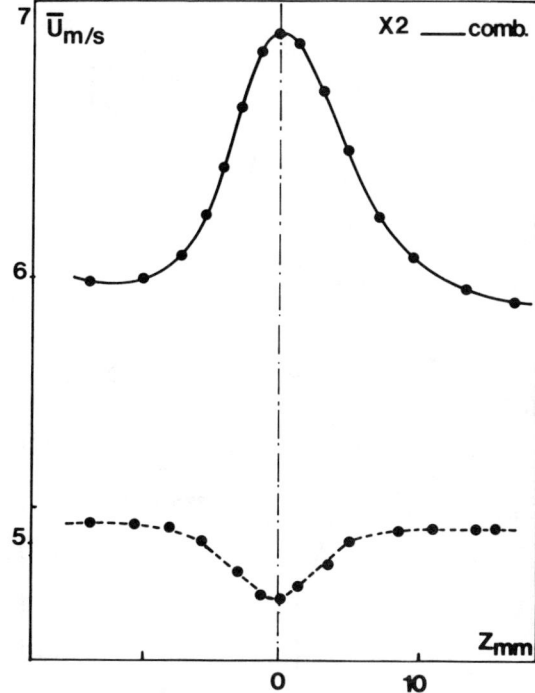

Fig. 3 Longitudinal mean velocity profiles with and without combustion in the x_2 section.

In the x_1 section, without combustion, a decay of mean velocity is observed in the wake of the wire; the minimum is located on the centerline. With combustion, the wake effect is somewhat smaller and the mean velocity increases in all the sections in respect to the noncombustion case.

In the x_2 section, the wake effect decreases without combustion. On the other hand, with combustion, an important increase in velocity within the flame can be observed.

The longitudinal evolution of mean velocity is summarized in Fig. 4.

Mean Pressure Gradient. Because of the confinement of the flame, a mean pressure gradient exists (Fig. 5). In the x_2 section, the gradient value is constant (-80 Pa/m in the combustion case).

Mean Temperature and Concentration Profiles. The corresponding profiles for temperature and concentration are shown in Figs. 6 and 7, respectively. An increase in the maximum temperature for small values of x is due to the

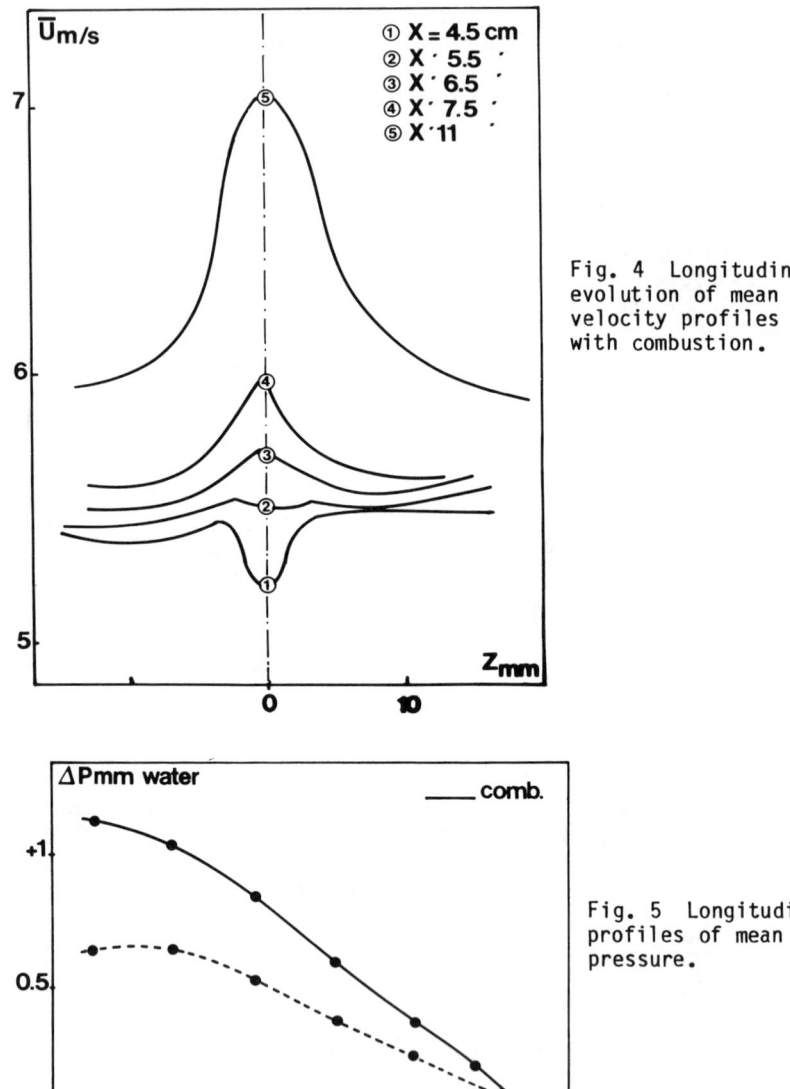

Fig. 4 Longitudinal evolution of mean velocity profiles with combustion.

Fig. 5 Longitudinal profiles of mean pressure.

presence of the catalytic wire; because of the low value of the hydrogen Schmidt number, the wire temperature is higher than the adiabatic temperature.

<u>Mean Deflection of the Streamlines</u>. The lateral mean velocity $\bar{U}_z$ is shown in Fig. 8: Without combustion, the

MODIFICATION OF FLOWFIELD BY AN OBLIQUE FLAME 153

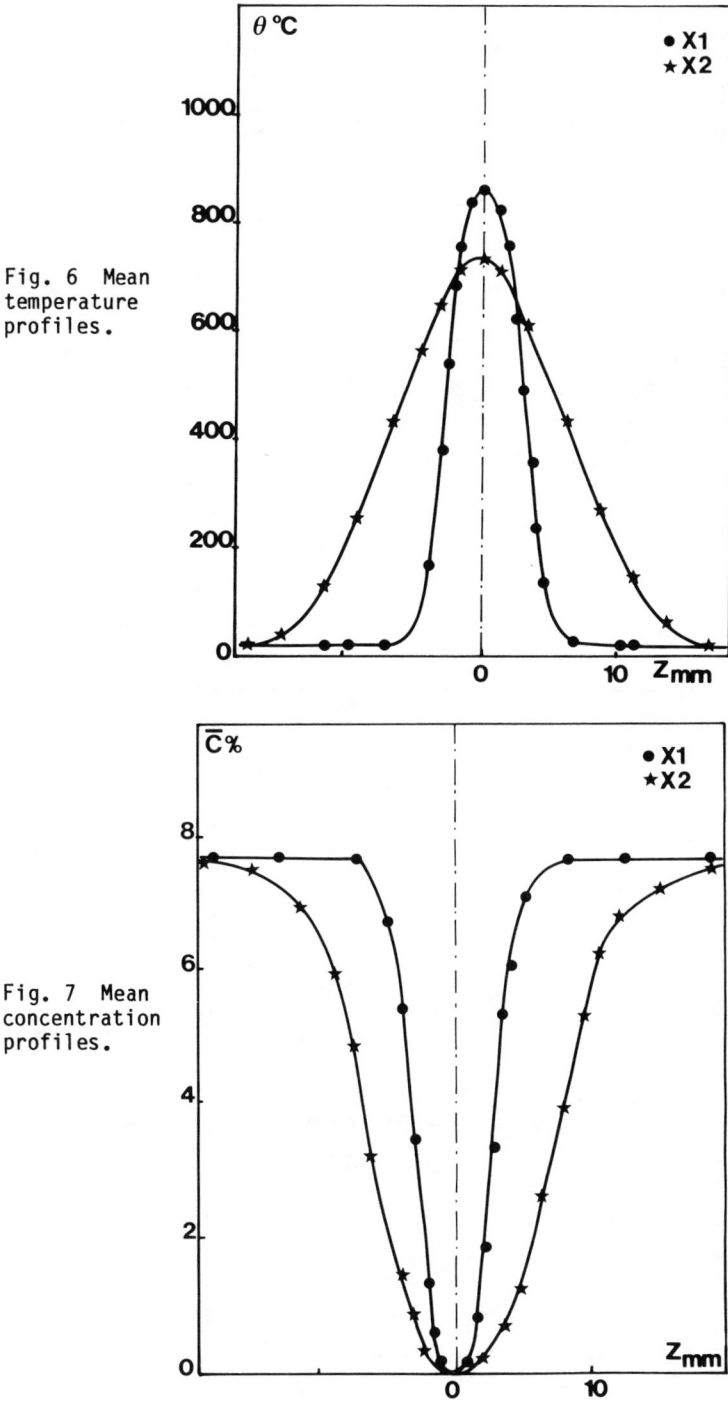

Fig. 6 Mean temperature profiles.

Fig. 7 Mean concentration profiles.

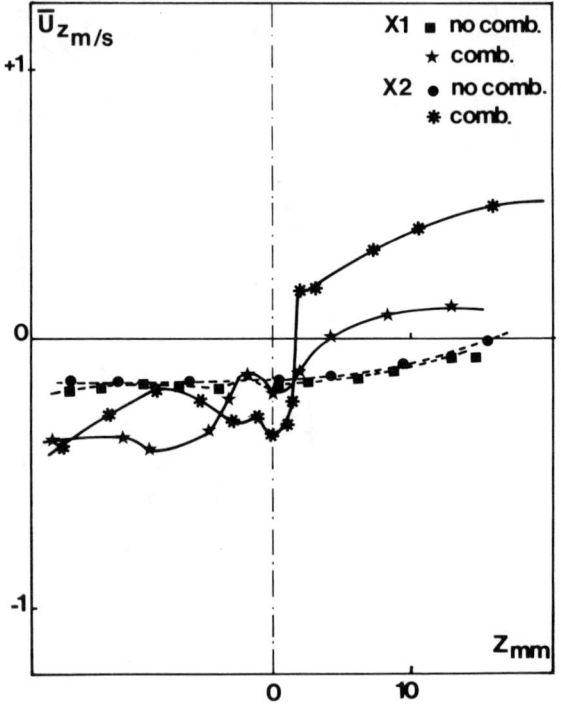

Fig. 8 Lateral mean velocity profiles.

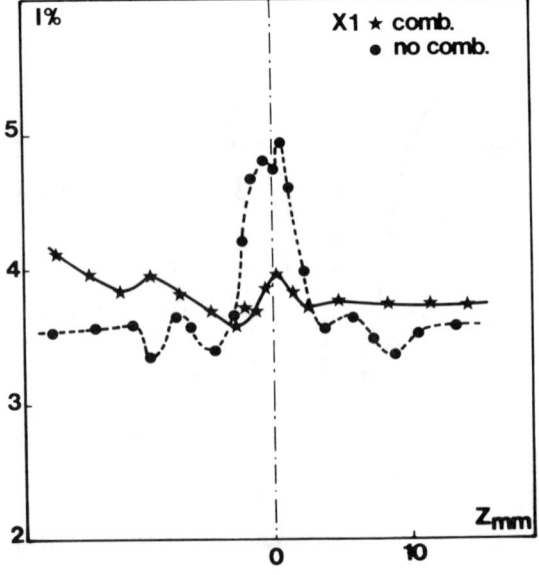

Fig. 9 Longitudinal turbulence intensity profiles with and without combustion in the x_1 section.

lateral velocity is constant and would be zero with a correct reference angle; with combustion, negative and positive components clearly appear.

Difficulties in obtaining more accurate results are due to the existence of gravity effects induced by the horizontal position of the combustion chamber.

Turbulence Measurements

Longitudinal Turbulence Intensity. Results are presented in Figs. 9 and 10 for sections x_1 and x_2, respectively.

In the x_1 section, without combustion, the maximum turbulent intensity in the centerline is due to the velocity gradient in the wake of the wire. With combustion, this maximum does not exist; it could be due to the increase of the dissipative terms with temperature. In the x_2 section, without combustion, the wake effect still remains, but with combustion, an important created turbulence can be observed. The two maximums are centered on the location of maximum velocity gradients.

This increase in turbulence intensity can be explained by considering the two productions terms: 1) the term of velocity gradient, $\overline{\rho\, u''v''}(\partial U/\partial y)$; and 2) the term of pressure gradient, $\overline{u''}\,(\partial p/\partial x)$.

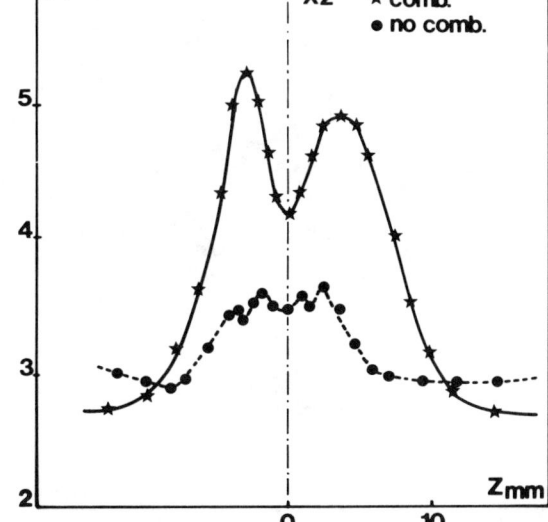

Fig. 10 Longitudinal turbulence intensity profiles with and without combustion in the x_2 section.

In the x_2 section the pressure gradient term predominates:

$$u'' \;\overline{\frac{\partial p}{\partial x}}/\rho \;\Big/\; \overline{u''v''}\;\frac{\partial \overline{U}}{\partial x} \sim 10$$

Transverse Turbulence Intensity. Profiles of the transverse component of fluctuation velocity measured within the flame are shown in Figs. 11 and 12. The turbulence decay observed at the x_1 location can be explained by an increase in dissipative terms for the longitudinal intensity. However, at the second location, transverse turbulent intensity is significantly greater within the flame than in the nonreactive flow. Furthermore, it is worth noting the dissymmetry of the profile due to the opposite thermal stratifications within the flame (stable stratification in the lower part of the flame and unstable stratification in the upper part of the flame).

Isotropy. In Figs. 11 and 12 longitudinal and lateral velocity components are compared with and without combustion.

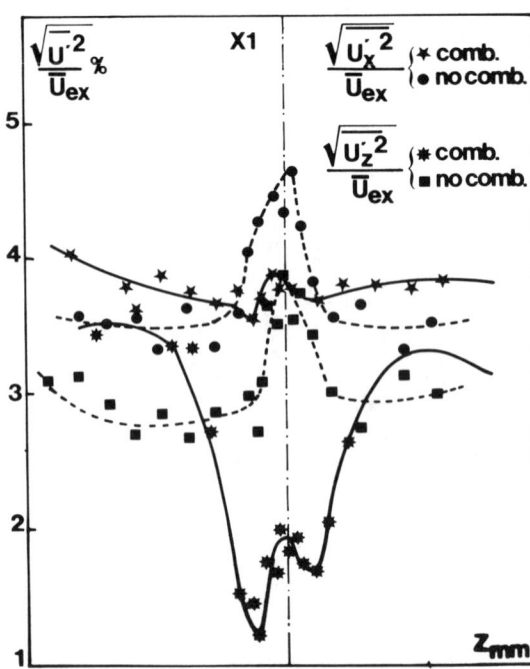

Fig. 11 Longitudinal and lateral fluctuation velocity profiles in the x_1 section.

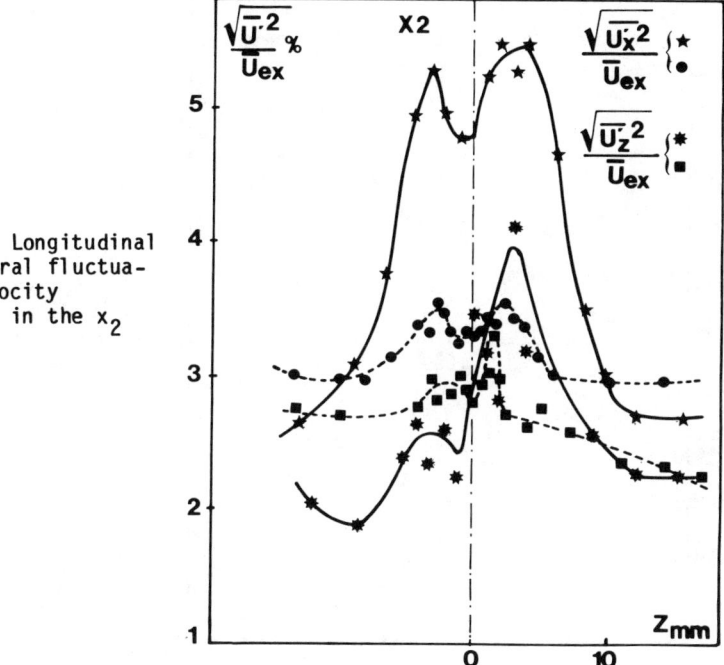

Fig. 12 Longitudinal and lateral fluctuating velocity profiles in the x_2 section.

In the first section, without combustion, a small anisotropy exists and remains constant in the wake of the wire. With combustion, an important decrease in the lateral fluctuation component is observed within the flame.

In the second section, this lack of isotropy is also demonstrated; increasing turbulence by combustion is less important for the lateral component than for the longitudinal one.

This effect was also noticed by Glass and Bilger (1978) in their study of hydrogen diffusion flame and by Bray et al. (1981b) in their analysis of confined turbulent flames predictions.

Skewness. Our data acquisition system was not sufficiently developed to obtain a direct probability density function, but an interesting indication has been obtained from the skewness of longitudinal velocity fluctuation presented in Figs. 13 and 14.

In the x_1 section, the two negative minimums correspond to the velocity profile in the wake without combustion. With combustion, a more complex profile is obtained through the mixing of the wake effect and the flame effect; negative values in the centerline are noted.

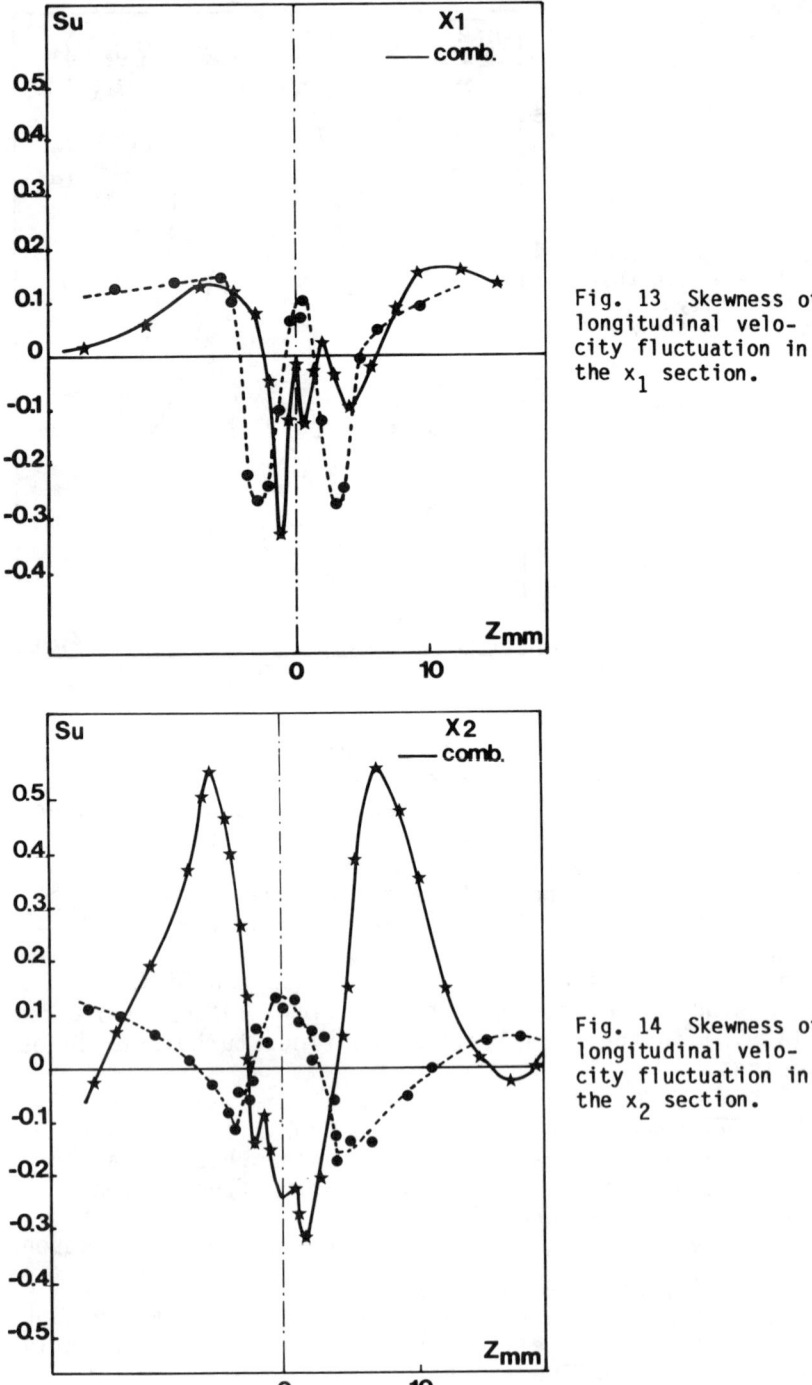

Fig. 13 Skewness of longitudinal velocity fluctuation in the x_1 section.

Fig. 14 Skewness of longitudinal velocity fluctuation in the x_2 section.

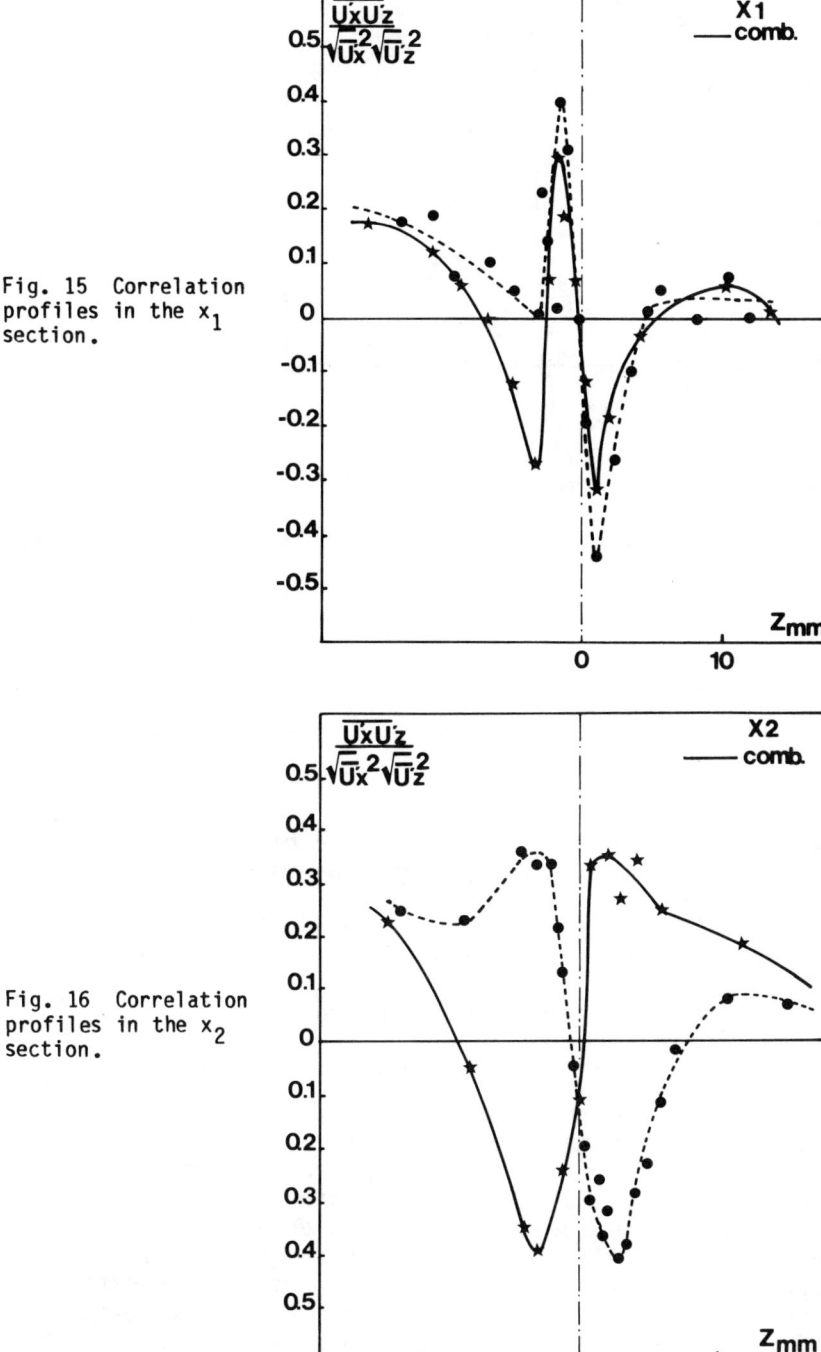

Fig. 15 Correlation profiles in the x_1 section.

Fig. 16 Correlation profiles in the x_2 section.

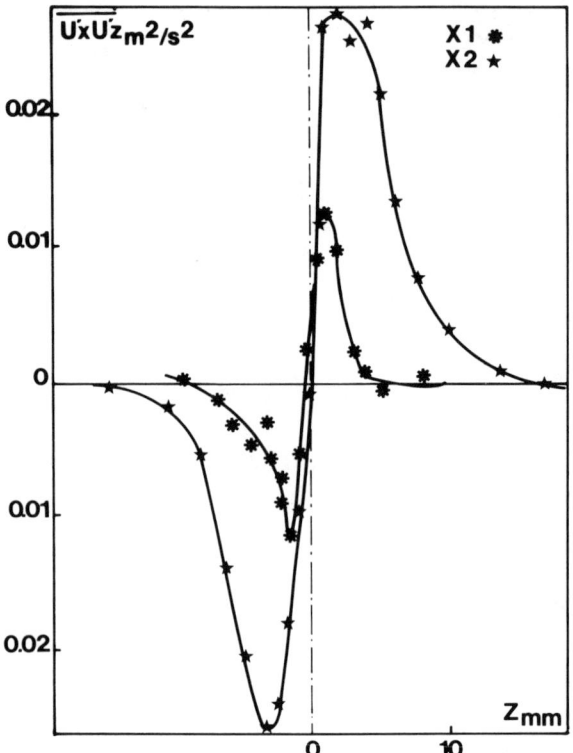

Fig. 17 Corrected correlation profiles.

In the x_2 section, the wake effect is less important and the combustion effect appears clearly: two positive maximums on the edge of the flame zone and a negative value in the centerline.

These profiles can be explained by the two states assumption: presence of all-burned or unburned packets in the mixing.

Correlation. the correlation coefficients $\overline{u'_x u'_z}/(\overline{u'^2_x})^{1/2}(\overline{u'^2_z})^{1/2}$ are shown in Figs. 15 and 16.

Without combustion, positive and negative extrema correspond to the maximum velocity gradients in the wake for the two sections. With combustion, new extrema appear with an opposite sign. In the first section these extrema do not correspond to the mean velocity gradients.

If the wake fluctuations generated by the wire are supposed to be independent of grid- or flame-generated fluctuations, it is possible to correct quadratically these profiles. The corrected correlations are presented in Fig.

17. For the first section, x_1, the assumption of the gradient transport theory seems doubtful, but the results can be explained by the two states assumption. In the presence of all-burned fluid, an increase in velocity exists corresponding to the production of correlated velocity fluctuations, and in the presence of unburned fluid, a decrease in velocity due to the wake exists corresponding to the production of correlated velocity fluctuations with the opposite sign.

Discussion and Conclusion

A satisfactory interpretation of some of these results can be obtained by considering, as did Bray et al. (1981a), a conditioned function, the concentration C, or the progress variable C (Fig. 18).

Only two values of C are possible: $\tilde{C} = 0$ in the all-burned gases (product), or $\tilde{C} = 1$ in the unburned gases (reactant).

This assumption is possible, in our case, because of the fast chemistry of hydrogen.

This analysis leads to consideration of the conditioned velocity within the reactant or the product: the mean

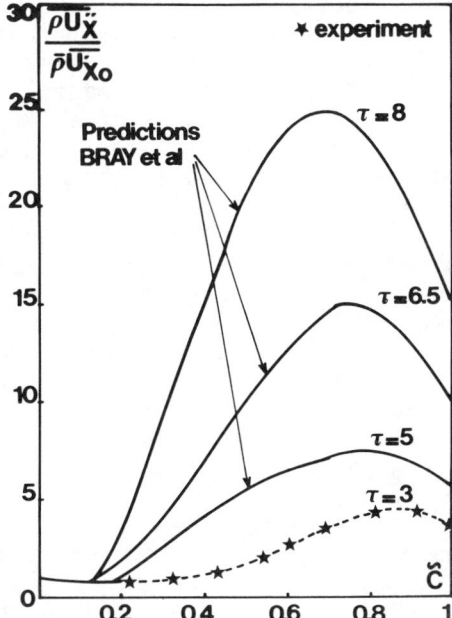

Fig. 18 Comparison with the Bray et al. prediction of the longitudinal turbulent energy.

velocities $\tilde{U}'_r$ and $\tilde{U}'_p$, and the fluctuations intensities $\widetilde{u'^2_r}$ and $\widetilde{u'^2_p}$.

The different equations can be expressed in terms of these four conditioned quantities. For example, the turbulence energy,

$$\overline{\rho u''^2}/\bar{\rho} = \tilde{C}(1-\tilde{C})(\tilde{U}_p - \tilde{U}_r)^2 + (1-\tilde{C})\widetilde{u'^2_r} + \tilde{C}\widetilde{u'^2_p}$$

The intensity of velocity fluctuations depends on the square of the difference between the two conditioned velocities and on the contributions of the two conditioned intensities.

A comparison between our results and the predictions of Bray et al. (1981a) has been made. These authors introduce the normalized temperature $\tau = T_\infty/T_0 - 1$ and predict the normalized turbulent energy $\overline{\rho u''^2_x}/\rho u'^2_{ox}$ vs the progress variable $\tilde{C}$, where u'^2_{ox} is the upstream turbulence. The predictions concern a plane flame analysis, and the corresponding value of τ is $\tau = 3$. Results are presented in Fig. 18. Only one variable is plotted for comparison with theory of Bray et al. (1981b) because these authors only consider the planar premixed flame perpendicular to the mean velocity. The surprising agreement can be explained by the large value of pressure gradient term in respect to the gradient term in our experiment, as already mentioned.

The production mechanisms of turbulence within the flame, proposed by Bray et al. (1981b) arises from the interaction between the self-induced mean pressure gradient and the fluctuations in the gas density within the turbulent flame.

The packets of reactant and product are not accelerated in the same way in the mean pressure gradient. The low-density burned gases are preferentially accelerated relative to the packets of reactants.

In the interest of obtaining conclusive proof, it seems more fruitful in this kind of experiment to perform conditioned measurements. Such measurements of the conditioned concentration were obtained in our flame and presented at Göttingen (Trinite and Beaudet 1981). Up to now, it has not been possible to connect the LDV measurements and the concentration measurements in our experiments. Moss (1980) has performed such measurements with an intensive seeding which is not possible with our apparatus.

Measurements presented in this paper show the influence of the longitudinal pressure gradient on the increase of the longitudinal component of fluctuating velocity. Gravity effects introduce a strong dissymmetry for transverse turbulent intensity profiles. Negative values of skewness of longitudinal velocity fluctuations seem to indicate the presence of unburned packets within the central zone of the flame.

In order to obtain a better knowledge of mechanism turbulence combustion interaction, it is essential to perform conditioned measurements in such experiments.

References

Beaudet, D. (1979) Interaction turbulence combustion dans le cas d'un prémélange air-hydrogène. Ph.D. Thesis, University of Rouen, France.

Bray, K. C. N., Libby, P. A., Masuya, G. and Moss, J. B. (1981a) Turbulence production in premixed turbulent flames. Combust. Sci. Technol. 25, 127-140.

Bray, K. C. N., Libby, P. A., and Masuya G. (1981b) Confined turbulent flames with premixed reactants. Paper presented at the First International Specialists Meeting of the Combustion Institute, Bordeaux, France.

Buchhave, P. and George, W. K. (1978) Bias corrections in turbulence measurements by the laser Doppler anemometer. Proc. 3rd International Workshop on Laser Velocimetry, Purdue University, Lafayette, Ind., pp. 80-85.

Glass, M. and Bilger, R. W. (1978) The turbulent jet diffusion flame in a co-flowing stream - some velocity measurements. Combust. Sci. Technol. 18, 165-177.

Moss, J. B. (1980) Simultaneous measurements of concentration and velocity in an open premixed turbulent flame. Combust. Sci. Technol. 20, 115.

Starner, S. H. and Bilger, R. W. (1980) LDA measurements in a turbulent diffusion flame with axial pressure gradient. Combust. Sci. Technol. 21, 259-276.

Trinite, M. and Beaudet, D. (1981) Measurement of the concentration probability density function in an air-hydrogen premixed turbulent flame. Combustion in Reactive Systems: AIAA Progress in Astronautics and Aeronautics (edited by Bowen, Manson, Oppenheim, and Soloukhin), Vol. 76, pp. 283-294, AIAA, New York.

Chapter III. Combustion of Solids

Combustion of Lithium Perchlorate, Ammonium Chloride-Ammonium Perchlorate Solid Mixtures

M.S. Al Fakir,* P. Joulain,† and J.M. Most†
Université de Poitiers, Poitiers, France

Abstract

This investigation is concerned with the search for the main parameters needed for the modeling of the burning of ammonium perchlorate solid mixtures in a turbulent flow of hot fuel. The influence of the additives (NH_4Cl; $LiClO_4$ $3H_2O$) is explained through the study of the reactive surface, gas, and condensed phase kinetics. The predictions of the mathematical modeling agree quite well with experimental results.

Nomenclature

A_s = overall gas phase frequency factor
A_D = overall solid phase (decomposition) frequency factor
B = mass transfer number
C_f = coefficient of friction without mass transfer
Cf^o = coefficient of friction with mass transfer
E_D = overall decomposition activation energy
E_s = overall gas phase activation energy
G = mass flow rate
h_D = heat of decomposition
h_g = heat of gasification
h_s = heat of sublimation
K_s = rate constant
K^s = constant equation (4)

Paper presented at the 8th ICOGER, Minsk, USSR, Aug. 23-26, 1981. Copyright © 1982 by M. S. Al Fakir. Published by the American Institute of Aeronautics and Astronautics, Inc. with permission.

*, Groupe de Recherches de Chimie Pysique de la Combustion. Presently at CERS, Damascus, Syria.

†, Groupe de Recherches de Chimie Physique de la Combustion.

m_g = 0.45 solid fraction which decomposes in the gas phase
m_s = $r \rho_s$ mass burning rate
n = pressure exponent
P = pressure
R = gas constant
r = linear burning rate
S/S_o = reacting surface/planar surface ratio
T_s = surface temperature
u_∞ = stream velocity
W = molar mass
Δn = variation of mole number during sublimation or dissociation
θ_D = surface adsorption ratio
ρ_s = solid density
τ_{s_o} = wall shear stress without mass transfer
τ_s = wall shear stress with mass transfer

Introduction

As a continuation of the study of combustion of composite propellants, the burning rate of ammonium perchlorate (AP) solid mixtures in a turbulent flow of gaseous fuel under pressure below the lower deflagration limit has been examined. The combustion is sustained by the heat released by the exothermic reaction between fuel and solid mixture decomposition products in a diffusion flame. Previous studies showed the hybrid character of these systems (Al Fakir et al. 1981; Al Fakir 1981).

Analysis of hybrid combustion must account for phenomena occurring in the solid phase (heat conduction, allotropic transformation, and decomposition); at the reactive surface (surface reaction and diffusion, adsorption-desorption, and sublimation); and in the gas phase (heat and mass transfer between the surface and the turbulent boundary layer and gas phase reactions: premixed and diffusion flame). This work is concerned with the evaluation of some of the most useful parameters for the modeling of these phenomena.

As previously shown (Al Fakir et al. 1981; Al Fakir 1981), the overall mass burning rate m_s can be obtained by simultaneous consideration of the diffusion and kinetic contribution:

$$m_s = A_s e^{-E_s/RT_s} = \frac{W_s}{m_g} \frac{S}{S_o} - \theta_D A_D e^{-E_D/Rt_s} \qquad (1)$$

and of an energy balance made on a control volume which encloses the solid surface:

$$m_s h_g = J_h + J_{rad} - J_{cond} \quad (2)$$

| gasification (solid phase) | convection (boundary layer including conduction and diffusion) | radiation | conduction (solid phase) |

Prior to the computation of m_s, determination of the unknown aerodynamic, thermodynamic, and kinetic parameters was attempted.

Reactive Surface

The structure of the solid propellant surface was studied by means of an electron scanning microscope. Pictures were taken after extinction (see Fig. 1). These pictures corroborate the assumptions about 1) the existence of a liquid layer on the surface; 2) a large reacting surface/planar surface ratio S/S_o; and 3) a small pore diameter. The existence of this liquid layer is very important. Without this layer it would be impossible to develop the following kinetic approach, which postulates [as did Guirao (1970)] that the reactions are initiated in the liquid layer.

Kinetic Study

Gas Phase

Ammonium Perchlorate. The analysis started with (Guirao 1970) previous kinetic schema for AP and includes the following equations for HCl and OH consumption:

$$ClO + HCl \rightarrow OH + Cl_2 \quad K = 2.12 \sqrt{T} \exp(-7050/RT)$$

$$OH + OH \rightarrow H_2O + O \quad K = 6.09 \times 10^{11} \sqrt{T}$$

and produced the results plotted on Fig. 2. The rate constant K_s and the activation energy E_s were determined to be

$$K_s = 4.65 \times 10^{14} \exp(-17,240/RT_s)$$

$$E_s \simeq 17,240 \text{ cal mole}^{-1}$$

This value agrees with data given by the literature (Friedman 1962; Levy and Friedman 1962; Jacobs and Whitehead 1969).

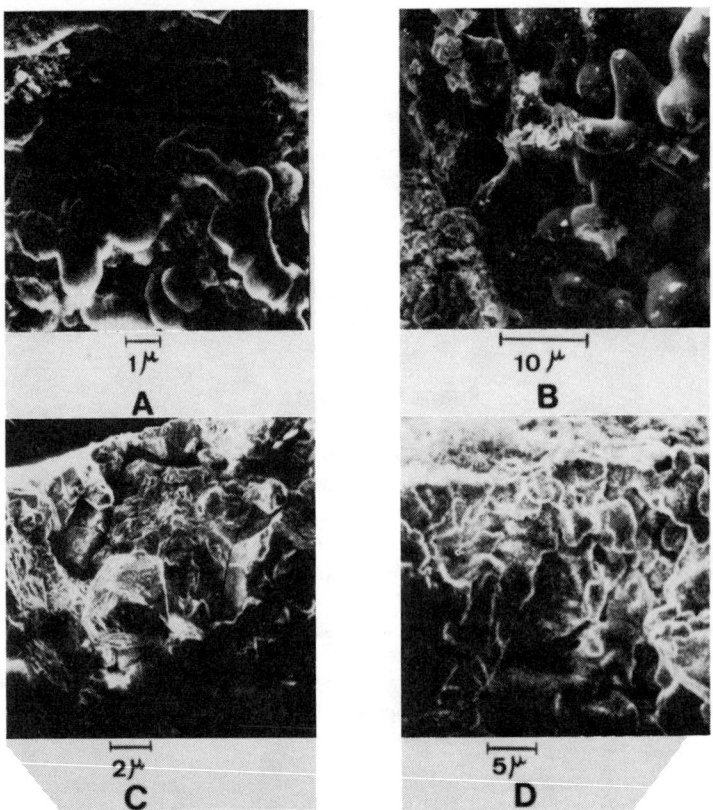

Fig. 1 Pictures of the reactive surface. a), b) Small pore diameters, viscous aspect of the surface: probably due to the existence of a liquid layer. c) Far from the surface: crystalline aspect, no pore. d) Near the surface: no more crystalline structure but pores and cavities.

Table 1 Pressure dependence of the stoichiometric coefficient

P bars	1 [a]	1	3	15	> 20 [b]
ν_{O_2}	0.675	0.525	0.558	0.744	1.015

[a] Levey and Friedman (1962).
[b] Guirao (1970).

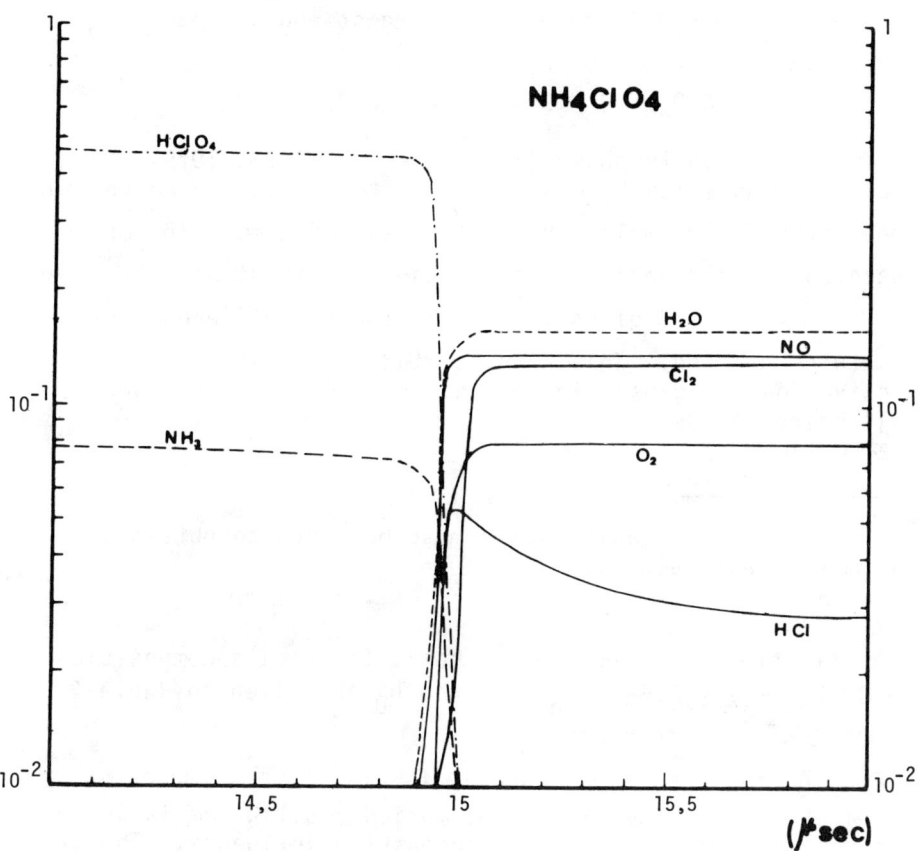

Fig. 2 NH_4ClO_4 decomposition. $G = 280$ kg-m^{-2}s^{-1} (propane); $P = 3$ bars; $T_s = 759$ K; $m_s = 0.96$ kg-m^{-2}s^{-1}.

The stoichiometric coefficient ν_i for O_2 and the main species were also determined. As shown in Table 1, ν_{O_2} increases with P and tends toward the value for the deflagration.

Solid mixtures. If NH_4Cl is a reducing agent and reduces the amount of O_2 available by the formula

$$(NH_3 + HCl) + 2\,O_2 \rightarrow 0.55\,NO + 0.1\,N_2O + 0.125\,N_2$$
$$+ 0.5\,Cl_2 + 2\,H_2O + 0.675\,O_2$$

then $ClO_4Li \cdot 3\ H_2O$ is an oxidizing agent and increases ν_{O_2} according to

$$ClO_4Li \cdot 3\ H_2O \rightarrow 3\ H_2O + LiCl + 2\ O_2$$

As previously shown (Joulain et al. 1978, 1979), the heat of combustion h_c must be known to obtain the convective heat flux at the wall. h_c is the heat of combustion of the decomposition products with propane. h_c is directly related to ν_{O_2}. Figure 3 gives ν_{O_2} and h_c for the different solid mixtures studied. This kinetic study gives also the activation energies (Fig. 4) and frequency factors for the different solids.

Condensed Phase

The heat of gasification must be known to obtain m_s from the heat balance. If

$$h_g = h_s + (1 - m_g) h_D \qquad (3)$$

gasification   sublimation   solid fraction decomposition

and $(1 - m_g) \approx 0.55$. h_g, h_s, and h_D are given in Table 2. Both additives increase h_g.

The surface temperature T_s is also a very important parameter. A monovariant sublimation equilibrium is assumed at the solid surface with no combustion influence. Then the

Table 2 Sublimation and gasification properties

Systems	W solid	ρ solid, $g\text{-}m^{-3}$	K	Δn	h_s solid, J/kg	h_D solid, J/kg	h_g solid, J/kg
5%LiP-95%AP	119.1	1944	18.93	2.2	2.101×10^6	-2.099×10^6	4.572×10^5
10%LiP-90%AP	120.7	1939	17.76	2.4	2.121×10^6	-2.792×10^6	5.851×10^5
15%LiP-85%AP	122.4	1933	16.77	2.6	2.140×10^6	-2.601×10^6	7.093×10^5
100%AP	117.5	1950	20.34	2.0	2.079×10^6	-3.188×10^6	3.257×10^5
5%NH$_4$Cl-95%AP	110.9	1923	20.05	2.0	2.172×10^6	-3.210×10^6	4.072×10^5
10%NH$_4$Cl-90%AP	104.9	1897	19.77	2.0	2.263×10^6	-3.213×10^6	4.959×10^5
15%NH$_4$Cl-85%AP	99.62	1872	19.48	2.0	2.349×10^6	-3.196×10^6	5.912×10^5
20%NH$_4$Cl-80%AP	94.81	1840	19.20	2.0	2.432×10^6	-3.161×10^6	6.938×10^5

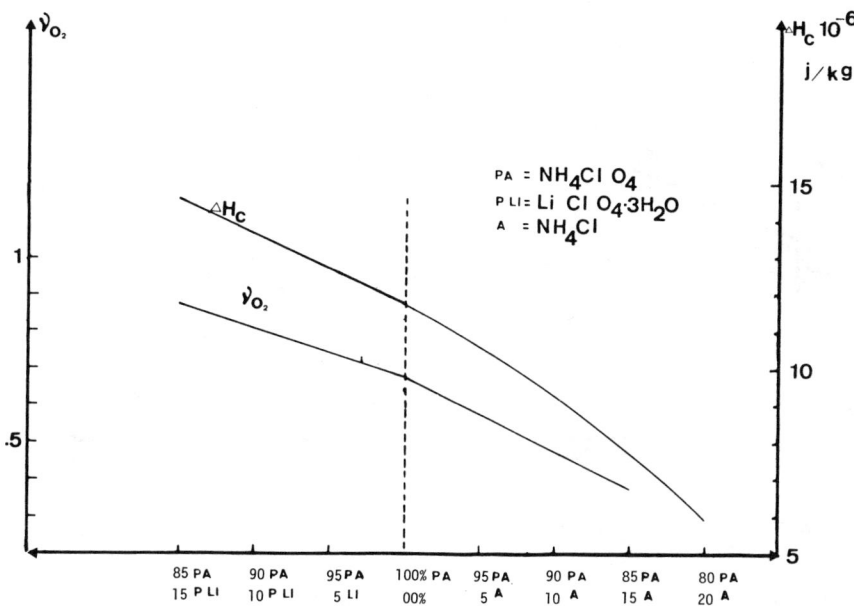

Fig. 3 Heat of combustion h_c and oxygen stoichiometric coefficient vs solid composition.

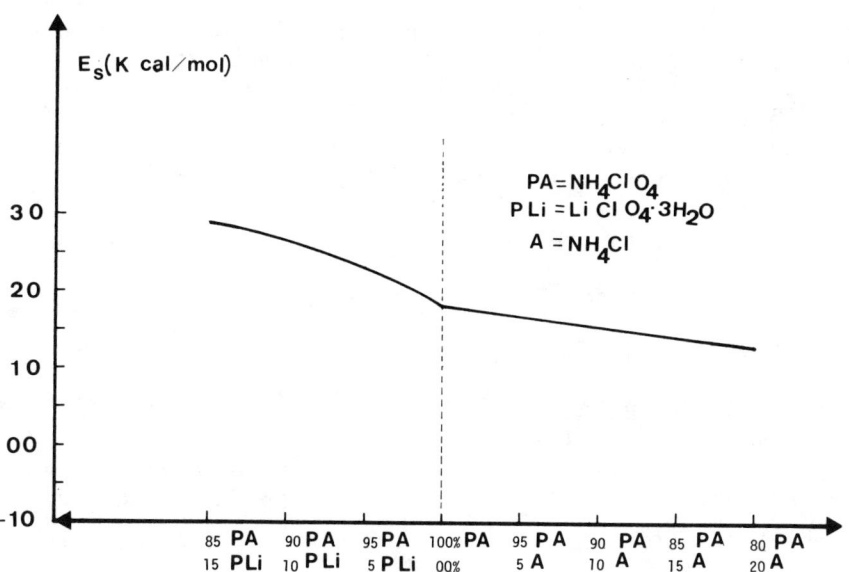

Fig. 4 Overall gas phase activation energy vs solid composition.

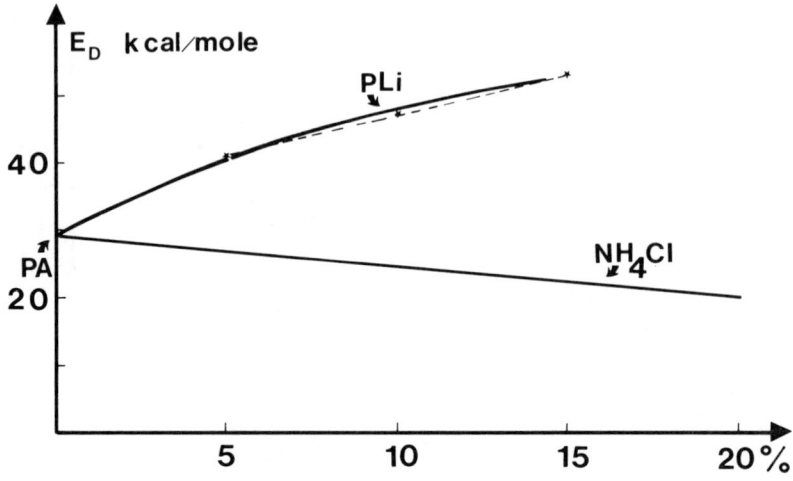

Fig. 5 Overall solid decomposition activation energy vs solid composition.

following relationship between P and T_s is obtained:

$$\ln P = K - h_s/\Delta n\, RT_s \tag{4}$$

where K is given by measurement at P = 1 or by the literature (Al Fakir 1981) and Δn is the variation of mole number during sublimation or dissociation. From the experimental results, E_s and A_D can be obtained if the overall kinetics in the gas and condensed phase are the same.

Addition of NH_4Cl to the solid moves the sublimation equilibrium to the solid phase and influences adsorption phenomena:

$$NH_4Cl \rightarrow NH_3 + HCl$$

solid phase   gas(adsorption-desorption)

Increasing the amount of NH_4Cl up to 20% tends to decrease E_D (Fig. 5) and A_D to the value obtained for the pyrolysis of pure NH_4Cl (Table 3).

This result can explain the extinction experimentally observed above 20% of NH_4Cl. With $ClO_4Li \cdot 3\, H_2O$, which favors liquid layer formation, m_s, E_D, and A_D increase with the amount of additive as previously shown (Al Fakir et al. 1981) after dehydration, the following kinetic schema can be

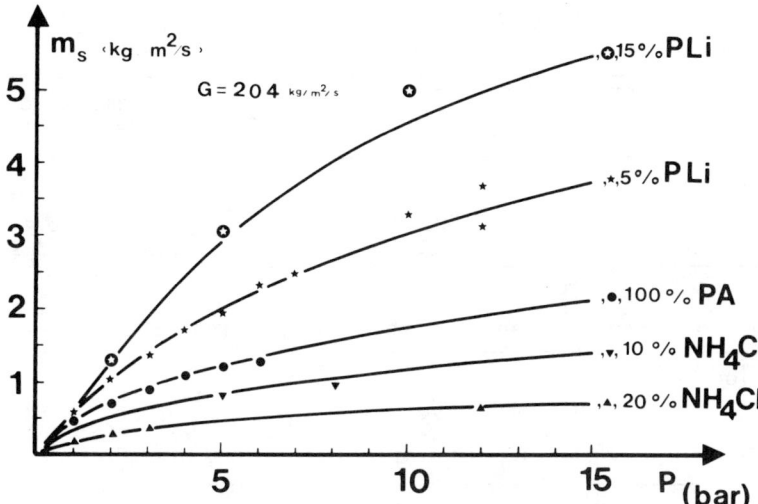

Fig. 6 Mass burning rate vs pressure for the different solids.

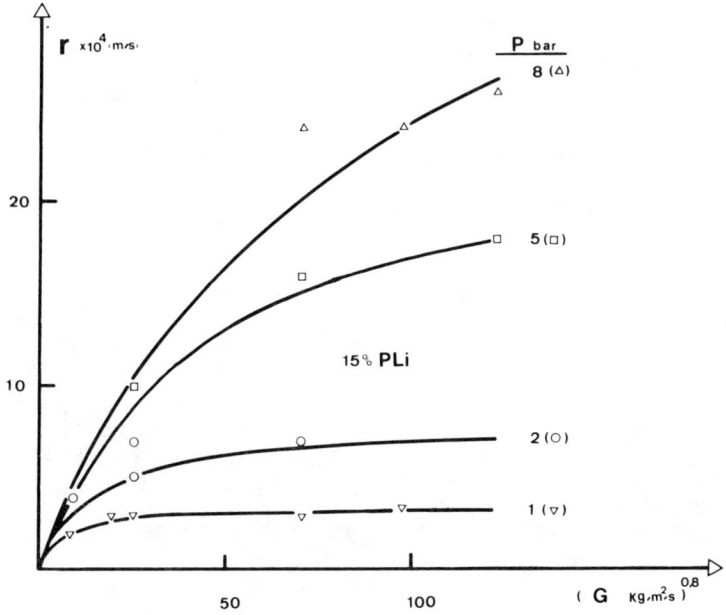

Fig. 7 Dependence of mass burning rate on flow rate and pressure. System: AP + 15% $LiClO_4 \cdot 3\, H_2O$.

Table 3 Effect of ammonium chloride on ammonium perchlorate overall combustion kinetic parameters

	Pyrolysis 100% NH_4Cl	Combustion 20% NH_4Cl + 80% AP
E_D kcal-mole^{-1}	13.5	13.2
A_D cm-s^{-1}	120.	119.

Table 4 Pressure exponent for the mass burning rate expression

Composition, %								
LiP	5	10	15					
AP	95	90	85	100	95	90	85	80
NH4Cl					5	10	15	20
n experimental	0.80	0.90	1.0	0.60	0.55	0.50	0.45	0.425
theoretical	0.85	1.04	1.20	0.62	0.59	0.54	0.52	0.47

assumed:

$$LiClO_4 \cdot 3\,H_2O \rightarrow LiClO_4,\ 2\,H_2O + H_2O \rightarrow LiClO_4 + 3\,H_2O$$

$$LiCl$$

$$LiClO_4 \rightarrow LiClO_3 + (1/2)\,O_2$$

$$LiClO_3 \rightarrow LiCl + (3/2)\,O_2$$

Then in the liquid layer at the surface,

$$NH_4^+ClO_4^- + Li^+Cl^- \rightarrow Li^+ClO4^- + NH_{3\,ads} + HCl_{ads}$$

According to this schema, E_D must tend toward the value corresponding to the breaking of one Cl-O bond (53 kcal mole^{-1}), as shown on Fig. 5.

As experimentally observed at high flow rate and low pressure, the burning rate is given by

$$M_s \cong b\,P^n \quad (5)$$

and assumed kinetically controlled.

Then as

$$n \simeq \frac{\partial \ln m_s}{\partial \ln P} \tag{6}$$

from Eqs. (1) and (4)

$$n = E_s \Delta n / h_g \tag{7}$$

As shown in Table 4 the agreement between the value for n given by experiment and that predicted from Eq. (7) is good.

The above results with their simultaneous solution of the conservation equations and Eq. (1), lead to a prediction of the mass burning rate vs flow rate, pressure, and solid composition (Al Fakir 1981). As is shown on Figs. 6 and 7, the agreement between experiment and theory is quite good.

Interaction Combustion - Boundary Layer

In the approximation solution of the momentum conservation equation, it is necessary to specify relationships between the main characteristics of the turbulent flow: wall shear stress τ, friction factor C_f, mass transfer number B. As in earlier calculations (Joulain et al. 1977; Joulain et al. 1978; Joulain et al. 1979), relationships given by Marxman (1965) and Spalding and Patankar (1970) were used:

$$\tau_{s_0} = 0.03 \, R_e^{-0.2} \tag{8}$$

$$\tau_s = \frac{1}{2} C_f \rho_\infty u_\infty^2 = \tau_{s_0} \frac{C_f}{C_{f_0}} \tag{9}$$

$$\frac{C_f}{C_{f_0}} = f(B) \quad \text{with} \quad B = \frac{m_s u_\infty}{\tau_s} \tag{10}$$

At atmospheric pressure and for low mass burning rate:

$$\frac{C_f}{C_{f_0}} = \frac{\ln(1 + B)}{B} \tag{11}$$

As Lengelle (1975) has shown that Eq. (11) is no longer valid for high mass burning rate and pressure the following expression was use

$$\frac{C_f}{C_{f_0}} = \beta \, (m_s) \, \frac{\ln(1 + B)}{B} \tag{12}$$

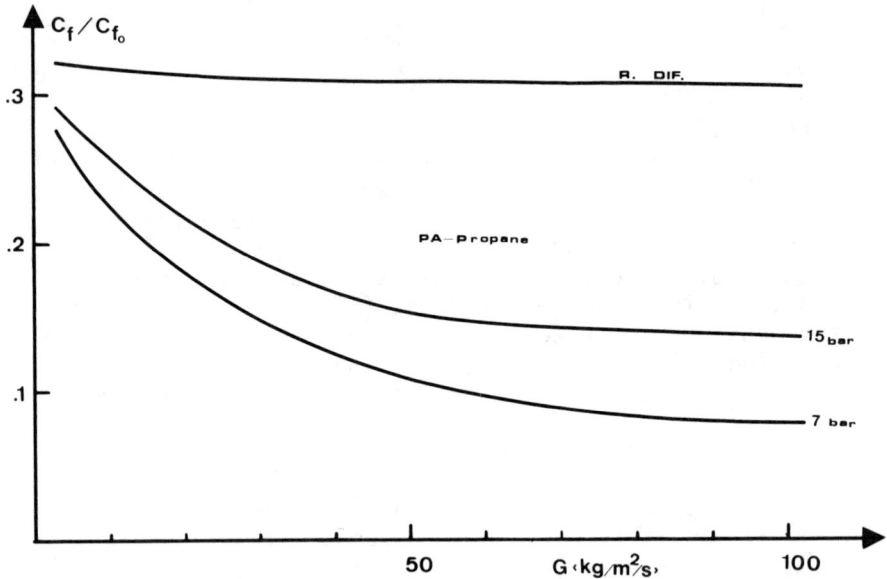

Fig. 8 Dependence of coefficient of friction on flow rate. System AP - propane.

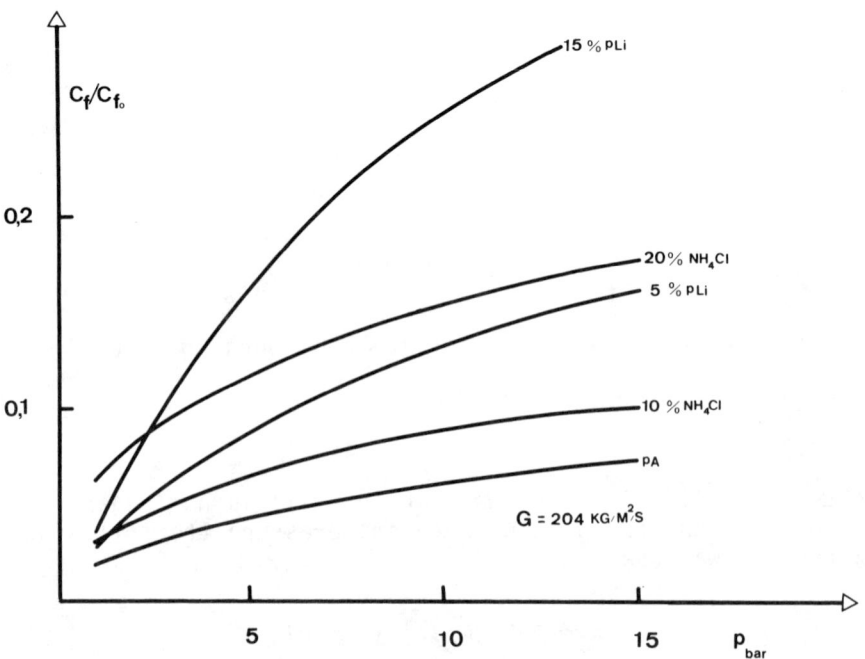

Fig. 9 Dependence of coefficient of friction of pressure.

COMBUSTION OF SOLID MIXTURES

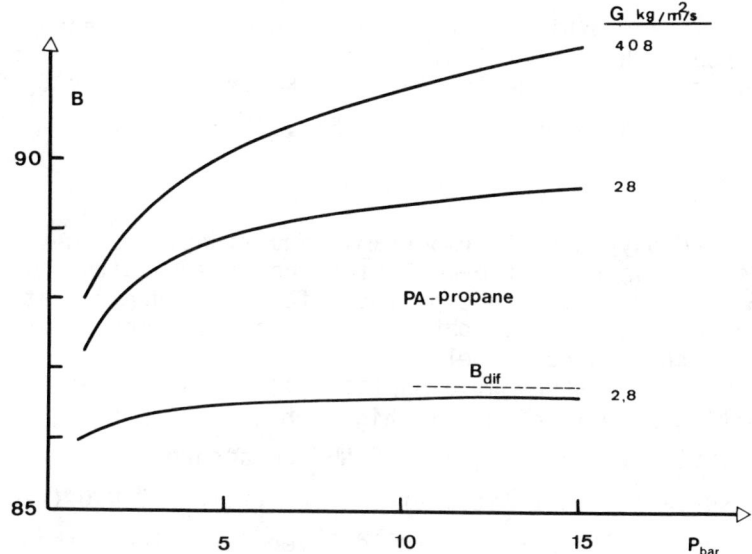

Fig. 11 Dependence of mass transfer number B on pressure.

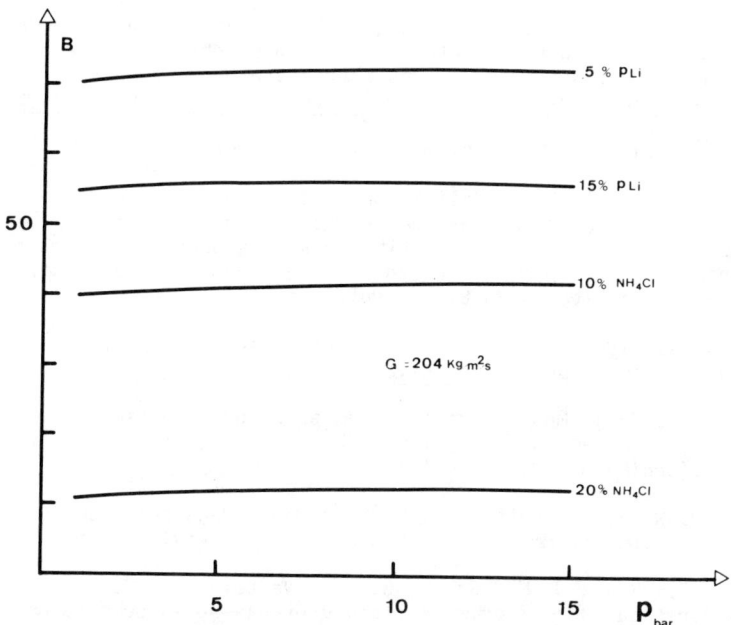

Fig. 10 Dependence of mass transfer number B on pressure and flow rate. System AP - propane.

All the systems exhibited the same flow rate and pressure dependence. These correlations are in good agreement with previous results (Lengelle, 1975; Al Fakir, 1981) for C_f/C_{f_0} (see Fig. 8 and 9) and for B (see Fig. 10 and 11).

Conclusions

The aerodynamic, thermodynamic and kinetic parameters obtained are in good agreement with previous results.

Good agreement with experiment for mass burning rate and pressure exponent is obtained if these parameters are used with theoretical models.

The kinetic study also explains 1) the so-called "inhibiting effect" of NH_4Cl (higher h_s, h_g; lower h_c, n; decomposition rate; influence of NH_3 on adsorption; etc; and 2) the so-called "catalytic effect" of $LiClO_4 \cdot 3\ H_2O$ (higher h_c, n; decomposition rate; liquid layer formation; etc.).

This study is being continued with composite solid propellants and applied to erosive burning phenomena.

References

Al Fakir, M. S., Joulain, P., Most, J. M., Sztal, B., and Bernard, M. L. (1981) Burning rate of ammonium perchlorate in a turbulent flow of propane under pressure below 20 bars. Combustion in Reactive Systems: Progress in Astronautics and Aeronautics (edited by Bowen, Manson, Oppenheim, and Soloukhin), Vol. 76, pp. 551-564. AIAA, New York.

Al Fakir, M. S. (1981) Modélisation de la combustion de propergols solides à base de perchlorate d'ammonium dans un écoulement turbulent combustible pour des pressions inférieures a 20 bars. Doctorat ès Sciences Physiques No. 339, Université de Poitiers, Poitiers, France.

Friedman, R. (1962) Mechanism of composite solid propellant combustion. Appl. Mech. Rev. 15(12), 935-937.

Guirao, C. (1970) Models for ammonium perchlorate dissociative sublimation and deflagration. Ph.D. Thesis, University of California, San Diego, Calif.

Jacobs, P. W. N. and Whitehead, H. M. (1969) Decomposition and combustion of ammonium perchlorate. Chem. Rev. (69) 551-590.

Joulain, P., Most, J. M., Sztal, B., and Vantelon, J. P. (1977) Theoretical and experimental study of gas-solid combustion in turbulent flow. Combust. Sci. and Technol. 15, 115-241.

Joulain, P., Most, J. M., Sztal, B., and Fuseau, Y. (1978) Combustion de materiaux solides dans un ecoulement gazeux turbulent. Acta Astron. 5(11-12), 1079-1093.

Joulain, P., Most, J. M., Fuseau, Y., and Sztal, B. (1979) Influence of coupled convection, conduction, radiation heat transfer on the burning of plastics. 17th Symposium (International) on Combustion, pp. 1041-1051. The Combustion Institute, Pittsburgh, Pa.

Lengelle, G. (1975) Model describing the erosive combustion and velocity response of composite propellants. AIAA J. 13, 315-322.

Levy, J. B. and Friedman, R. (1962) Further studies of pure ammonium perchlorate deflagration. 8th Symposium (International) on Combustion, pp. 1257-1269. Williams and Wilkins, Baltimore, Md.

Marxman, G. A. (1965) Combustion in the turbulent boundary layer on a vaporizing surface. 10th Symposium (International) on Combustion, pp. 1337-1349. The Combustion Institute, Pittsburgh, Pa.

Spalding, D. B. and Patankar, S. V. (1970) Heat and Mass Transfer in Boundary Layers. A General Calculation Procedure. 2nd Ed., Intertext Books, London.

Radiation from Polyurethane Pile Fires

J.M. Souil,* H. Azov,† and P. Joulain‡
Université de Poitiers, St. Julien L'Ars, France
and
S. Galant§
Bertin et Cie, Plaisir, France

Abstract

An experimental and theoretical study of radiative transfer from free burning polyurethane piles for four fuel arrangements is reported. Temporal flux variations at different locations were recorded by 14 radiometers distributed around the fire to determine the effects of the orientation and distance of the gages from the fire. These measurements were needed to calculate radiative fluxes from pool or crib fires with assumptions of a mean absorption coefficient, effective radiation temperature, and given flame shape. Values of the total radiative power output of the fire were compared with those inferred from flux measurements.

Introduction

The study of free diffusion flames is important to promote increased protection of people and property. To this end, information about heat transfer from burning objects to the surroundings is essential. While convective and radiative heat transfer are significant in fires, radiative heat transfer is dominant in moderate and large-scale fires. Radiative transfer is an important mechanism in the spread of building fires. In the industrial field, studies have been performed for human safety on offshore drilling rigs which have radiating flares.

Paper presented at the 8th ICOGER, Minsk, USSR, Aug. 23-26, 1981. Copyright ©American Institute of Aeronautics and Astronautics, Inc., 1983. All rights reserved.
*Chargé de Researche, du Centre Scientifique et Technique du Batiment, Poitiers, France.
+Research Assistant.
‡Director-Adjoint de 1-ERA 160 and Maître de Recherche, CNRS.

RADIATION FROM POLYURETHANE PILE FIRES

For a better understanding of radiative transfer, experimental and theoretical work is necessary; many essential quantities, such as temperature and absorption coefficient, are needed for use in theoretical modeling. For this scope, an experimental and theoretical study of polyurethane pile fires is proposed, based on radiation flux measurements performed at the Société Bertin facilities.

Experimental Features

Radiative fluxes from burning piles were measured with 14 wide-angle, water-cooled Medtherm radiometers, each viewing the entire flame volume, in the configuration showed on Fig. 1. The radiometers were located at horizontal distances between 3 and 7 m from the pile center, and at heights between 0.6 and 1.05 m above the ground. Radiative

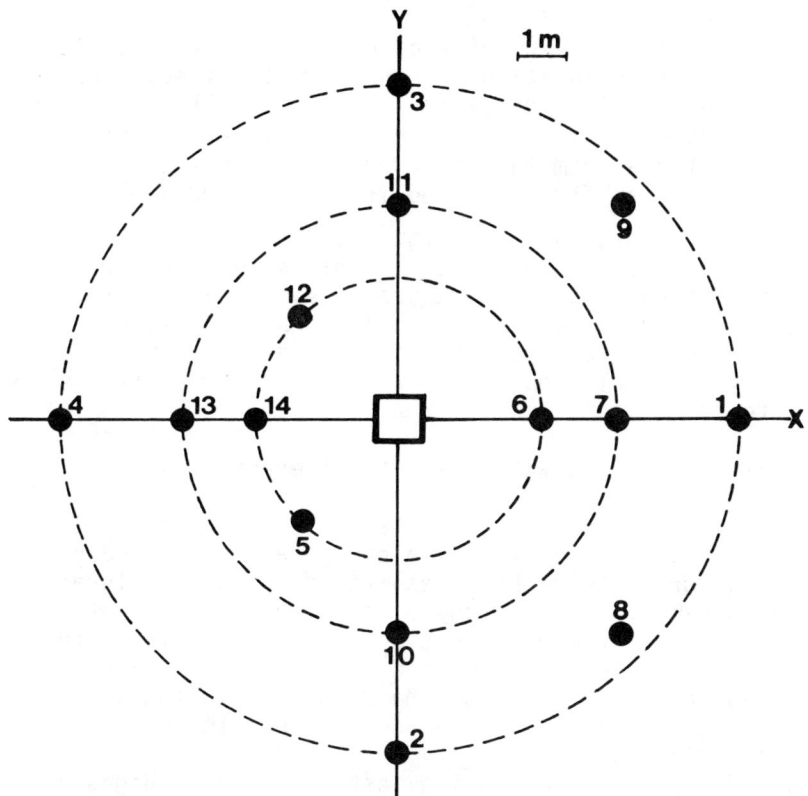

Fig. 1 Gage locations around the fire in the X-Y plane.

Table 1 Test parameters

Test	Pile dimensions, m			Total surface area, m^2	S/V, m^2/m^3	Test duration, min
	a	b	c			
1	0.7	0.7	0.6	2.66	9.05	7
2	0.9	0.9	0.6	3.78	7.78	3
3	0.7	1.4	0.3	3.22	10.95	5
4	0.6	1.8	0.45	4.32	8.89	5

fluxes were measured every 2 s on all radiometers; video records were made of the tests. Four arrangements of polyurethane were burned (Table 1); piles 1 and 3 and piles 2 and 4 have the same weight.

Analysis of Experiments

These experiments were made to obtain flux distributions at different locations around the pile fire. The contribution of solar energy to the measured fluxes was taken into account by subtraction of the initial flux (before the test) from the recorded values on each radiometer. Solar fluxes ranged from about 100 W/m^2 (gage 7) to 500 W/m^2 (gage 14). As the measurements were performed in an open air area, the influence of the wind cannot be ignored. This influence appeared on the flux recordings which exhibited many brief peak values; moreover, the difference between values recorded on radiometers located symmetrically on the X or Y axis were mainly due to wind effects.

Flux Variation with Time at Different Locations

The fluxes recorded during test 1 on gages 5, 6, 12, and 14 located 3 m from the pile center are presented on Fig. 2. Unfortunately these results do not give relevant information about the isotropy of the radiative transfer to the surroundings at a given distance from the pile center. This remark is also valid for the recordings of gages 7, 10, 11, and 13 located 4.5 m from the fire; as mentioned before, the influence of crosswind produced higher fluxes in particular directions.

The fluxes recorded during tests 1 and 3 on gages 1-4, 7-13, and 6-14, located symmetrically by pair on the X axis,

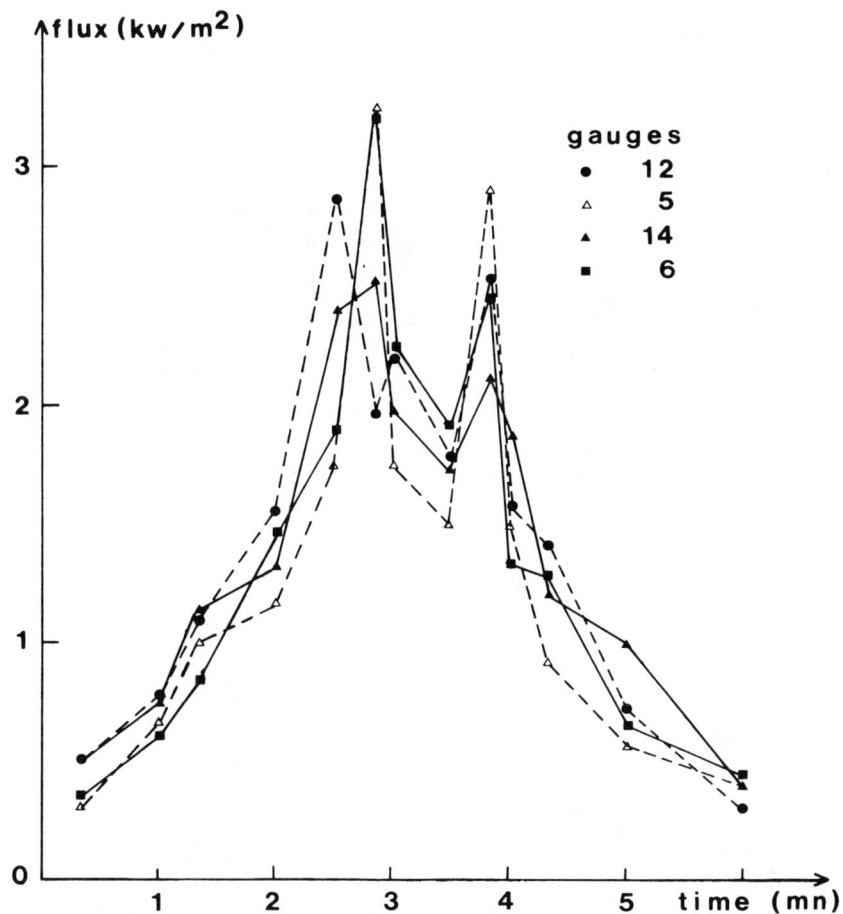

Fig. 2 Flux variations with time for equidistant gages in fire test 1.

are shown on Figs. 3 and 4. The influence of crosswind on these measurements appeared again very clearly and consistently with the previous observations. The fluxes recorded on gages 1, 6, and 7 located on the same side of the fire were higher, which corresponded with the wind main direction during the tests. The mean values of flux for the three different pairs of radiometers 1-4, 7-13, and 6-14 were used to determine the flux variation vs distance between the radiometer and pile base center, for different times after ignition, as shown on Figs. 5 and 6.

Mean Flux Variation with Time

If for each time interval an arithmetical mean of the experimental values given by all the radiometers is computed,

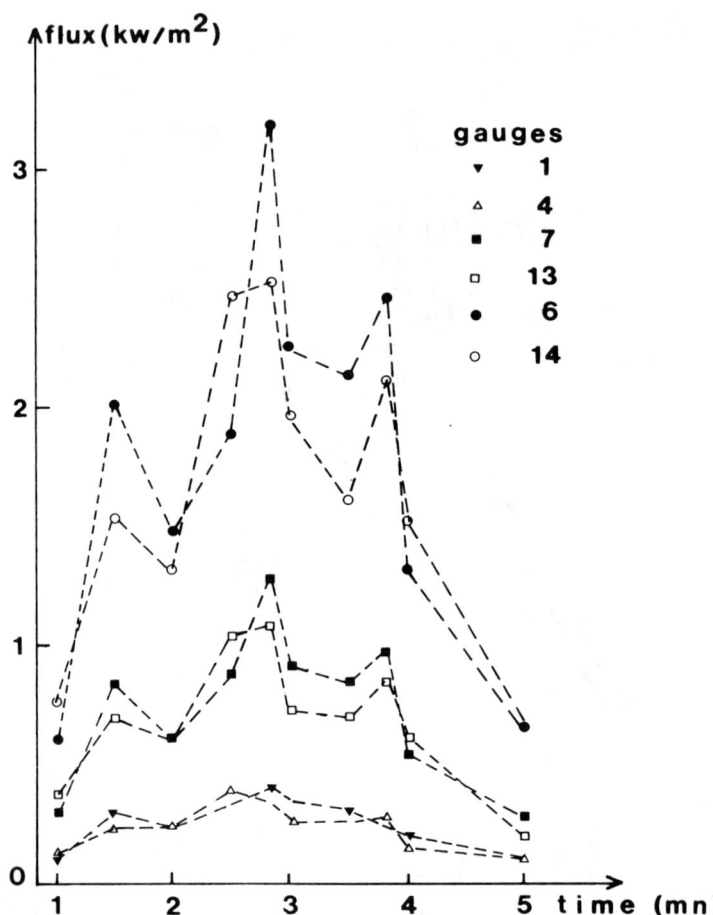

Fig. 3 Flux variations with time for gages on X axis in fire test 1.

an average time distribution for this mean flux which corresponds to a hypothetical radiometer located 4.4 m from the burning pile center can be predicted. The obtained mean flux variations with time for tests 1, 3, and 4 are shown on Figs. 7-9. The solid lines represent the radiative transfer for the three different tests corresponding to three different pile configurations. Only the solid curve on Fig. 7, which corresponds to a square base pile, is symmetrical. The two other curves, relative to rectangular base piles, are asymmetrical. The maximum flux is reached earlier for a rectangular base pile fire.

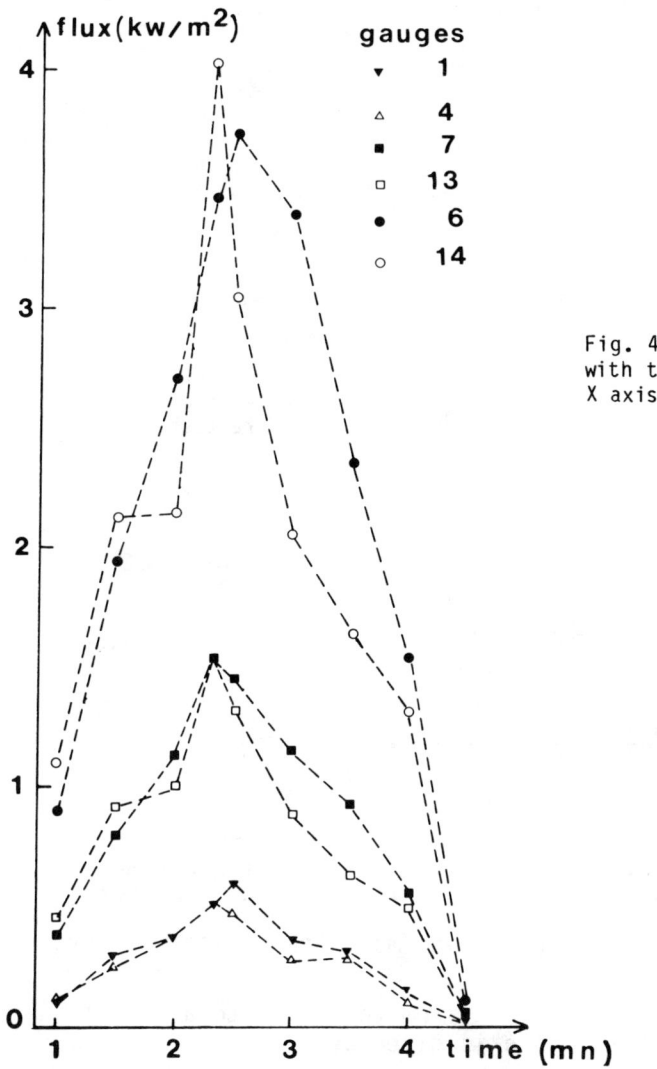

Fig. 4 Flux variations with time for gages on X axis in fire test 3.

Theoretical Study

A method given by Modak (1977) for radiative flux calculation in pool fires was used to correlate the experimental results. Modak's expressions in their most general form were applied to compute radiative fluxes on targets of arbitrary orientation. The physical parameters needed for calculation are an effective radiation temperature T_f, a mean absorption coefficient k, and a

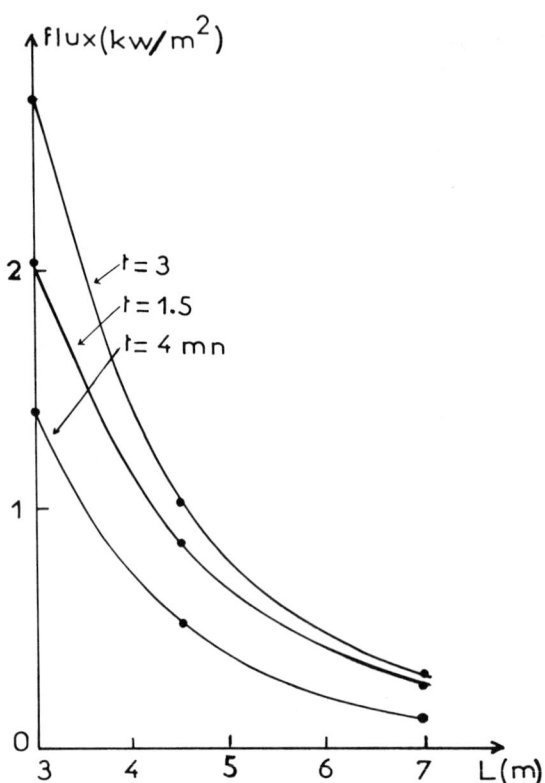

Fig. 5 Flux variations with distance for fire test 1.

specified axisymmetric flame shape. For polyurethane flames, Markstein's measurements (1979) give values of 1400 K for T_f and 1.3 m^{-1} for k. The main theoretical features of the method are given below.

The radiative heat flux from the flame to a target element with unit normal $\vec{n}$ is given by

$$\dot{q}'' = \int_\Omega N \vec{n} \cdot \vec{r} \, d\Omega$$

where $\vec{r}$ is the unit vector along a ray to the flame and $d\Omega$ is the elemental solid angle subtended by the flame at the target element.

The flame radiance N is given by

$$N = (\sigma T_f^4/\pi) [1 - \exp(-kl_f)] \qquad (1)$$

where l_f is the path length through the flame (Fig. 10).

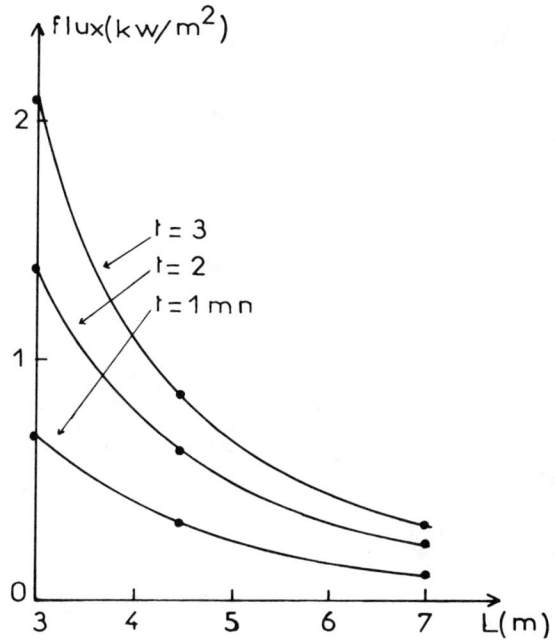

Fig. 6 Flux variations with distance for fire test 3.

In a spherical coordinate system (r, θ, ϕ) about the target element, the radiative flux is expressed as

$$\dot{q}'' = (\sigma T_f^4/\pi) \int_\theta \int_\phi [1 - \exp(-kl_f)] (u \sin\theta \cos\phi + v \sin\theta \sin\phi + w \cos\theta) \sin\theta \, d\phi \, d\theta \qquad (2)$$

where u, v, and w are the direction cosines of the target element.

The path length through the flame l_f is given by

$$l_f = (z_o - z_i)/\cos\theta \qquad (3)$$

where z_i and z_o are the vertical coordinates where the ray respectively enters and leaves the flame.

It is more convenient to compute the value of the radiative flux $\dot{q}''$ in the fire centered cylindrical coordinate system (z_o, ψ) for upward $(\theta \leq \pi/2)$, and (z_i, ψ) for downward $(\theta > \pi/2)$ directed rays.

In the (z_o, ψ) cylindrical coordinate system, the radiative flux from the flames to a target element is

$$\dot{q}'' = \frac{2\sigma T_f^4}{\pi} \int_{z_o=z_p}^{z_t} dz_o \int_{\psi=0}^{\psi_o} \frac{1}{r_o^4} \{1 - \exp[-kr_o(1 - \frac{z_i}{z_o})]\}$$

(equation continued on p. 191)

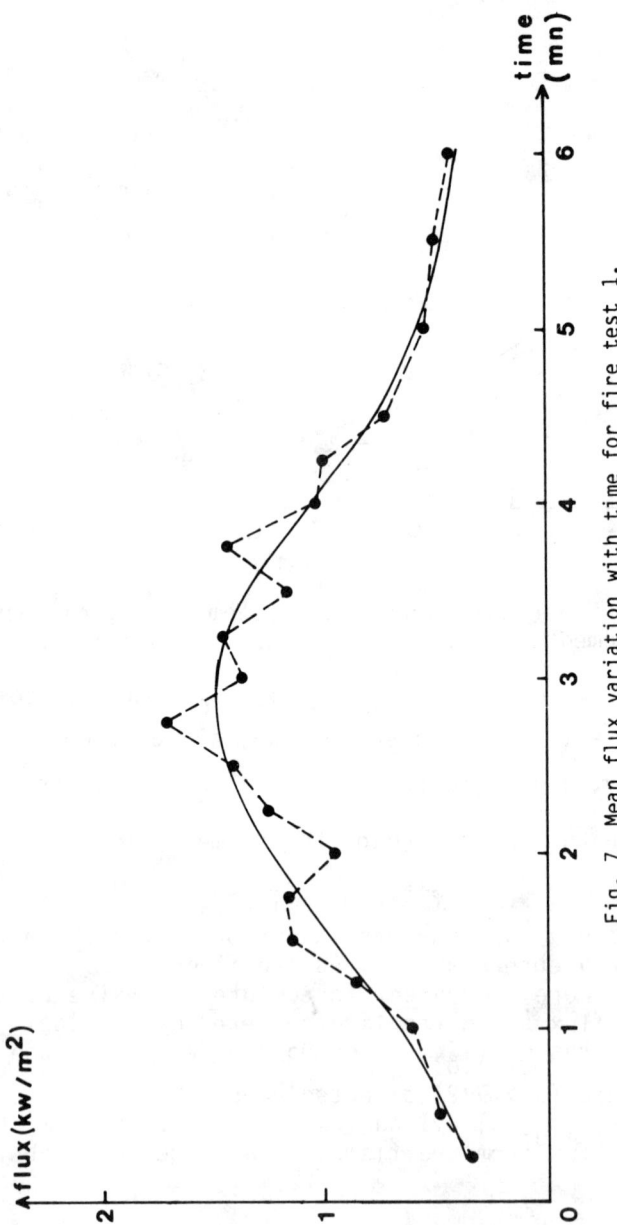

Fig. 7 Mean flux variation with time for fire test 1.

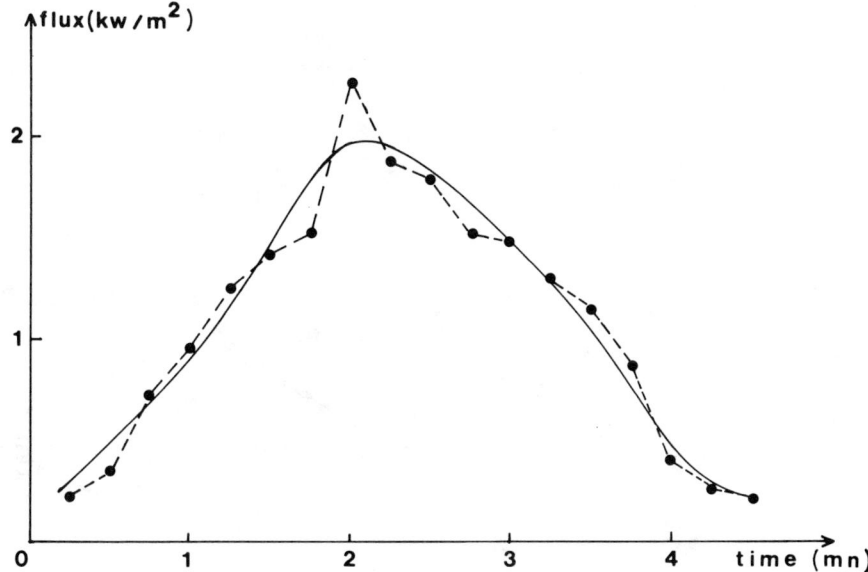

Fig. 8 Mean flux variation with time for fire test 3.

$$\times \{u\ S(z_o)\ \sin \psi + v\ [L+ S(z_o)\ \cos \psi] + w\ z_o\}\ S(z_o)$$
$$\times \{S(z_o) + L\ \cos \psi - z_o (\frac{dS}{dz})_{z_o}\}\ d\psi \quad (4)$$

The upper limit of integration for the azimuthal angle is

$$\psi_o = \cos^{-1} \{[z_o(\frac{dS}{dz})_{z_o} - S(z_o)]/L\} \quad 0 \leq \psi_o \leq \pi \quad (5)$$

where $S(z)$ is the radius of the axisymmetric flame profile at height z. This parameter can be derived from photographs of flames, but in the present work, this procedure was not feasible because of the strong wind. From a theoretical point of view, there is insufficient information about correlations between flame heights, burning rates, and the size of pools or cribs.

With averaged time distributions for the radiative fluxes of the fire tests, the heights of simple conical shaped flames which produce the measured flux distributions were computed. The mean values of measured fluxes $\dot{q}''$ were used to obtain a time-dependent variation for the flame height with the assumption of a conical shape whose hydraulic radius is equivalent to the base of the fire. Flame heights in column 3 of Table 2 were obtained with this

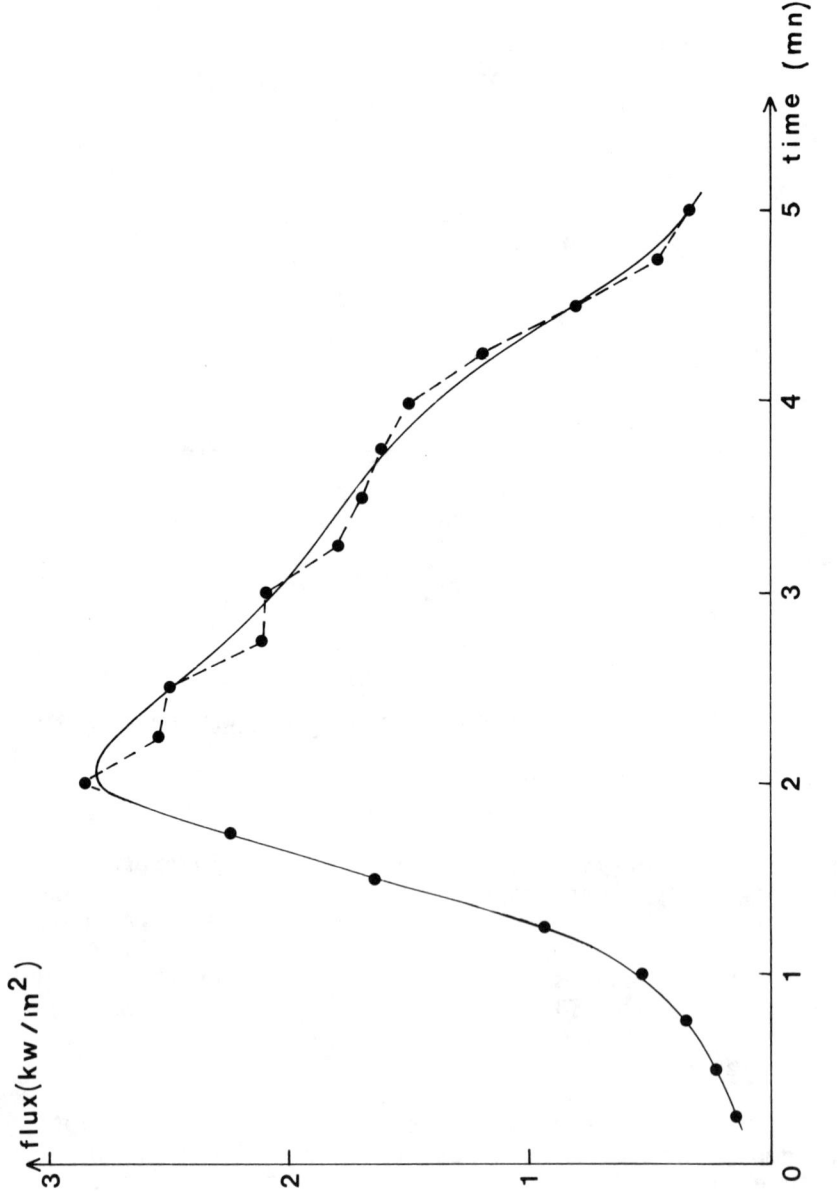

Fig. 9 Mean flux variation with time for fire test 4.

procedure for fire test 1. The total radiative power output of a fire to its surroundings is an important factor in the evaluation of fuel fire hazard. For a convex flame volume V_f surrounded by a flame surface A_f, the total radiation leaving the flame volume is

$$\dot{Q}_R = A_f \sigma T_f^4 (1 - e^{-kl_m}) \qquad (6)$$

where the mean beam length l_m is given by Hottel and Sarofim (1967):

$$l_m = 3.6 \, V_f/A_f \qquad (7)$$

For isotropic radiation, the total radiative power of the flame is

$$\dot{Q}_R = 4\pi R_o^2 \dot{q}_o''$$

where $\dot{q}_o''$ is the measured flux at the distance R_o (here $R_o = 4.4$ m). The values of $\dot{Q}_R$ obtained by these two methods reported in columns 4 and 5 of Table 2 do not differ greatly.

As the main concern of these tests was the measurement of radiation, mass loss rate measurements were not made. As a consequence, the total heat release rate of the fires

Table 2 Fire test 1: mean fluxes, $R_o = 4.4$ m, conical flame shape

Time, s	Mean flux $\dot{q}_o''$, W/m^2	Theoretical flame height H_f, m	Total radiative power output[a] $\dot{Q}_R$, kW	Total radiative power output[b] $\dot{Q}_R$, kW
15	200	0.45	52	49
45	340	0.75	83	84
60	460	1.03	111.6	113.4
75	650	1.45	154.3	160
90	760	1.7	180	188
105	900	2.05	215	222
120	1000	2.3	240	247
135	1140	2.66	277	281
165	1300	3.1	323	321
195	1240	2.95	306	306
255	700	1.55	165	170
345	250	0.55	61.8	61.6

[a] $\dot{Q}_R = A_f \sigma T_f^4 [1 - \exp(-3.6 \, k \, V_f/A_f)]$. [b] $\dot{Q}_R = 4\pi R_o^2 \dot{q}_o''$.

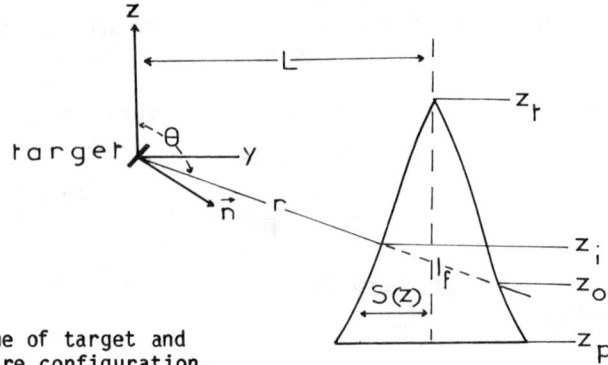

Fig. 10 Basic scheme of target and axisymmetric pool fire configuration.

cannot be determined directly. Nevertheless, with the total radiative power output calculation for fire test 1 (Table 2), an estimation of the radiated fraction of the total heat released can be obtained after an integration of $\dot{Q}_R$ over time. For a mean value of 6000 cal/g for the heat of combustion of polyurethane (Tewarson and Pion 1977), the fraction of heat transferred as radiation to the surroundings is roughly 20%, which is consistent with other observations of free diffusion flames and especially polyurethane flames (Bard et al. 1979).

The theoretical variation of flame height, as derived from radiation flux measurements, is plotted as a function of total radiative power (Fig. 11). The correlation $H_f \propto \dot{Q}_R^{1.05}$ should be compared with some predictions available in the literature of buoyant flame height vs energy input ($\dot{Q}_T$), since it can be assumed that $\dot{Q}_R$ is directly proportional to $\dot{Q}_T$. In their work on buoyant diffusion flames, Tomas et al. (1961) proposed the correlation

$$\frac{H_f}{D} \propto \left(\frac{\dot{Q}_T^2}{D^5}\right)^{1/(2n+1)}$$

where D is a characteristic dimension of the fuel source and the index n is expected to increase from 0 to 2 with increasing H_f/D. In the present work, n = 0.45, which is a realistic value for a conical flame, since for this geometry, n is expected to increase from 0 when $H_f/d \ll 1$ to 1 when $H_f/d > 1$.

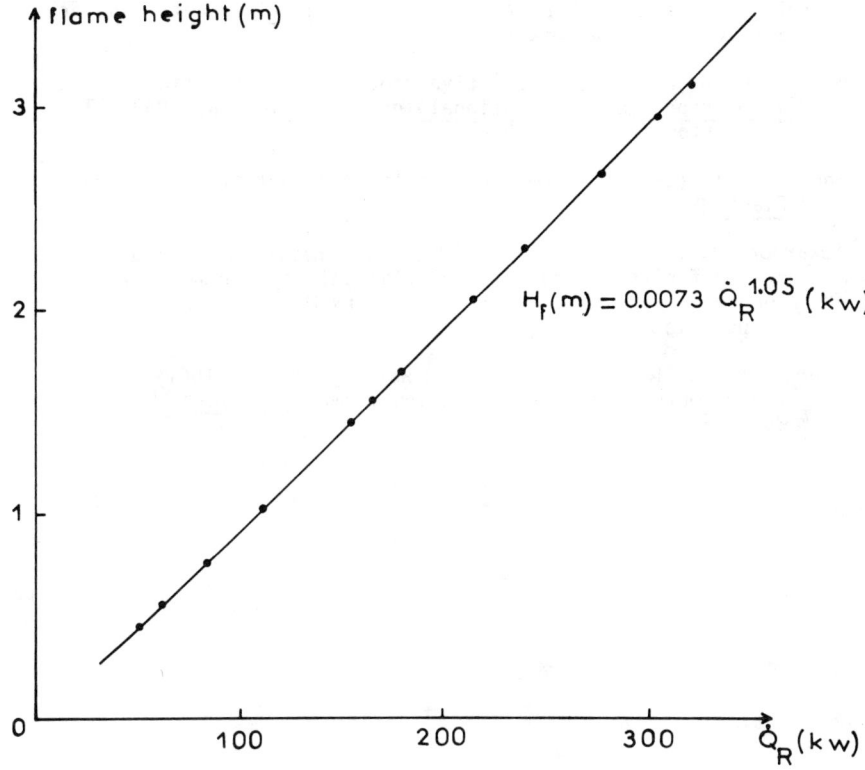

Fig. 11 Theoretical flame height variation vs $\dot{Q}_R$.

Conclusions

This work is a first step in a general study which is intended to describe the main features of the radiative transfer between free burning fires and their surroundings. The pile fire configuration is very interesting and convenient for the completion of such a study. This configuration allows an exhaustive theoretical parametric study: influence of main parameters, such as fire size and geometry, nature of fuel, surrounding conditions, etc. These results can be easily checked by conducting some pile fire tests. Despite the rough experimental results, the agreement with the theoretical evaluation is quite good for this first approach. This study is being completed through new experimental measurements and extended theoretical modeling.

References

Bard, S., Clow, K.H. and Pagni, P. J. (1979) Combustion of cellular urethane. <u>Combust. Sci. Technol</u>. 19, 141.

Hottel, H. C. and Sarofim, A. F. (1967) <u>Radiative Heat Transfer</u>. McGraw Hill, New York.

Markstein, G. H. (1979) Radiative properties of plastic fires. <u>17th Symposium (International) on Combustion</u>, p. 1053. The Combustion Institute, Pittsburgh, Pa.

Modak, A. T. (1977) Thermal radiation from pool fires. <u>Combust. Flame</u> 29, 177.

Tewarson, A. and Pion, R. F. (1977) A laboratory-scale test method for the measurement of flammability parameters. Technical Report FMRC 22524. Factory Mutual Research Corporation, Norwood, Mass.

Thomas, P. H., Webster, C. W., and Raftery, M. M. (1961) Some experiments on buoyant diffusion flames. <u>Combust. Flame</u> 5, 359.

Study of Condensed System Flames by Molecular Beam Mass Spectrometry

O.P. Korobeinichev* and L.V. Kuibida†
Academy of Sciences, Novosibirsk, USSR

Abstract

To study the microstructure (species concentration and temperature profiles) of condensed system flames, an experimental technique has been developed including: 1) a molecular beam system of flame sampling, a time-of-flight (TOF) mass spectrometer as a detector of the beam, a combustion chamber, and a system of combustion zone scanning; 2) a system for automation of, and data collection for, mass-spectrometry experiments based on CAMAC apparatus and a minicomputer SARATOV-2. The concentration profiles of atoms and radicals, as well as stable species, can be measured in flames with the molecular beam sampling technique. The system of automation of experiments provides automatic mass-spectrometry (TOF) data acquisition, accumulation, primary processing, and control over the TOF and experimental apparatus. Data on the microstructure investigations of flames of ammonium perchlorate (AP) and its mixtures are reported as an example. For the first time, to the authors' knowledge, perchloric acid (highly reactive stable molecules) has been detected in AP flames as the basic product of the condensed phase gasification. This fact confirms experimentally the hypothesis used as the basis for a number of mathematical models of combustion of AP and its mixtures. Analysis of the mass peak intensity profiles for $HClO_4$, ClO_2, $ClOH$, and ClO has enabled us to infer that the concentration of ClO radicals in AP flame is appreciable.

Presented at the 8th ICOGER, Minsk,USSR, Aug. 23-26, 1981. Copyright © American Institute of Aeronautics and Astronautics, Inc., 1982. All rights reserved.
  *Senior Research Scientist, Institute of Chemical Kinetics and Combustion.
  †Junior Research Scientist, Institute of Chemical Kinetics and Combustion.

Introduction

Mass spectrometry is one of the most popular and effective methods for studying combustion processes (Fristrom and Westenberg 1965; Korobeinichev 1980). This fact results from the universality of the method which can be used 1) to identify the species of the flame, 2) to determine their concentration, 3) to study the flame structure through measurement of the concentration distribution in the combustion zones, and 4) to analyze flame processes in terms of species fluxes and chemical reaction rates.

The most important problem of the flame mass spectrometry is that of sampling the species from the combustion zone and transferring them to the ion source of a mass spectrometer. The species sampling from flames is accomplished by a microprobe which is a cone with an orifice at the apex. The gas expansion in the cone "freezes" the chemical reactions. When taken, the sample must be transported without further reactions to the ion source of the mass spectrometer. Molecular beams are the most effective system to transport the sample from the flames to the ion source of a mass spectrometer since collisions of the beam species with one another and with units are minimized and ensure minimal change of the chemical composition of the sample due to homogeneous chemical reactions. Moreover, since the molecular beam stems from the supersonic jet core free of wall collisions, heterogeneous-catalytic wall reactions do not affect the sample composition. The progress in the experimental technique achieved recently by some laboratories [see, for example, Stearns et al. (1979)] made it possible to design apparati with molecular beam sampling from flames stablized by a flat burner and subsequent mass-spectrometry analysis for atoms and free radicals in the flames. An apparatus based on a time-of-flight (TOF) mass spectrometer for sampling condensed systems flames is described by Fristrom and Westenberg (1965). A special mechanism moves a fuel strand with a velocity exceeding the strand burning rate. The sample, taken by a quartz probe, is transported to the ion source of the TOF by a molecular beam flow.

The mass spectrum was displayed on an oscillograph screen and recorded with a camera. On the basis of this apparatus and a device for automatic detection of (STROB) TOF mass spectra (Korobeinichev et al. 1977) more elaborate facilities have been designed for investigations of condensed system flames.

In the present communication the apparatus is described and species data from sampling condensed system flames are reported.

Apparatus

The system, shown in Fig. 1, schematically consists of 1) a device for sampling species from condensed system (CS) flames with an inlet molecular beam system, TOF mass spectrometer as a beam detector, a combustion chamber, and a device for scanning the combustion zones; 2) a control and mass-spectrometric data acquisition system based on a CAMAC apparatus and a minicomputer SARATOV-2.

The apparatus for containing and sampling CS flames is shown schematically in Fig. 2. It consists of combustion chamber (1) connected to the scanning device (2) which moves the burning strand relative probe (3). The ion source (4) of the MSKh-4 TOF serves as a detector of the molecular beam sample of the CS flame.

The flame is sampled by a quartz probe, a 25-mm-high cone with a 40-deg external angle, a 30-deg internal angle, and a 0.1-mm-diam orifice at the apex. Gas expansion in the cone results in a supersonic jet directed into the skimmer chamber (5), which is evacuated by a booster pump BN-3 (300 l/s). Behind the sampling cone is a stainless steel skimmer (8), a 50-mm-high cone with a 60-deg external angle, a

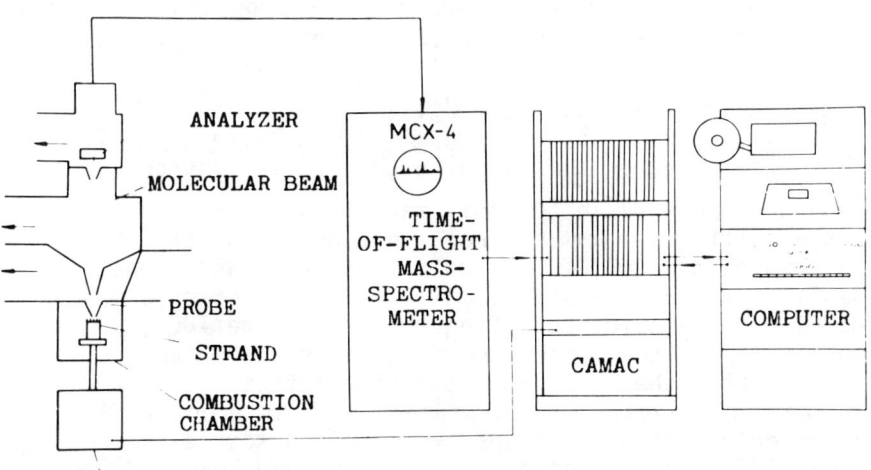

Fig. 1 Schematic of experimental apparatus and control and data acquisition system.

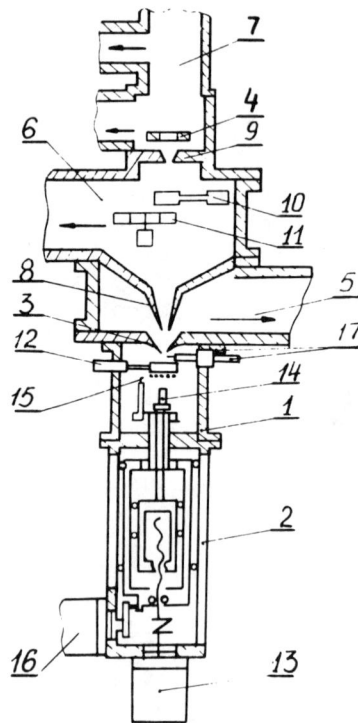

Fig. 2 Schematic of apparatus for mass-spectrometric observations of flame structure.

40-deg internal angle, and a 2-mm-diam orifice at the apex. The skimmer is designed so that only the supersonic jet core enters the collimator chamber (6), which is evacuated by an oil diffusion pump NO-5 (500 l/s). The pressure in the skimmer chamber is $5 \cdot 10^{-2}$ mm Hg. The molecular beam from the skimmer passes through a 4-mm-diam collimator orifice (9) to the detector chamber with the TOF ion source. The phone gas pressure in the collimator chamber is 10^{-5} mm Hg. In the collimator chamber are: an electromagnetic chopper (10) which can cut off the molecular beam, and a beam modulator which is a slotted disk (11) rotated by a DG-2TA engine with an adjustable frequency from 1 to 100 Hz. Subtraction of the phone (the detector signal with the beam cut off) from the beam+phone sum (the detector signal with the beam on) gives a useful signal which is proportional to the molecular beam intensity. The ion source of the MSKh-4 TOF is evacuated by a heteroion pump NORD-250, and the drift tube MSKh-4, by a turbomolecular pump TMN-200. The pressure in the ion source is 10^{-6} mm Hg. The distance between the orifices varies: the skimmer-collimator is 310 mm; the collimator-detector is 50 mm; and the skimmer-sampling cone is varied from 5 to 25 mm by changing the flange to which the probe is attached. In the combustion chamber the ignition spiral (12), after ignition, is automatically removed from the combustion zone. To increase the accuracy of concentration distribution measurements in CS flames, a new scanning device has been designed (Korobeinichev et al. 1980) which is more elaborate than that reported by Korobeinichev and Tereshenko (1977). To scan a flame supported by condensed reactants, a control system and two step-by-step motors (13) and (16) are required. With the first motor (13) (0.00125-mm-step), the burning strand (14) is moved relative to the thermocouple (15) so that the thermocouple junction is always at a fixed temperature inside the flame zone. As a result, the spatial location of the burning surface is stabilized. The second motor (16) moves the strand-microthermocouple system relative to the microprobe and fixed thermocouple (17) as directed by the control system. The phase commutation frequency being $\leq$ 4 kHz, the strand speed is $\leq$ 5 mm/s. By varying the phase commutation frequency, one can continuously control the strand speed.

The experimental control and data acquisition system consists of a SARATOV-2 minicomputer, a CAMAC stand with two crates and a control desk, a double-trace oscillograph, and a unit of the scanning device control. The data acquisition system automatically detects mass spectra, i.e., measures the mass peak intensities and corresponding mass numbers, as well as temperatures as functions of time. During a single mass-spectrum scan, realized by time gating, the intensities of 12 mass peaks of 60-100 ns can be measured and the data stored in the computer. The time lag range of a gate is 0-160 µs, its duration being 35-640 ns. The time intervals can be set at 10 ns. The mass peak intensities are measured within the dynamic range of 1-1000. The starting frequency of the mass spectrometer is 100 kHz. The variable amplification factors are 4, 8, 16, and 32. The accuracy of the measurement channel is not less than 0.1%. The intensities of 40 mass peaks of a spectrum can be measured at a frequency of some tens of hertz. With a complex of programs, the system can measure and store peak intensities at a required frequency (up to 500 spectra/s), calibrate the mass scale semiautomatically, realize selective filtering of experimental data (increasing thereby the signal-to-noise ratio), control and correct the zero line, and display the data with a plotter. This system allows more than an order increase in the accuracy of detecting the mass peak inten-

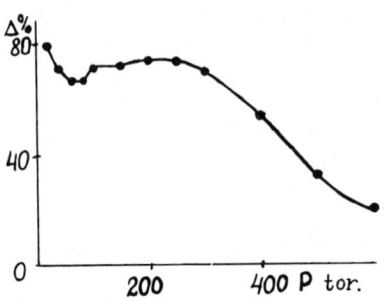

Fig. 3 Influence of combustion chamber pressure on molecular beam efficiency.

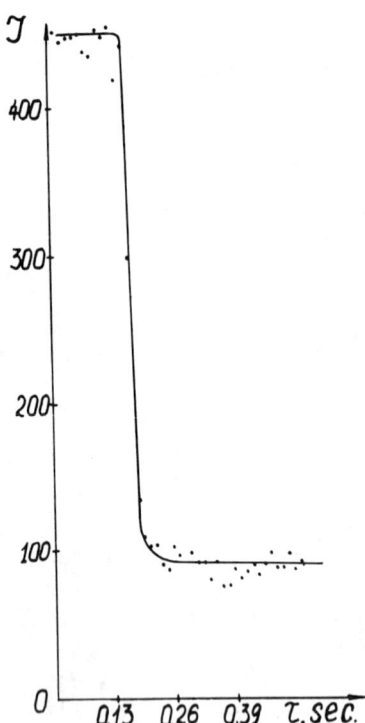

Fig. 4 Response time of the sampling system as observed through decay of m/e 36 (HCl) peak intensity.

sities and mass numbers, in the sensitivity of the apparatus, and reduces the time of primary data processing.

System Characterization

The molecular beam efficiency Δ was measured at various distances L_1 between the skimmer and the sampling cone (intercone distance) and at various pressures in the combustion chamber. Here $\Delta = (I_{op} - I_{cl})/(I_{op} - I_0)$, where I_{op}, I_{cl}, I_0 are nitrogen peak intensities with the chopper opened, closed, and at zero pressure in the combustion chamber, respectively. With these experiments the optimum intercone distance was determined to be $L_1/d = 100$. Figure 3 shows Δ as a function of air pressure in the combustion chamber at $L_1 = 23$ mm. Figure 4 depicts the time evolution of the peak intensity with m/e 36 (HCl) in the mass spectrum of the combustion product, the beam chopper closed and opened, which allows the time of response of the sampling system to be determined.

AP-PMMA Composite Flames

The sampling system without a molecular beam was used to study the chemical structure of AP-based flames (Korobeinichev and Tereshenko 1977). Ammonia and chlorine dioxide, rather than ammonia and perchloric acid as predicted by conventional models of solid AP-base fuel combustion (Korobeinichev et al. 1980), were detected as the basic oxidant gasification products maintaining the flame. However, in the sampling system used, chlorine dioxide could originate from perchloric acid by the reaction $HClO_4 \rightarrow ClO_2 + H_2O$ on the hot walls of the sampler. With the molecular beam sampling system this reaction does not occur. Perchloric acid was detected in the flame of AP and PMMA composite (stoichiometric composition) at 100 mm Hg. Figure 5 shows the mass peak intensity profiles with m/e 36 (HCl), 83 ($HClO_4$), 67 (ClO_2, $HClO_4$), 52 (ClOH), 51 (ClO, ClOH, $HClO_4$, ClO_2), 41 (methyl methacrylate $C_5H_8O_2$), 44 (CO_2, N_2O), 30 (NO, N_2O), 18 (H_2O), 17 (H_2O, NH_3) measured in the flame zone (L > 0) and in the reaction C-phase (L < 0). L = 0 corresponds to the burning surface. Analysis of the mass peak intensity profiles in the flame leads to the conclusion that the zone of perchloric acid decomposition in the flame (m/e = 83) and that of methyl methacrylate (41) oxidation practically coincide. In both cases the zone widths are about 0.3 mm. The zone of ammonia oxidation (obtained from the analysis of 17 and 18 mass peak profiles) is much wider (0.6-0.7 mm); i.e., ammonia is oxidized at a longer distance from the burning surface. The widest zone is that of CO_2 generation (m/e = 44).

As seen from Fig. 5, the distributions of the mass peak intensities m/e 83, 67, 52 are significantly different:

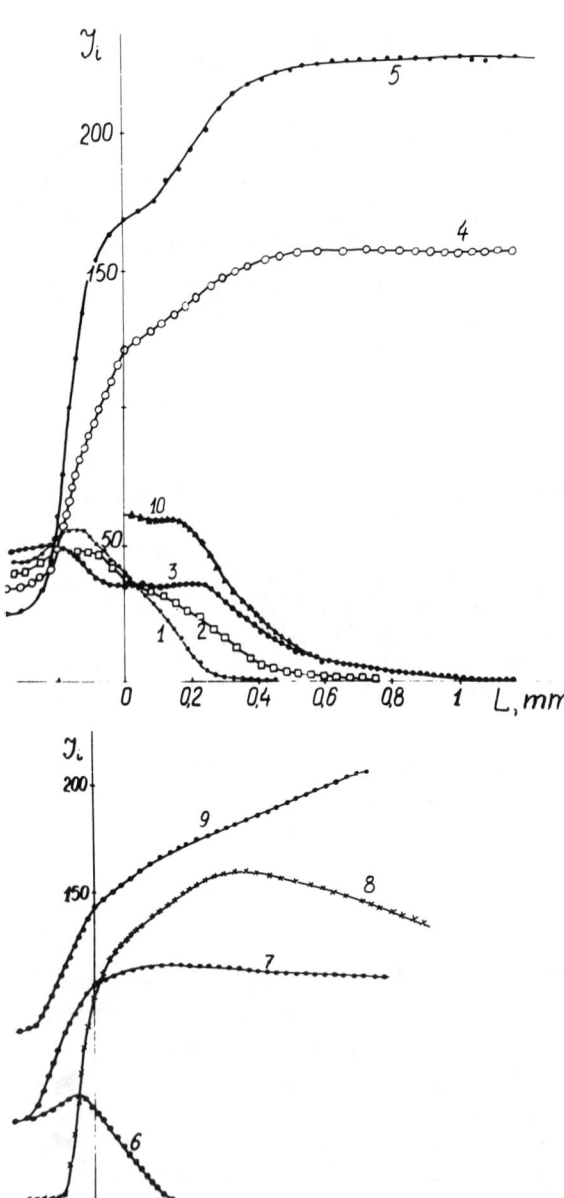

Fig. 5 Distribution of mass peak intensities in stoichiometric AP-PMMA flames. Initial pressure = 100 Torr; burning surface, L = 0; flame zone, L > 0. (1) $HClO_4(I_{83})$; (2) $HClO_4$, $ClO_2(I_{67})$; (3) $ClOH$ (I_{52}); (4) H_2O $(2I_{18})$; (5) $HCl(2I_{36})$; (6) $C_5H_8O_2(I_{41})$; (7) NH_3, $H_2O(I_{17})$; (8) NO, N_2O (I_{30}); (9) CO_2, N_2O $(2I_{44})$; (10) ClO, $ClOH$, ClO_2, $HClO_4(I_{51})$.

first, the m/e 83 ($HClO_4$) peak falls, then 67 (ClO_2, $HClO_4$), and, at last, 52 (ClOH) follows. The plateau observed near the burning surface for m/e 52 (ClOH) is additional evidence that ClOH is generated at a certain distance from the burning surface in the interior of the high-temperature flame zone. Taken together, these observations support the following transformation mechanism of chlorides in the flame: $HClO_4 \to ClO_2 \to ClOH \to HCl$.

The mass peak intensity profiles for m/e 83, 67, 52, 51 (Fig. 5) were analyzed because of the overlap from the individual components $HClO_4$, ClO_2, and $ClOH$. The m/e 51 peak intensity is equal to the sum of the contributions from the above components (neglecting the contribution of ClO radicals) if $I_{51}/I_{52} = 1$ of the ClOH spectrum. There are no data on this latter point in the literature. Analysis of experiments made under conditions when ClO radicals cannot be detected [in the system without a molecular beam (Korobeinichev and Tereschenko 1977)] for the ClOH spectrum $I_{51}/I_{52} \leq 0.5$. This contradiction can be explained by the presence of ClO radicals in the flame, their concentration being approximately half that of ClOH.

As the distance from the burning surface increases, the m/e 30 (NO, N_2O) peak first grows (owing to nitrogen oxide generation in ammonia oxidation), then falls because of the nitrogen oxide reduction to molecular nitrogen in the high-temperature zone. The chemical transformation of nitrogen proceeds through the following scheme:

$$NH_3 \to NO, \quad N_2O \to N_2$$

Comparison of reaction zone widths for perchloric acid and chlorine dioxide obtained a 0.03-mm orifice sampler (Korobeinichev and Tereshenko, 1977) with those obtained in the present work for the same composite system with 0.1-0.3-mm orifice samplers shows that a three-to-tenfold change in the sampler orifice has a negligible effect ($\sim$ 10-20%) on the observed flame reaction zone width. While this observation merits further investigation, the present results prove the applicability of this method to structure investigations of flames whose zone widths are comparable to the sampler orifice diameter.

AP Flames

Figure 6 shows the mass peak intensity profiles with m/e 36, 83, 67, 52, 51 in ammonium perchlorate flames at 1 atm pressure and 260°C initial temperature of the strand. These data indicate that perchloric acid is the primary product of AP gasification. At the same time, ClO_2 and ClOH are also generated in measurable concentrations. Analysis of the m/e 51 peak indicates that the ClO radical concentra-

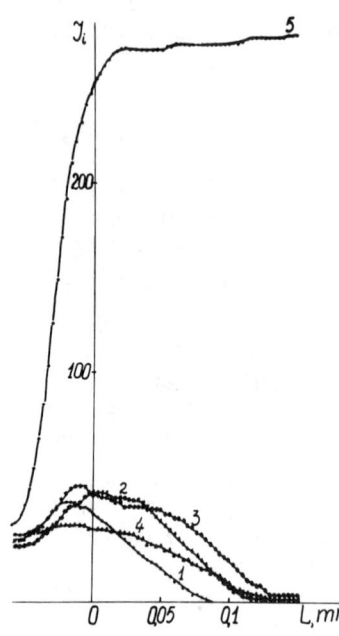

Fig. 6 Distribution of mass peak intensities in ammonium perchlorate flames. Initial pressure = 760 Torr; initial temperature = 260°C; burning surface, L = 0; flame zone, L > 0. (1) $HClO_4(I_{83})$; (2) $ClO_2(I_{67})$; (3) $ClOOH(I_{52})$; (4) ClO, $ClOH$, ClO_2, $HClO_4(I_{51})$; (5) $HCl(2I_{36})$.

tion in the AP flame is smaller than that in the AP-PMMA composite flame. This observed difference probably is a consequence of the higher temperatures of the composite flame. The observed zone widths and the distribution of the components in the flame are in good agreement with those reported by Korobeinichev and Tereshenko (1977).

Conclusions

The data on the gasification products of AP and composite systems obtained in the present work confirm the hypotheses which underlie models of combustion of AP and AP-based composite systems. These data make it possible to specify the chemical transformation mechanisms in the flames. The apparatus developed can be used to study condensed system flames as well as chemical transformation due to shock waves.

References

Fristrom, R. M. and Westenberg, A. A. (1965) <u>Flame Structure</u>. McGraw-Hill, New York.

Korobeinichev, O. P. (1980) Primenenie mass-spectrometrii dlya izucheniya struktury plamen i protsessov goreniya. <u>Usp. Khim.</u> 49, 946-965.

Korobeinichev, O. P., Scovorodin, I. N., Emeliyanov, E. L., Kascheev, K. P., and Polozov, S. V., Tereschenko, A. G., Kuibida, L. V., and Ivanov, V. V. (1980) Ustroistvo dlya issledovaniya protcessa goreniya tverdykh topliv. Copyright No. 756279. Bull. Izobreten. Otkytii. No. 30.

Korobeinichev, O. P., Scovorodin, I. N., Polozov, S. V., Postoenko, U. K., Astrahantsev, I. E., Safonov, P. G., Kuzin, V. A., and Skomorohov, V. B. (1977) Avtomatizirovannaiya ustanovka dlya izucheniya dinamiki bystryh khimicheskih prevrascheniy s pomoschiu vremya proletnogo mass spectrometra. Zh. Fiz. Khim. 51, 1542-1547.

Korobeinichev, O. P. and Tereshenko, A. G. (1977) Issledovanie structury zon goreniya condensirovannyh sistem s pomoschiu mass spectrometrii. Khimicheskaiya phisica protsessov goreniya i vzryva. Gorenie condensirovannyh sistem. Chernogolovka, 73-75.

Stearns, C. A., Kohl, F. J., Fryburq, G. C., and Miller, R. A. (1979) High pressure molecular beam mass spectrometric sampling of high temperature molecules. Proc. 10th Materials Res. Symp. Charact. High Temp. Vapors and Gases. pp. 303-355. NBS Special Publication 561, No. 1.

Unsteady Burning of Double-Base Propellants

V.E. Zarko,* V.N. Simonenko,† and A.B. Kiskin†
Academy of Sciences, Novosibirsk, USSR

Abstract

The dynamic burning rate response of a modified double-base propellant (with carbon black and lead oxide added) to an externally exposed radiant heat flux was measured directly by detecting the propellant recoil. The unsteady burning rate was measured continuously at a sharply decreasing external heat flux and under the conditions of ignition by a combustion wave. An attempt was made to simulate numerically the burning rate behavior under conditions corresponding to experimental ones. Experimental data were found to qualitatively agree with the Zeldovich-Novozhilov phenomenological theory. However, the accuracy and reliability of the available experimental data on steady-state combustion are insufficient for correct verification of theoretical predictions.

Introduction

At present the phenomenological theory (PT) of Zeldovich (1942) and Novozhilov (1973) is widely used to describe a propellant unsteady combustion. A further development of this theory must be based on objective experimental data obtained by up-to-date methods. Unfortunately, the available methods of experimental measuring of instantaneous burning rates have insufficient time and space resolutions, the only exception being the

---

Presented at the 8th ICOGER, Minsk, USSR, Aug. 23-26, 1981. Copyright © American Institute of Aeronautics and Astronautics, Inc., 1982. All rights reserved.

*Senior Research Assistant. Institute of Chemical Kinetics and Combustion.

†Junior Research Assistant. Institute of Chemical Kinetics and Combustion.

microwave technique (Levine and Andrepont 1979). Scanty information can be found in the literature on systematic experimental studies of unsteady combustion of condensed substances with direct dynamic measurements of burning rate levels.

The technique of propellant recoil measurement (Simonenko and Zarko 1981) is a convenient and simple method of detecting the instantaneous burning rate. Since a reactive force is related to a mass burning rate (ρU) as

$$F = (RT_g/PM)(\rho U)^2 S \qquad (1)$$

the steady-state relation $F = F(\rho U)$ determined beforehand suffices to obtain (ρU). Here, U is the propellant burning rate, ρ is the condensed phase density, S is the cross section of the strand, M is the molecular weight of the combustion products, P is the pressure, T_g is the maximum gas temperature, and R is a universal gas constant. This technique can be used to detect the regression of surface layers with a thickness of some 10-20 μ. At moderate pressures it equals approximately the reaction zone width in the condensed phase of a propellant.

It is necessary to remember, however, that Eq. (1) is valid only in the case of complete gasification of a condensed substance and when (ρU) does not vary too quickly. Useful properties of this method are a high clarity of data presentation and the lack of necessity of differentiating the initial signal, in contrast to the case of continuous length and mass measurements.

The present paper reports experimental data on the instantaneous burning rate of a double-base propellant under periodical and pulse action of a radiant flux, as well as under the conditions of ignition by a combustion wave. The experimental results are discussed in PT terms.

Experimental Technique

The reactive force was measured by a capacity transducer (Mikheev and Borin 1973) with 10^{-3} g sensitivity and a 0-400 Hz working frequency range. The reactive force data were converted to those of instantaneous burning rate using the steady-state dependence $F = F(\rho U)$. The propellant strands were 4-mm-long cylinders, 10 mm in diameter, protected with a thin mica envelope. The envelope also ensured a one-dimensional escape of combustion products and prevented an inflow of external oxidizer. We investigated two modifications of the double-base propellant N: with 1% of carbon black (N + CB, the effective coefficient of radiation absorption $\sigma \cong 450$ cm^{-1}) and with 1% of lead oxide (N + Cat, $\sigma \cong 40$ cm^{-1}).

A 10-kW xenon lamp was used as a radiation source (see Fig. 1). The light flux amplitude was modulated by a slotted disk. A step-by-step variation of the radiation flux intensity was realized with the help of shutters. The flux was completely cut off by a central photoshutter with 10^{-3}-s operation time. A partial cutting of the flux was realized by a blind-type shutter with 10^{-2}-s operation time.

In experiments with high initial temperatures the propellant samples were heated in a tube electrical furnace, the sample temperature at the center and periphery being measured by thermocouples.

Experimental Results

Figure 2 depicts typical oscillograms of U(t) at a step-by-step change of the light flux level from q_1 to q_2. Under fixed initial conditions a drop in U(t) increases with the relative value of $A = q/c\rho U(T_s - T_0)$ (curves 1, 2, Fig. 2, N + CB propellant).

However, at different starting steady-state burning rates, the same perturbation ($A \cong 0.2$) leads to qualitatively different results. In the case of a low original burning rate, the combustion "fails" (curve 3, N + CB propellant) followed by reignition due to the radiation flux. In experiments with N + Cat propellant, a transition to stationary conditions at a complete cutoff of the radiation flux occurs practically without reduction of U(t) below a new stationary level (curve 4, Fig. 2).

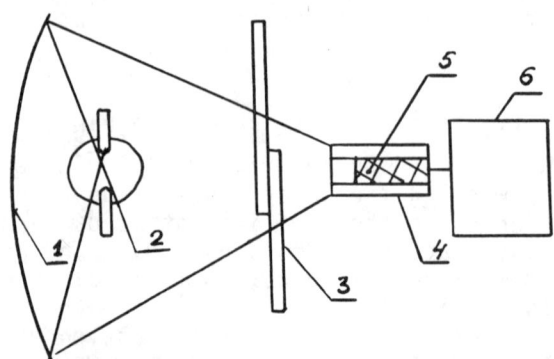

Fig. 1 Experimental setup for measuring propellant recoil. (1) reflector, (2) Xenon lamp, (3) shutter, (4) mica envelope, (5) propellant sample, (6) capacitance transducer.

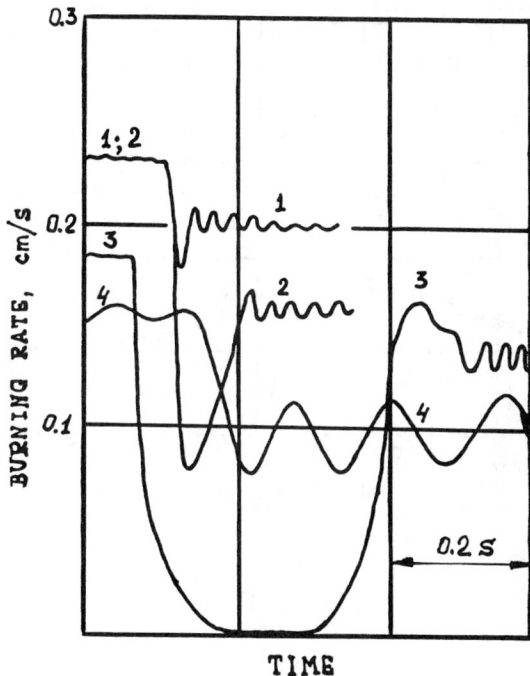

Fig. 2 Burning rate behavior (experimental) after a sharp decrease in the light flux. (1) N + CB, $T_o = 80°C$, $q_1 = 15$, $q_2 = 10$ cal/cm^{2}s; (2) N + Cb, $T_o = 80°C$, $q_1 = 15$, $q_2 = 5$ cal/cm^{2}s; (3) N + CB, $T_o = 20°C$, $q_1 = 10$, $q_2 = 5$ cal/cm^{2}s; (4) N + Cat, $T_o = 80°C$, $q_1 = 15$, $q_2 = 0$.

The experiments on a burning wave transition in compound strands were performed at $T_0 = 100°C$. Figure 3 shows oscillograms of direct and back transitions in N + CB and N + Cat samples glued together. Note that a transition from a high to low burning rate (N + CB → N + Cat) is accompanied by a deep fall in U(t) trace, its character resembling that observed during the combustion of opaque N + CB propellant with a radiation cutoff.

Measurements performed at self-sustaining burning conditions show that the propellant combustion is accompanied by burning rate oscillations, these being especially pronounced in the case of an N + Cat propellant. An increase in the burning rate, induced by either growing pressure, or rising initial temperature T_0, or by an external radiation flux q, results in an increase of the burning rate oscillation frequency. Special experiments with a reacting surface irradiated periodically by low-intensive radiation have proved the resonance response frequency of the burning rate to coincide with the oscillation frequency under the self-sustaining conditions. Hence the latter can be identified with the eigenvalue of the burning rate frequency

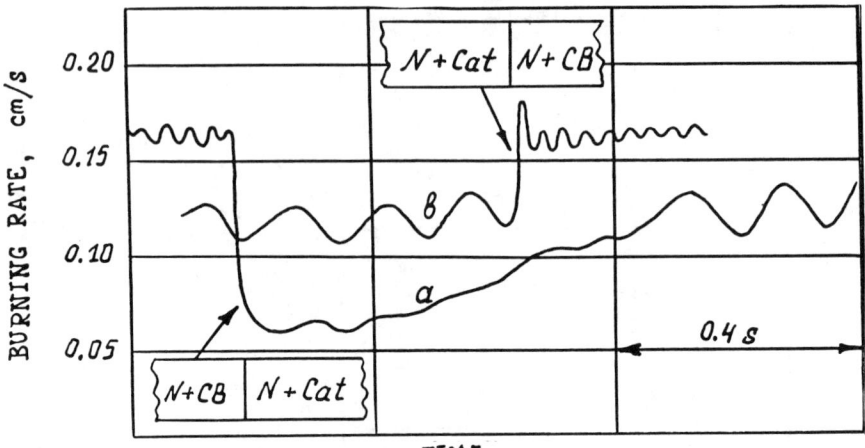

Fig. 3 Burning rate behavior (experimental) during combustion wave propagation in compound samples. $T_o = 100°C$, $P = 1$ atm. a) N + CB → N + Cat; b) N + Cat → N + CB.

ν, which obeys the expression (Novozhilov 1973):

$$\nu = \frac{U^2}{2\pi a}\sqrt{\frac{k}{r^2} - \lambda^2} \qquad \lambda = \frac{(k+1)r - (k-1)^2}{2r^2}$$

$$k = \beta(T_s - T_o) \qquad \beta = \left(\frac{\partial \ln U}{\partial T_o}\right)_P \qquad r = \left(\frac{\partial T_s}{\partial T_o}\right)_P$$

(2)

Here T_s is the surface temperature, and a is the thermal diffusivity. Solving Eq. (2) with respect to r, we obtain an ordinary differential equation (P = const):

$$\frac{dT_s}{dT_o} = r(T_s, T_o, \nu, U, \beta) \qquad (3)$$

The functions $T_s(T_o)$ for N + CB and N + Cat propellants are determined from experimental dependences $\nu(T_o)$ and $U(T_o)$ (Table 1) by numerical integration of Eq. (3). Table 1 lists the values of $T_s(T_o)$ obtained.

Analysis of Experimental Data

The behavior of the burning rate in passing through the site of gluing can be explained by taking into account the surface temperature, the gradient at the surface, and the

Table 1 Values of T_s (T_0) obtained for N + CB and N + C at propellants

Composition	T_0, K	U, cm/s	$\beta \times 10^3$, (deg)$^{-1}$	ν, Hz	T_s, K	k	z
N + CB	293	0.08	6	15	518	1.31	0.06
	323	0.10	7	17	520.2	1.40	0.08
	353	0.12	9	21	522.9	1.48	0.11
	373	0.14	10.5	27	525.2	1.55	0.12
	393	0.18	12	35	527.9	1.63	0.15
N + Cat	293	0.075	4	5.0	563.0	1.05	0.04
	323	0.084	5	5.5	565.2	1.21	0.11
	353	0.103	6.5	6.0	569.8	1.43	0.20
	373	0.120	8	7.0	574.6	1.57	0.27
	393	0.140	9.5	8.5	589.7	1.73	0.34

Table 2 Initial data and calculated values of ϕ and ΔH

Composition	P, atm	T_0, °C	U, cm/s	$a \times 10^3$, cm^2/s	$\lambda \times 10^4$, cal/cm-s-deg	T_s, °C	$\phi \times 10^4$, deg/cm	ΔH, cal/cm^2
N + CB	1	100	0.14	1.25	5.6	252.2	1.7	0.61
N + Cat	1	100	0.12	1.25	5.6	301.6	1.94	0.94

solid heat storage. The value of ϕ and ΔH for steady-state waves can be calculated by the formulas

$$\phi = (U/a)(T_s - T_0) \qquad (4)$$

$$\Delta H = c\rho \int_{-\infty}^{0} (T - T_0)dx = c\rho(a/U)(T_s - T_0) = (\lambda/U)(T_s - T_0) \qquad (5)$$

Table 2 lists initial data and calculated values of ϕ and ΔH.

Table 2 shows that for the N + Cat propellant the heat storage and the surface temperature in steady-state combustion are high compared to those for the N + CB propellant. Hence, when the combustion wave propagates from the N + CB to the catalyzed propellant, the heat must be accumulated in the solid, and the surface temperature must rise. As follows from PT analysis (Zarko and Kiskin 1980), this situation is possible providing the burning rate reduces below the level of the steady-state burning rate of the catalyzed propellant, which corresponds to the curve U(t) with a minimum. A substantial drop in the instantaneous burning rate can lead to the propellant extinction.

Under the back transition, from N + Cat to N + CB, the solid heat storage and the surface temperature must decrease. Using similar reasoning, one can show that in this case the burning rate curve must have a maximum. However, in the experiment the N + CB burning rate reaches a new stationary level in fact instantaneously, subsequent to the burning wave transition from the N + Cat strand. Taking into account the chemical kinetics of the propellant studied, we can assume that in the N + Cat → N + CB transition the temperature profile of the catalyzed propellant cannot be entirely the same as that of the N + CB sample, since the surface temperature of the former exceeds that of the latter. Under this assumption, as the burning wave approaches the samples contact, N + CB is ignited at a lower temperature than that of the burning surface of N + Cat, and the finite-thickness layer of the catalyzed propellant must be shot off. In this case, the temperature profile in the N + CB strand is close to that observed in steady-state combustion. However, since the thickness of the shot-off layer does not exceed 20×10^{-6} m, which is comparable with the characteristic size of inhomogeneities on the reaction surface, the details of this transition cannot be detected experimentally.

Numerical Simulation of the Transition Processes

Nonstationary burning rates have been calculated numerically in PT terms under conventional assumptions of a one-dimensional process, temperature-independent thermophysical properties being the same in both parts of a compound strand. In the general case, the problem of nonstationary combustion is reduced to solving the heat equation with certain initial and boundary conditions, and a phenomenological relation for the burning rate, the gradient, and the surface temperature:

$$\frac{\partial T}{\partial t} + U \frac{\partial T}{\partial x} = a \frac{\partial^2 T}{\partial x^2} \quad -\infty < x < 0 \quad (6)$$

$$x = 0 \quad T = T_s(U) \quad \left.\frac{\partial T}{\partial x}\right|_{x=0} = \frac{U}{a}[T_s - T_0(U)] \quad (7a)$$

$$x = -\infty \quad T = T_0 \quad (7b)$$

$$t = 0 \quad T(x) = T_0 + (T_{si} - T_0)\exp(V/a)(x-C) \quad (7c)$$

Here T_{si} is the burning surface temperature of the igniter; U and V are the burning rates of the ignited sample and the igniter, respectively. The function $T_0(U)$ can be calculated either from theoretical or empirical dependence $U(T_0)$.

For the initial temperature distribution $T(x,0)$ it is necessary to take into account that at the ignition moment only a part of the temperature profile of the igniter can be realized. It is assumed that the unburnt b-thickness layer of the first sample is carried away (shot off) by the gaseous products of pyrolysis of the second strand at the moment of its ignition. The value of b is determined from a relation analogous with the ignition criterion (Averson et al. 1968):

$$\lambda \frac{\partial T}{\partial x}\bigg|_{x=0} = c\rho V(T_{si}-T_0)e^{-VC/a} = \alpha Qk_0 \int_{-\infty}^{0} (e^{-E/RT} - e^{-E/RT_0})dx \tag{8}$$

The coefficient α is chosen so that Eq. (8) holds identically (with b = 0) when the combustion wave front passes through the gluing of the samples with the same composition. For a particular composition, in the general case, $\alpha = \alpha(T_0)$.

An important point of the PT is that at any moment U, ϕ, and T_s must have correlated values satisfying Eq. (7a). In order to satisfy the condition (7a) at the initial moment

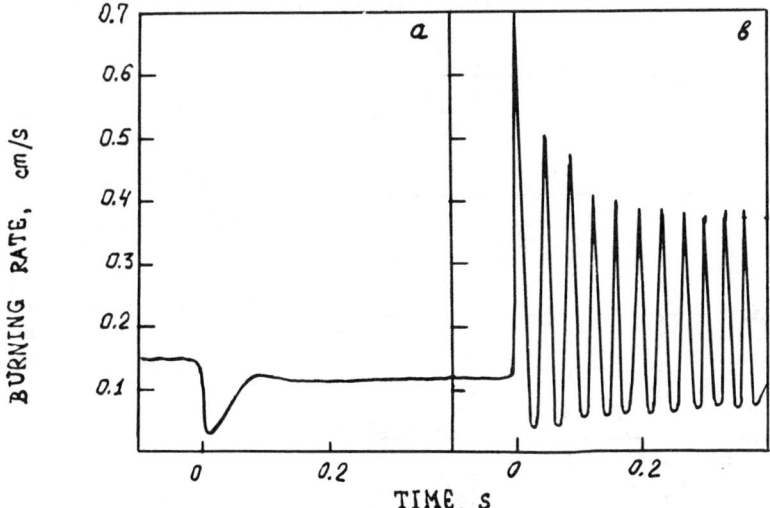

Fig. 4 Burning rate behavior (calculated) during combustion wave propagation in compound samples. $T_0 = 100°C$, P = 1 atm. a) N + CB N + Cat; b) N + Cat N + CB.

(t = 0), the temperature profile in a narrow subsurface layer ($\Delta x = L$) of the ignited sample must be transformed in a certain manner. In this case, the value of $T_s = T_o + (T_{si} - T_o)e^{-V_c/a}$ is fixed on the surface, the gradient $\partial T/\partial x$ at $x = 0$ is set equal to $\phi = \phi[T_s(U)]$, the transformed profile and the original one are "stuck" together at $x = L$ (the first and the second partial space derivations being the same for both profiles at $x = L$). Numerical computations have shown that variations of the layer width L ($L < a/U$) negligibly affect the time and amplitude characteristics of the nonstationary burning rate.

To calculate U(t) for combustion under irradiation, a $(c\rho)^{-1} q\sigma e^{\sigma x}$ type term must be added to the right side of Eq. (6), and the temperature distribution of the steady-state combustion under a light flux q_o serves as the initial distribution.

In numerical computations we considered the following nonstationary conditions: 1) direct and opposite propagation of the burning wave in (N + Cat) - (N + CB) compound strands (Fig. 4); 2) N + Cat burning subsequent to the light flux cutoff (Fig. 5).

Discussion

It should be noted that, according to Eqs. (6) and (7), the combustion wave to the left of the contact surface in a

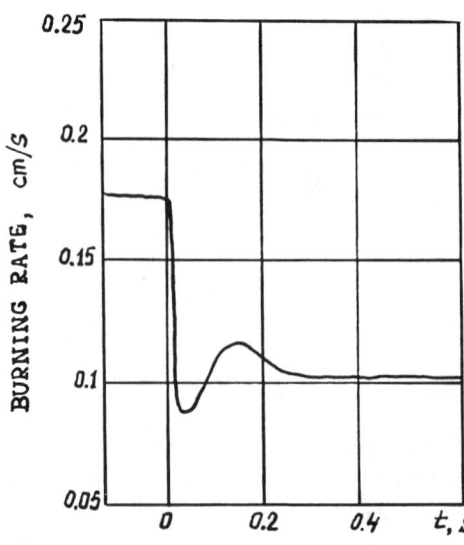

Fig. 5 Burning rate behavior (calculated) of the N + Cat propellant after a sharp decrease in the light flux. $T_o = 80°C$, $q_1 = 15$ cal/cm^{2}s; $q_2 = 0$.

compound sample is assumed to have strictly steady-state behavior without any oscillations of the burning rate. Therefore the calculated burning rate behavior should be analyzed only after the combustion wave has passed through the gluing site. The comparison of calculated and experimental dependences U(t) (Figs. 3 and 4) demonstrates the following difference. The N + CB → N + Cat transition of the burning wave occurs in calculations faster and with a deeper decrease in the burning rate level. Moreover, on reaching a new steady state, the calculated U(t) shows no regular oscillations typical of the experimental curve. In the case of N + Cat → N + CB transition the calculated burning rate sharply increases at the initial moment which is not observed in experiment.

The above discrepancies seem to be caused by the following factors. Apparently, the PT approach is a rough approximation of the combustion phenomenon. On the other hand, the initial data used in calculations are set with a certain inaccuracy; these data must be extrapolated far beyond the limits of experimental determination. Indeed, in the case of N + CB → N + Cat transition, the initial T_s assumed is 250°C for the N + Cat surface, which is substantially lower than the minimum T_s observed in experiment (see Table 1). Therefore numerical calculations yield almost zero level of U(t) at the moment of transition.

The fact that U(t) reaches a steady-state level comparatively fast can result from neglecting the laws of N + Cat ignition at such low T_s values. In particular, as in experiments with N + CB under pulse light, we can assume that the N + Cat surface is ignited by the burning wave gradually, in a local manner. Therefore the combustion wave stabilization time would be longer compared to the case of uniform combustion throughout the surface.

The sharp initial increase in the burning rate obtained in numerical simulation of a N + Cat → N + CB transition can be accounted for by peculiarities of igniting a composition with a lower surface temperature. Since $T_s(N + CB) < T_s(N + Cat)$, the simulated N + CB ignition is accompanied by shooting off the unburnt 3×10^{-5} N + Cat layer. In this case, the entalphy excess in the surface layer increases the burning rate. Note that it is rather difficult to observe this situation in detail experimentally: the combustion front must be parallel to the gluing plane accurate to $(10-20) \times 10^{-6}$ m.

The absence of auto-oscillations in simulated U(t) curves for N + Cat strands results from the fact that at set T_s, k, U, r (see Table 1), and $T_0 = 100°C$, a self-sustaining propellant combustion is far from the stability boundary, i.e., in the region determined by linear analysis (Novozhilov 1973), as that of burning rate perturbations decaying in time. According to the integration procedure of Eq. (3), the burning process would approach the calculated stability boundary with auto-oscillations of the burning rate if $T_s(T_0)$ is within 350-400°C. A similar effect can be achieved if the thermal diffusivity coefficient is assumed to exceed the nominal one by several times. At present there is no sufficient reason for such a modification of the original data.

Conclusions

Analysis of numerical calculations has shown the accuracy of the available experimental data on the burning rate sensitivity to the initial and surface temperatures to be insufficient for a reliable numerical simulation of transient combustion. In particular, variations of $U(T_0)$ within the range of experimental data may result in changing the calculated unsteady burning rate behavior to the steady-state one and vice versa. To increase the reliability of numerical calculations, it is necessary to widen the range of determining the parameters of steady-state burning laws in order to cover the whole interval of transient burning rates.

A nonstationary combustion model must be developed on the principle of a more adequate description of heat- and mass-transfer processes in a heat wave. To take into account the real temperature dependence of thermophysical coefficients and phase transformations (melting, boiling, etc.) presents no special problem. It is much more difficult to describe chemical transformations in condensed and gas phases. However, the question of whether or not a phenomenological approach should take into account these processes can be answered only after a careful comparison of experimental data on nonstationary burning rate with numerical estimations by the above model using a full set of reliable empirical dependences.

References

Averson, A. E., Barzykin, V. V., and Merzhanov, A. G. (1968) The approximate method of solution of thermal theory iginition problems. Dokl. Akad. Nauk SSSR 178, 131-134.

Levine, J. N. and Andrepont, W. C. (1979) Measurement methods of transient combustion response characteristics of solid propellant--an assessment. AIAA Paper 79-1209, 36 pp.

Mikheev, V. F. and Borin, S. M. (1973) Mass velocity measurements in solid explosive burning. Fiz. Goren. Vzryva 9, 327-329.

Novozhilov, B. V. (1973) Nonstationary Combustion of Solid Rocket Fuels, p. 176. Nauka, Moscow, USSR.

Simonenko, V. N. and Zarko, V. E. (1981) Reactive force of combustion products as a measure of unsteady velocity of powder combustion. Fiz. Goren. Vzryva 17, 129-132.

Zarko, V. E. and Kiskin, A. B. (1980) Numerical simulation of unsteady powder combustion at a light flux action. Fiz. Goren. Vzryva 16, 54-59.

Zeldovich, Y. B. (1942) On theory of combustion of powder and explosives. Zh. Eksper. Teor. Fiz. 12, 498-524.

Surface Layer Destruction during Combustion of Homogeneous Powders

V.E. Zarko* and V.Y. Zyryanov†
Academy of Sciences, Novosibirsk, USSR

Abstract

Under certain conditions the combustion of homogeneous powders (HP) is accompanied by dispersion of the reaction layer of the condensed phase (C phase). As a theoretical analysis shows dispersion can affect both the combustion rate and its dependence on the external conditions; studies on the dispersion mechanism are of particular importance. In the present work, HP combustion is considered as simultaneous processes of a chemical reaction in the C-phase reaction layer and destruction of this layer by the issued gases. The dispersion induced by gasification in the subsurface layers has been shown to result from the aerodynamic forces acting on the liquified reaction layer. A phenomenological dispersion model has been proposed for HP combustion with a liquid-viscous reaction layer in the condensed phase. The model fairly well describes the available experimental data: 1) a decrease in the dispersion rate with increasing pressure and decreasing portion of the volatile components in the HP, and 2) a decrease in the mean size of dispersed particles with increasing pressure.

Introduction

The mechanical destruction of the reacting surface layer following particle ejection into the gas phase is

---

Presented at the 8th ICOGER, Minsk, USSR, Aug. 23-26, 1981. Copyright © American Institute of Aeronautics and Astronautics, Inc., 1983. All rights reserved.
*Senior Research Assistant, Institute of Chemical Kinetics and Combustion.
†Junior Research Assistant, Institute of Chemical Kinetics and Combustion.

conventionally called dispersion. The dispersion of
homogeneous powders (propellants) results from the volume
gas evolution in the reaction layer of the condensed phase
(C phase). It follows from the physical grounds that the
dispersion must result in incomplete heat release in the
C-phase and reduce the reacting surface temperature.
Theoretical analysis (Merzhanov 1960, 1969; Khaikin 1975)
has shown that the degree of dispersion can essentially
affect the burning rate, whose dependence upon the external
conditions may be determined from a model of the variations
of the degree of dispersion. Observation of the dispersed
phase during combustion is difficult except in model
experiments (Ginsburgh and Stepanov 1978; Pokhil 1954;
Zyryanov 1980). As a consequence, studies on the dispersion
mechanism and its simulation acquire especial significance.

Formulation of the Problem

Observations of the structure of burning and
extinguished propellant samples prove that the reaction
C-phase layer is a viscous liquid. Gas release in the
interior of the liquid reaction layer causes dispersion of
the liquid layer. In this case, two fundamentally different
mechanisms of droplet generation are possible: 1) gas
bubble bursting at the surface and 2) aerodynamic
interaction of the gas jet with the liquid layer when the
gas issues from the surface layers. The bursting of gas
bubbles at the surface does not appear to result in a
substantial dispersion of the propellant. According to the
analysis of the experimental data on the droplet formation
from the liquid gas-bubbled layer, not more than 10^{-4} g of
droplets per gram of gas is formed by this mechanism
(Kutateladse and Styrikovich 1976; Datta et al. 1972).

The situation is quite different when the gas stream
issues from the surface layer. In this case, the mass flow
of droplets can be comparable, in magnitude, to the mass
flow rate of the gas. This dispersion will be further
referred to as entrainment (Kafarov 1962).

Although the possibility of entrainment has already
been mentioned by Khaikin (1975), so far it has not been
confirmed experimentally. Analysis of the longitudinal cuts
of extinguished double-base propellant N strands (burning at
$P \sim 10^{-3}$ atm) has shown that, at least, at low pressures the
reaction layer has a system of long pores through which the
volatile propellant component vapors are blown (Zyryanov
1980; Zarko et al. 1979). The pores, i.e., channels, arise

because the propellant viscosity ($\mu \sim e^{E'/RT}$, $E' = 28,000-30,000$ cal/mole) (Grinute et al. 1966) falls sharply toward the burning surface and thus the gas cavities grow in this direction. When issuing by these channels, the gas partially destroys the reaction layer and entrains a certain amount of the C phase.

Model of Injection of the C-Phase Surface Layer

Physical considerations suggest that a liquid element in the reaction layer on the burning surface follows, with a finite probability, either the channel of chemical transformation (C-phase reactions → gaseous products of pyrolysis) or the channel of mechanical transformation (solid destruction → aerosol) (viz. e.g., Abramson 1975). The degree of dispersion, associated with mechanical destruction, is augmented when the surface layer is crushed faster and the rate of chemical transformation is slower. This qualitative statement can be expressed quantitatively as follows.

The dispersion degree is defined as

$$\eta_d = j_d/(\rho_c U_1 + j_d) \tag{1}$$

where the numerator is the mass flux of the aerosol transferred from the burning surface; the denominator is the overall mass flux of the gas; U_1 is a component of the observed burning rate U contributed from the chemical decomposition; and ρ_c is the C-phase density. Division of the numerator and denominator of Eq. (1) by the characteristic size (δ) of the chemical reaction zone and by the density of the substance in the reaction zone (ρ) gives the value of η_d in terms of the characteristic times of dispersion τ_d and chemical decomposition τ_1:

$$\eta_d = \tau_d^{-1}/(\tau_1^{-1} + \tau_d^{-1}) \tag{2}$$

The dispersion of the δ layer is associated with the conditions of instability in a liquid interface. Instability of an interface is by the rheological characteristics of the liquid (mainly, surface tension), the characteristic scale of the liquid element involved into dispersion, inertial factors, and the time of dynamic interactions between the gas and the liquid. With a propellant reaction layer dispersion, the collective factor of interacting jets can also be of importance, since the same liquid element can be influenced by various gas microjets.

The critical deformation of the reaction layer which results in the liquid dispersion is assumed to be induced mainly by aerodynamic forces. In this case, Raushenbakh and Belyi (1964) propose that the time necessary to reach the critical deformation, or the dispersion time, is

$$\tau_d = \xi'(\delta/\overline{U}_g)(\sqrt{\rho_c/\overline{\rho}_g}) \qquad \xi' = \text{const} \qquad (3)$$

Here $\overline{U}_g$ is the mean relative velocity of the gas and the liquid in the δ-thick reaction layer, and $\overline{\rho}_g$ is the mean gas density. Since $\tau_1 \sim (\delta/U_1) \simeq (\delta/\overline{U}_g)(\rho_c/\overline{\rho}_g)$, $\overline{\rho}_g = \overline{\rho}_g(T_s)$, where T_s is the surface temperature, Eq. (2) can be written as

$$\eta_d = [1 + \xi \sqrt{\overline{\rho}_g(T_s)/\rho_c}]^{-1} \qquad \text{const} \qquad (4)$$

In the case of a double-base propellant combustion at a low pressure (P < 300 mm Hg), it must be recognized that the reaction layer is also blown through by the volatile component vapors (Zyryanov 1980; Zarko et al. 1979). It can be readily shown that in this case,

$$\overline{U}_g \simeq (\rho_c U_1/\rho_*)(1 + 2\alpha)/2$$

where α is the content of the volatile components in the propellant (α = 0.4 in N propellant), and ρ_* is the mean density of the vapor-gas mixture in the reaction layer. In this case,

$$\eta_d \simeq (1 + \frac{1}{1 + 2\alpha} \xi \sqrt{\rho_*(T_s)/\rho_c})^{-1} \qquad (5)$$

Figure 1 shows the data on the degree of dispersion, $\eta_d(P)$, of N-propellant (Zarko et al. 1979; Zyryanov and Zarko 1980) and the predictions of Eq. (5) at P < 300 mm Hg and Eq. (4) at P > 300 mm Hg. For these calculations, experimental values of $T_s(P)$ and the relation for the mean gas-vapor mixture density,

$$\rho_* = (P/RT_s)[M_g M_v/M_g + (M_v - M_g)\frac{1-\alpha}{1+\alpha}]$$

are used. Here $M_v \simeq 200$, $M_g \simeq 30$ are molecular weights of the volatile N-propellant components and the C-phase

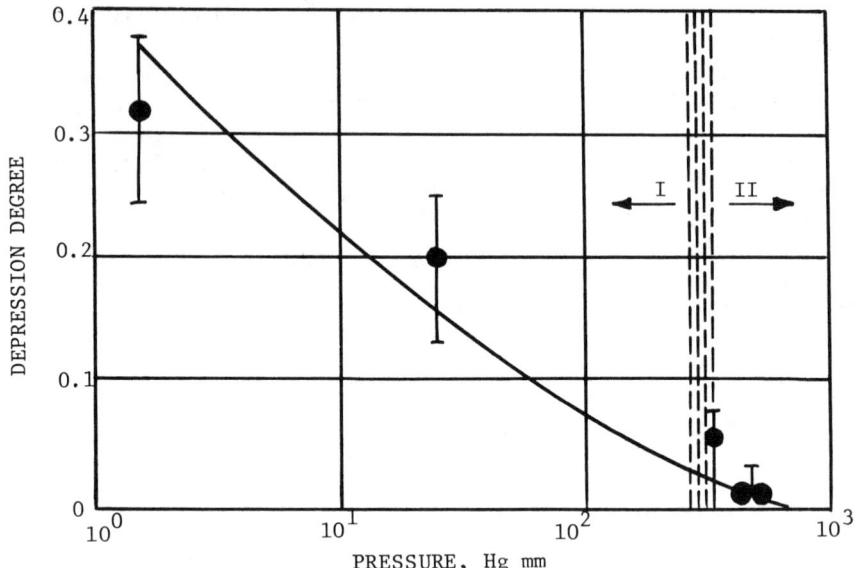

Fig. 1 Degree of dispersion for N propellant. $T_o = 110°C$. Region I: separated (spatially) gasification fronts. Region II: single gasification front region. (———): calculation, Eq. (5) at $P < 300$ mm Hg; Eq. (4) at $P > 300$ mm Hg. ●: experiments.

decomposition products, respectively. The constant ξ is assigned the value of 1500 to ensure fit to the data.

As follows from Eq. (4), η_d is reduced with increasing pressure as $\eta_d \sim 1/(1+\xi''\sqrt{\rho_g})$, $\xi'' = $ const, and for N propellant it does not exceed fractions of a percent at $P \geq 1$ atm.

Dispersion Energy

The expression for $\eta_d(P)$ can be also obtained from the mechanical energy balance in the reaction layer. During burning, a propellant sample portion $(1 - \eta_d)$ of initial mass m_o forms gases with the kinetic energy $W=(1-\eta_d)m_o U_g^2/2$. This energy must be partly spent to make a new interface, i.e., dispersion. The work W_Σ expended to create a new surface Σ is $W_\Sigma = \sigma\Sigma$, where σ is the surface tension. If the energy of the gas escaping from the reaction zone is spent for dispersion with a certain efficiency ϕ,

$$W_\Sigma = \sigma\Sigma = \phi(1 - \eta_d)m_o U_g^2/2 \qquad (6)$$

Here Σ is the total surface of particles per $n_d m_o$ propellant. It can be easily shown that n_d, d_o (particle size), m_o, and Σ are related as

$$n_d = \rho_c d_o \Sigma / 6 m_o \qquad (7)$$

Equations (6) and (7) yield

$$n_d = 1/(1 + \frac{a}{\phi d_o} P^{2(1-\nu)}) \qquad (8)$$

where

$$a = (\frac{M}{RT_s U_0 \rho_c})^2 \times \frac{12\sigma}{\rho_c} \quad U_0 = U P^{-\nu} = \text{const}$$

From Eq. (8) it follows that $n_d \to 0$ at $d_o \to 0$, meaning that the escaping gas energy does not suffice to generate any substantial amount of very small droplets.

Discussion

Analysis of the physical model of a liquid layer entrainment by a gas flow has shown that the aerodynamic mechanism must be of primary importance in the formation of a dispersed phase over the burning surface. As noted above, under usual conditions the condensed-phase transfer due to bursting bubbles is negligible. Similar data for reacting liquids are unavailable in the literature; however, there is reason to belive that in this case the contribution to the dispersion from bursting bubbles is also negligible. Holography and high-speed movies have revealed that local clusters of bubbles of $(5-50) \times 10^{-6}$ m in diameter appear periodically (Ginsburgh and Stepanov 1978). The calculated degree of dispersion due to the bursting of these bubbles does not exceed fractions of percent.

Experimental data indicate that dispersion degree decreases with increasing pressure, in good agreement with the semiempirical model proposed in the present paper. Indeed, in accordance with Eq. (3), the dispersion degree falls with pressure as $(1 + \text{const } \sqrt{P})^{-1}$. From the physical viewpoint this trend of $n_d(P)$ results from a substantial increase in the C-phase reaction due to rising temperature of the surface. As a result, the major part of the condensed substance burns to form gaseous products rather than being entrained as an aerosol.

A decrease in the dispersion degree with increasing pressure can be associated also with structure changes in the subsurface propellant zones. The initial spatial separation (at pressures 10^{-3}-10^{-1} atm) of the zones of evaporation and C-phase reactions vanishes with increasing pressure; the two zones are replaced by a single narrow $[(20\text{-}30) \times 10^{-6}$ m] zone. With a further increase in the pressure, the zone of chemical reactions in the propellant reaches $(5\text{-}10) \times 10^{-6}$ m, when gaseous decomposition products can be successfully removed by a diffusion mechanism. At the same time, the dispersed fragments of the liquid-viscous layer must have sizes comparable with the layer thickness, i.e., approximately 10^{-5}-10^{-6} m. For very small particles, however, simple energy estimations show [see Eq. (8)] that the mass transfer from the surface is extremely low, since the kinetic energy of the gaseous decomposition products is limited in magnitude and is insufficient to cover the energy deficit needed for generation of liquid aerosol particles.

Conclusions

1) The analysis of the available experimental data and the semiempirical model of burning surface destruction have shown that in the case of propellants with liquid-viscous reaction layers, dispersion is of essential importance at low pressures and practically vanishes at atmospheric and higher pressures.

2) Most of conventional theoretical models of double-base propellant combustion dealt with comparatively high values of the (constant) thermal effect in the C phase and the variable dispersion degree. The absence of dispersion at high pressures results in a change of the magnitude of the C-phase thermal effect and its dependence upon the burning conditions.

3) The reported concepts of the burning surface destruction of homogeneous propellants can prove useful for physical simulation of heterogeneous mixtures burning. The picture of their combustion is complicated by heterogeneous transformations and different reactivities of separate components. In some cases, these factors can result in a strong dispersion, and hence must be included in appropriate combustion models.

References

Abramson, A. A. (1975) <u>Surface-Active Substances. Properties and Applications.</u> Khimiya Press, Leningrad, USSR.

Datta, A., Nam, P. S., and McMillan, H. K. (1972) The role of impurity particles in the combustion of double-base propellants. Combust. Sci. Technol. 6, 37-46.

Ginsburgh, V. M. and Stepanov, B. M. (1978) Optical Holography Practice, Soviet Radio Press, Moscow, USSR.

Grinute, G. A., Zubov, P. I., and Sangarovski, A. T. (1966) The temperature influence on the nitrocellulose film strength time dependence. The Mechanism of Film Formation From the Polymer Solution and Dispersion, pp. 165-170. Nauka, Moscow, USSR.

Kafarov, V. V. (1962) The Principles of Mass Transfer, p. 404. High School Press, Moscow, USSR.

Khaikin, B. I. (1975) The investigation on the heterogeneous systems combustion theory. Ph.D. Thesis, Institute of Chemical Physics, Moscow, USSR.

Kutateladse, S. S. and Styrikovich, M. A. (1976) Gas-Liquid Systems Hydrodynamics, Energiya Press, Moscow, USSR.

Merzhanov, A. G. (1960) The dispersion influence on the propellant combustion. Dokl. Acad. Nauk USSR 135, 1439-1441.

Merzhanov, A. G. (1969) Theory of stable homogeneous combustion of condensed substances. Combust. Flame 13, 143-156.

Pokhil, P. P. (1954) The combustion mechanism of colloid powders. Ph.D. Thesis, Institute of Chemical Physics, Moscow, USSR.

Raushenbakh, B. V., Belyi, S. A. (1964) Physical Principles of Air-Jet Combustion Chamber Working Process, Mashinostroenie Publishing, Moscow, USSR.

Zarko, V. E., Zyryanov, V. Y., and Koutzenogii, K. P. (1979) Combustion mechanism of double-base propellant at subatmospheric pressures. Sixth International Symposium on Combustion Processes, pp. 56-60. Polish Academy of Science, Karpacz, Poland.

Zyryanov, V. Y. (1980) The investigation of reaction zone structure in the double-base propellant combustion. Candidate Thesis, Institute of Chemical Kinetics and Combustion, Novosibirsk, USSR.

Zyryanov, V. Y. and Zarko, V. E. (1980) The dispersion in the homogeneous powders combustion. Preprint No. 4, Institute of Chemical Kinetics and Combustion, Novosibirsk, USSR.

Heat Transfer Ahead of Flame Spreading over a Cured Epoxy Resin Surface in an Opposed Flow

B.Y. Kolesnikov,* V.L. Efremov,† N.S. Umarbekov,‡ and A.B. Kolesnikov§
Kazakh State University, Alma-Ata, USSR

Abstract

The significance of the solid-phase heat conduction ahead of the flame, spreading horizontally, is investigated. The experimental method employed in this study permits the determination of the delay period of the flame skipping over an inert spacer which is flush with the sample surface and which is embedded in the epoxy sample normal to the flame propagation. An approximate solution of the one-dimensional solid-phase heat-transfer equation indicates that the delay period is directly proportional to the squared spacer thickness and inversely proportional to the thermal diffusivity of the spacer material. The experimental data obtained are in good agreement with the theoretical results. A mechanism of the flame skipping is proposed.

Introduction

The flame spreading across the surface of a solid fuel is of interest both theoretically and practically. The complete mathematical description of such flames involves the simultaneous solution of the system of nonlinear differential equations for chemical kinetics and heat and mass transfer. Solution of this system requires various assumptions and approximations whose validity can be checked only by experiment.

Presented at the 8th ICOGER, Minsk, USSR, Aug. 23-25, 1981.
Copyright © 1982 by B. Y. Kolesnikov, V. L. Efremov, N. S. Umarbekov, and A. B. Kolesnikov. Published by the American Institute of Aeronautics and Astronautics, Inc. with permission.
*Assistant Professor, Chemical Faculty.
†Junior Associate, Chemical Faculty.
‡Research Engineering, Chemical Faculty.
§Graduate, Chemical Faculty.

HEAT TRANSFER AHEAD OF FLAME IN AN OPPOSED FLOW

A rigorous description of heat transfer in a spreading flame must account for all modes of energy transfer ahead of the flame to the layer of unburnt solid fuel. Heat transfer occurs by gas-phase or solid-phase heat conduction, convection, and radiation. Heat release in situ in the course of exothermic reactions which result from oxygen diffusion into the surface layer of fuel may be important. However, it should be noted that the radiative heat transfer is significant only for large-scale, highly turbulent flames. While the effect of exothermic oxidation reactions on the energy balance in a flame has been studied by Stuetz et al. (1975), this phenomenon merits further study.

Heat conduction, either through gas-phase or through solid-phase, is the principal mechanism of heat transfer. Whether gas-phase or solid-phase conduction is considered to be the predominant mode in flame models depends on a number of factors, the most important of which is the thickness of a combustible sample. For thin samples, gas-phase heat conduction is predominant. The mathematical model for this case has been developed, and the analytical expression for the flame-spread rate has been obtained by Frey and T'ien (1979) and Lastrina et al. (1971). Experiments by Fernandez-Pello and Williams (1975) have shown that gas-phase heat conduction during flame spread across the surface of PMMA contributes only 10% of the total heat required to heat the region upstream of the point of the flame attachment. The effect of opposed flow with different concentrations of oxygen on heat transfer during flame spread across the surface of PMMA has also been investigated experimentally Lalayan et al. (1979). This work indicated that at low oxygen concentrations, the two heat flux components are both of importance during the preliminary heating of a sample. Lalayan et al.(1979) suggest that at high oxygen concentrations, solid-phase conduction does not contribute heat significantly, and that the mechanism of heat transfer changes from the flame to the preflame zone.

It is evident that the mechanism of heat transfer to the preflame zone is not fully understood. In this paper, a study of the horizontal propagation of the flame across the surface of cured epoxy resin is presented (see Fig. 1). An attempt has been made to determine the contribution of condensed-phase heat conduction to heating and gasification of unburnt polymer material. The flame spreading across the surface of a solid is stopped before a noncombustible source (see Fig. 1) and heat flux is conveyed from the flame to the sample by two routes. For a heating of the sample and its ignition the heat flux through the solid (Q_s) is more significant than the heat flux through the gas (Q_g). The

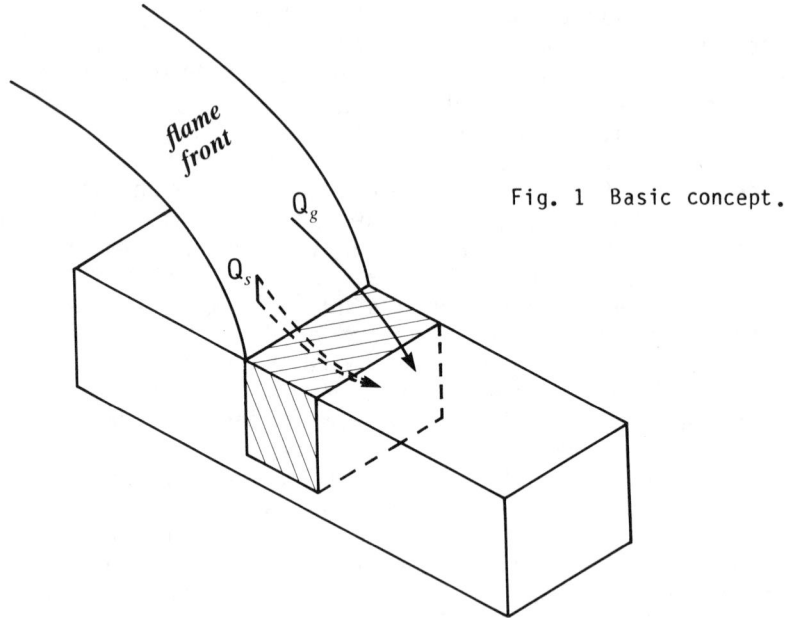

Fig. 1 Basic concept.

essential measurements were a flame delay period defined as the elapsed time between time of flame arrival at one side of an inert spacer and the time of flame ignition at the other side of the spacer, and the temperature on each side of the spacer. This phenomenon was termed flame skipping.

Experiments

Combustion of the epoxy polymer, resin ED-20 cured with m-phenylenediamine, was observed in the opposed flow of a gas mixture (80% of O_2 and 20% of N_2) at a linear velocity of 3 cm/s. The schematic of the apparatus is shown in Fig. 2. The cross section of housing (1) was equal to 40 x 40 mm. Sample (3) was mounted inside on a metal frame (2). The upper part of the device was provided with a small optical window for the photorecording of flame propagation. The ignition of samples was realized by means of a nichrome spiral (6).

The samples were shaped as plates 20 x 3 x 100 mm in size. These were provided with cross-cut grooves of different widths where polished spacers of alumina or silica were affixed with the same epoxy resin to ensure reliable heat contact. The upper end faces of the spacers were ground to the sample surface levels and then the whole surface was polished.

HEAT TRANSFER AHEAD OF FLAME IN AN OPPOSED FLOW

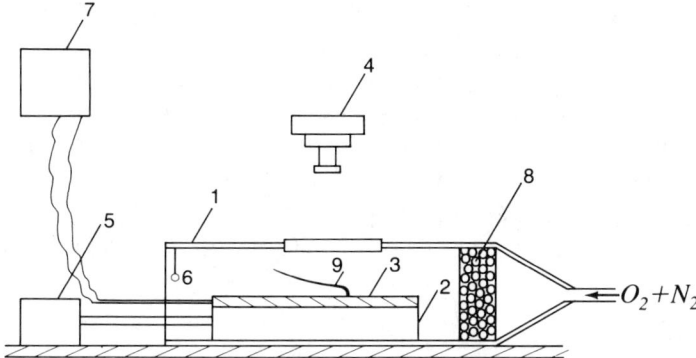

Fig. 2 Experimental system: 1) housing of apparatus, 2) metal holder of epoxy sample, 3) epoxy sample, 4) camera, 5) mechanism for moving sample, 6) nichrome ignition spiral, 7) temperature recorder, 8) silica gel, 9) flame front.

To measure the sample surface temperature at the moment of flame skipping, in some experiments chromel-alumel thermocouples with filaments 50 µ in diameter were implanted into samples at the edges of the spacers.

The flame delay periods were determined by two independent methods, visual and photographic; the latter also served to determine a flame spread velocity in each experiment. Temperatures were recorded by the mirror-galvanometer oscillograph H-105.

In the selection of the spacer material, the following considerations were important. If gas-phase heat conduction predominates the flame delay period should not depend on the inert spacer material. When the solid-phase heat conduction predominates, the delay period should increase as thermal diffusity of the inert space material increases. Silica (thermal diffusivity 0.055 cm^2/s) and alumina (thermal diffusivity 0.0114 cm^2/s) were selected as inert spacer materials.

Model Predictions

As a first approximation to the analytical expression for a flame delay period, the one-dimensional equation of solid-phase heat conduction is used:

$$\frac{\partial H}{\partial t} = \alpha \cdot \frac{\partial^2 H}{\partial x^2} \qquad (1)$$

where $H = (T-T_i)/(T_s-T_i)$. Its solution is obtained under the following constraints:
1) The spacer is a semi-infinite solid bounded with a flat surface $x = 0$; 2) the initial temperature H_i is the same everywhere and is equal to 0; and 3) the surface temperature H_s ($x = 0$) at $t = 0^+$ increases instantaneously to unity and remains constant.

The solution is of the form (Belyaev and Ryadno 1978):

$$T(\eta) = T_s \, \mathrm{erfc}\, \eta \tag{2}$$

$$\mathrm{erfc}\, \eta = \frac{2}{\pi} \int_\eta^\infty e^{-\xi^2} d\xi \tag{3}$$

and

$$\eta = x/\sqrt{4\alpha t}$$

The series expansion gives

$$\mathrm{erfc}\, \frac{x}{2\sqrt{\eta t}} = 1 - \frac{2\eta}{\sqrt{\pi}} + \ldots \tag{4}$$

Solution of Eq. (2) subject to Eq. (4) for t yields

$$t = \frac{1}{\pi(1-H)^2} \cdot \frac{1}{a} x^2 \tag{5}$$

If the flame skips over an inert spacer at the moment when its rear side reaches a temperature T_o, then, for spacer thickness $x = d$, the flame delay period is approximately

$$T = \frac{1}{\pi(1-H_0)^2} \cdot \frac{1}{a} \cdot d^2 = K_o \cdot \frac{d^2}{a} \tag{6}$$

Within the limitations of the approximations, the flame delay period is predicted to be a linear function of the square of the spacer thickness. The slope of the linear plot is inversely proportional to the thermal diffusivity, and the ratio of the slopes for the ($t - d^2$) dependencies on different materials is inversely proportional to the ratios of the thermal diffusivities of those materials.

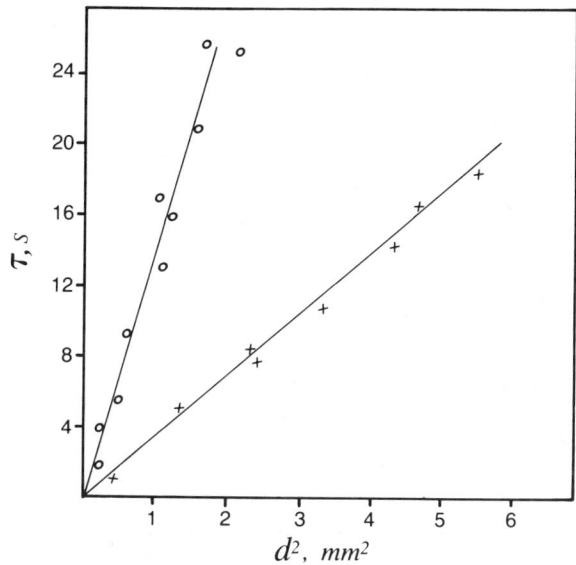

Fig. 3 Reignition times for flames propagating over inert spacer.
o - alumina = 0.0114 cm^2/s; x - quartz = 0.055 cm^2/s.

Results and Discussion

Flame delay periods for alumina and quartz spacers are shown in Fig. 3. The straight lines were fitted to the experimental points by the least-squares methods. The ratio of the slopes of these lines ($\sim$4.5) is very close to the inverse ratio of thermal diffusivities for alumina and quartz (4.82).

Equation (6) is valid provided tha the temperatuare T_s does not depend on t, and that the temperature T_o at the moment of the flame skipping depends neither on the spacer material nor thickness. Sample surface temperatures were measured during the process of flame skipping, to determine the validity of these assumptions.

Results of temperature measurements are shown in Fig. 4. The subscripts s and o refer to the burning surface temperature before an inert spacer and to a nonburning surface temperature behind it, respectively. Dashed lines indicate the time at which flame skipping occurs. Zero time corresponds to the time of arrival of the flame at the spacer. In all experiments, reignition was observed in the center of the sample at a spot about 1 mm in diameter. The temperature T_o at that time was between 380 and 405 degree C (somewhat lower for quartz) and, within the limit of the

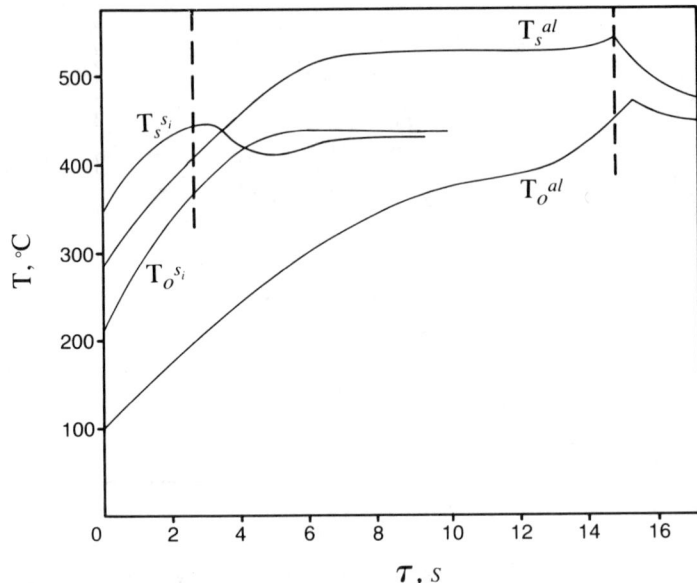

Fig. 4 Temperature of spacer surfaces: superscript Al stands for alumina, 0.85 mm in diameter, superscript Si stands for quartz, 1.03 mm in diameter.

uncertainty of measurement, was constant (Efremov and Kolesnikov 1980). The instantaneous adjustment of T_s to a steady value is not observed (see Fig. 4). However, T_s may be approximated as $T_s = m + nt$, where m and n are constant coefficients. For this case, the approximate solution of Eq. (1) may be shown to have the same form as Eq. (6), though the proportionality factor K_o has a different form.

Experiment observations indicate that the reignition phenomenon is not one-dimensional, since the flame, as a rule, skips over the center, not across the whole surface, owing to heat losses from the edges of the metal frame. Additionally, account should be taken of the heating of a sample in depth. Even though the simple analytic dependence of a flame delay period is fairly consistent with the experimental findings, in general, the problem must be considered as three-dimensional heat transfer.

Reignition occurs in the center of the sample because the temperature, and gasification rate has a maximum there. A cloud of evolved pyrolizate is mixed with oxidant due to diffusion and convective flow of the latter. The formed mixture is transferred to the flame zone before the spacer. As soon as the gas fuel ahead of the flame, i.e., just over the spacer, reaches a concentration sufficient for ignition, reaction and flame skipping follow. A more appropriate concept, then is not the ignition temperature but the minimum quantity of the evolved pyrolizate sufficient to maintain combustion. This quantity depends on the quantity of heat absorbed by the sample and, consequently, on the temperature and the depth of a heated layer. The latter should be deeper obviously for the quartz spacer than for the alumina one with the same thickness. This results in a greater sample gasification rate and hence a reduced flame delay period.

For the above reasons, T_0 at the moment of the flame skipping is variable. The rate of gasification increases with an increase of temperature and/or depth of the heated layer. As a consequence a gasification rate sufficient for the flame skipping may be realized at a lower temperature with a quartz spacer than with alumina. The experimental measurements of the temperature at the moment of the flame skipping (Fig. 4) indicate that the temperature T_0 for the quartz spacer is lower than for the alumina spacer, while the non-one-dimensionality is of significance, and the theory within a first approximation gives a good agreement with the experimental results.

Conclusion

The representation of the dependence of flame delay periods on the square of spacer thickness in the form of straight lines as a first approximation apparently correlates well with the experimental data for the range of conditions studied. Thus, when the flame spreads over the 3-mm-thick epoxy resin samples in the flow with high oxygen concentration, the preliminary heating of a sample is reached mainly by condensed phase heat conduction.

References

Belyaev, N. M. and Ryadno, A. A. (1978) Metody Nestatsionarnoi Teploprovodnosti, 328 pp. Vysshaya Shkola, Moscow, USSR.

Efremov, V. L. and Kolesnikov, B. Y. (1980) Issledovaniye respredeleniya temperatury pri rasprostranenii plameni po poverkhnosti otverzhdennoi epoksidnoi smoly. Goreniye Kondensirovannykh i Geterogennykh Sistem, Materialy VI Vsesoyuznogo Simpoziuma po Goreniyu i Vzryvu, pp. 38-41. Akademia Nauk SSSR, Chernogolovka.

Fernadez-Pello, A. and Williams, F. A. (1975) Laminar fire spread over PMMA surface. 15th Symposium (International) on Combustion, pp. 217-231. The Combustion Institute, Pittsburgh, Pa.

Frey, A. E. and T'ien, J. S. (1979) A theory of flame spread over a solid fuel including finite-rate chemical kinetics. Combust. Flame 36, 263-289.

Lalayan, V. M., Khalturinsky, N. A. and Berlin, A. A. (1979) Teploperenos pri rasprostranenii plameni po poverkhnosti polimetilmetakrilata. Vysokomol. Sodenin. Ser. A 21, 1139-1143.

Lastrina, F. A., Magee, R. S., and McAlevy, R. F. III (1971) Flame spread over the fuel beds: solid phase energy considerations. 13th Symposium (International) on Combustion, pp. 935-946. The Combustion Institute, Pittsburgh, Pa.

Stuetz, D. E., DiEdwardo, A. H., Zitomer, F., and Barnes, B. P. (1975) Polymer combustion. J. Polym. Sci. Polym. Chem. Ed., 13, 585-621.

Chapter IV. Ignition and Extinction

Self-Ignition of Atomized Liquid Fuel in Gaseous Medium

A.A. Borisov,* B.E. Gel'fand,† E.I. Timofeev,‡ S.A. Tsyganov,*
and S.V. Khomik§
Academy of Sciences, Moscow, USSR

Abstract

The available and newly obtained data on the values of self-ignition delays of atomized liquid fuel in air and oxygen in the submillisecond region have been analyzed. A significant dependence on temperature and pressure, consistent with the chemico-kinetic nature of self-ignition, was found at delays above 100 µs. The significant role of chemico-kinetic factors in the ignition of atomized fuel is further supported by the direct demonstration of promotion and inhibition of ignition by some fuel additives. The phenomenon of failure ignition at a certain critical temperature characteristic of each fuel was observed for atomized fuel. In some previous studies of ignition of atomized fuel it was shown that the results observed may be distorted because of the interaction of shock waves with closely spaced droplets of liquid.

Introduction

The study of combustion and ignition of atomized liquid fuels in a gaseous environment (quiescent or moving at high velocities) is of practical importance. Self-ignition delay is the critical parameter describing the combustion of liquid fuel in diesel engines, and combustion chambers of various types, as well as in optimizing external combustion.

Presented at the 8th ICOGER, Minsk, USSR, Aug. 23-26, 1981. Copyright © American Institute of Aeronautics and Astronautics, Inc., 1982. All rights reserved.
*Head of Laboratory, Institute of Chemical Physics.
†Senior Researcher, Institute of Chemical Physics.
‡Institute of Chemical Physics.
§Junior Researcher, Institute of Chemical Physics.

The most important consideration in the optimization of combustion performance is the dependence of the self-ignition delay on the ambient parameters, namely, temperature, pressure, and composition. The modification of self-ignition delay by occasional impurities or by deliberately added promotors or inhibitors of combustion can be of practical significance. The magnitude of self-ignition delay can determine the dimensions of the combustion chamber, while the temperature dependence of self-ignition delay controls the stability of the combustion process with respect to variation in pressure and temperature. In the analysis of explosion hazards, the value of self-ignition delay determines the length and the structure of the reaction zone and the possibility of the detonation itself. The relationship between breakup times of droplets or jets of liquid in transverse gaseous flows and the time lag of chemical reactions govern the limiting stages of heterogeneous combustion. The self-ignition delay changes with the transition from oxygen mixtures to air mixtures, as well as with the change of one type of liquid fuel to another. Estimation of the combustion efficiency and an accurate understanding of combustion chamber performance are rather difficult without knowledge of the self-ignition delays of a liquid fuel and their relation to characteristic times of other elementary stages. In the investigation of self-ignition of liquid fuel, special attention should be paid to the determination of characteristics of these processes under conditions close to those realized at present in practical installations.

There is little information on the characteristics of self-ignition at temperatures 800 K < T < 2000 K and pressures 1.0 < P < 10 MPa. Ignition in most power devices occurs in the submillisecond region, while most measurements have been made for self-ignition delays $\tau_i > 10^{-3}$ s or even $\tau_i \gg 10^{-3}$ s. Direct extrapolation of the data from these time intervals to shorter times can lead to appreciable error because the physical and chemical processes involved are not linear in temperature or pressure. Also, the data obtained for homogenous mixtures, especially for those highly diluted with an inert gas, are in most of the cases quite useless for the determination of the values of τ_i in two-phase gas-droplet systems.

The scarcity of experimental data on self-ignition delay in the literature, as well as disagreement between the results obtained by different techniques, make it difficult to develop a comprehensive model of the ignition of droplets in moving gaseous flows and the dependence of this process upon various physical, gasdynamic, and chemical factors, or, for that matter, to develop correlations for self-ignition delay of drops.

The analysis of self-ignition of motionless droplets in a quiescent oxidizing medium (Kadota et al. 1976) indicates that in such a case self-ignition delays are 10-20 times higher than those for drops moving in an oxidizing gaseous medium (Mullins 1953; Kadota et al. 1970). As a consequence, the applicability of measurements of self-ignition delay of fixed drops in a stagnant medium is limited, since in cases of practical interest, the atomized fuel moves relative to the gas flow. Analysis of the data (Kadota et al. 1970) also shows a considerable effect of gas pressure on the delay. In this connection, detailed measurements of ignition delays by Mullins (1953) are useful only for $P = 0.1$ MPa and $\tau_i > 10^{-3}$ s. Thus measurement of self-ignition parameters requires a consistent systematic study of the ignition of droplets of various liquid fuels in submillisecond time intervals.

Experimental Technique

Extended measurements of self-ignition delays behind the reflected wave front have been carried out to provide experimental information on self-ignition of various liquid fuels for conditions of practical interest. A single drop of a test liquid is placed in a low-pressure section of a shock tube at a distance $\Delta l = 10$ mm from the tube end plate so that droplet-wall collision does not occur prior to ignition. This technique is preferable to that used by Kauffman et al. (1971), Kauffman and Nicholls (1971) and Miyasaka and Mizutani (1975a, 1975b) in which a column of fuel is injected into the tube. The single-droplet technique provides better control over the equivalence ratio at the observation point and minimizes the disturbances from interaction of the shock wave with the fuel droplets.

In this investigation the time Δt, during which the flow parameters behind the wave did not change appreciably, was $\Delta t \cong 10^{-3}$ s. While the shock wave traveled from the droplet to the end plate and returned, the droplet underwent atomization in the gas flow. While the gas velocity behind the reflected wave front is zero, the drop core continued

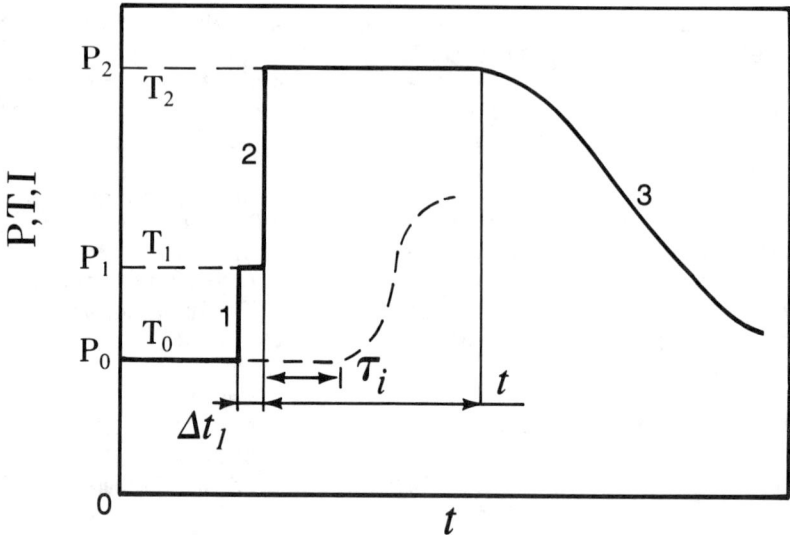

Fig. 1 Schematic of ignition delay of fuel droplet in a shock tube: pressure P, temperature T, and emission I history; subscripts 0, 1, and 2 indicate the undisturbed, postincident wave and post-reflected wave, respectively. Discontinuities 1 and 2 indicate passage of the incident and reflected wave. Smooth transition 3 indicates rarefaction wave. $\Delta t_1 \equiv$ elapsed time between passage of the incident wave and the return of the reflected wave. τ_i $\equiv$ self-ignition delay of the atomized fuel droplet. (- - -) represents light emission I.

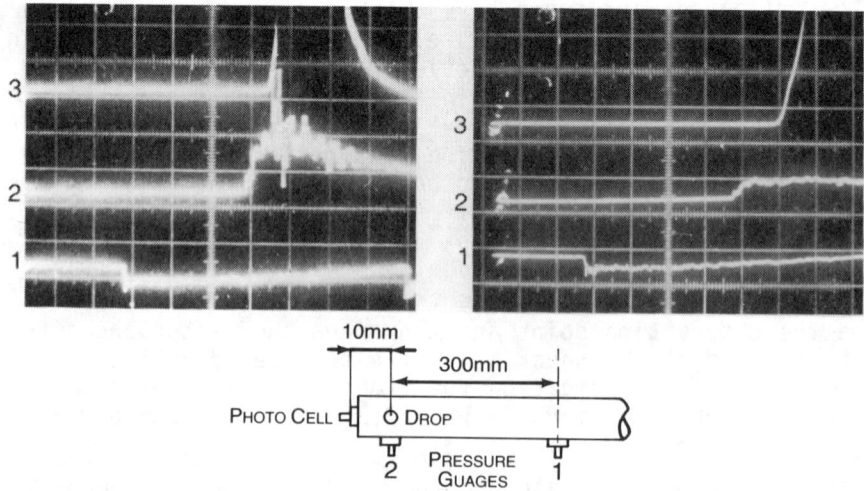

Fig. 2 Oscillograms for ignition delay experiments for kerosene droplet: a) in oxygen, T_2 = 1260 k, P_2 = 1 MPa; b) in air, $\overline{T}_2$ = 1300 K, P_2 = 3.6 MPa. Traces 1 and 2 - pressure; trace 3 - emission; time divisions = 250 μs.

moving to the end plate because of its inertia. The change in the pressure, temperature, and light emission with time is shown in Fig. 1. The oscillograms of pressure and light emission for kerosene droplet self-ignition in oxygen and in air are given in Fig. 2. The oscillograms show characteristic features of hydrocarbon fuel ignition in a gas. The ignition of drops in oxygen always involves generation of strong secondary waves, while the ignition in air is characterized by smooth combustion of atomized fuel without generation of secondary waves.

Self-Ignition of Hydrocarbons

The main features of self-ignition of an atomized fuel drop are illustrated by the ignition of kerosene droplets. The correlations of the ignition delay measurements, for kerosene drops in air and oxygen are presented in Fig. 3. These results indicate that an increase in pressure or concentration of oxygen leads to a considerable decrease in the ignition delays. The temperature dependence of the ignition delay become weaker as the pressure approaches $P_2 \simeq 10$ MPa, which is confirmed by the smaller slope of line 3, as compared to that of lines 1 and 2. The study of self-ignition behind reflected shock waves reveals the existence of a certain critical temperature, T^*, below which self-ignition is not possible and cannot be achieved by changes in shock tube design which extend the duration of the conditions behind the reflected shock wave. Self-ignition of kerosene at $P_2 = 1.0$-4.0 MPa is not possible below $T^* \simeq 1300$ K in air or $T^* \simeq 950$ K in oxygen. An increase in the pressure expands the temperature range of self-ignition to $T \simeq 1050$ K at $P_2 = 10$ MPa in air. A similar result was obtained earlier by Kauffman and Nicholls (1971) for self-ignition of diethylcyclohexane in oxygen, for which $T^* \simeq 900$ K. The impossibility of self-ignition of kerosene at $T < T^*$ is consistent with the observations of Laurent et al. (1973) that an artificial ignition source is required for a kerosene fueled supersonic chamber when $T^* < 1450$ K.

Another distinctive feature of atomized fuel self-ignition is apparent from the comparison of ignition delays of gaseous and liquid fuels in air under similar conditions given in Fig. 4. The shaded area between 1 and 5 indicates the range of self-ignition delays for various liquid hydrocarbons in air at $P_2 = 4$ MPa. Ignition delays for stoichiometric mixtures of gaseous methane (2), butane

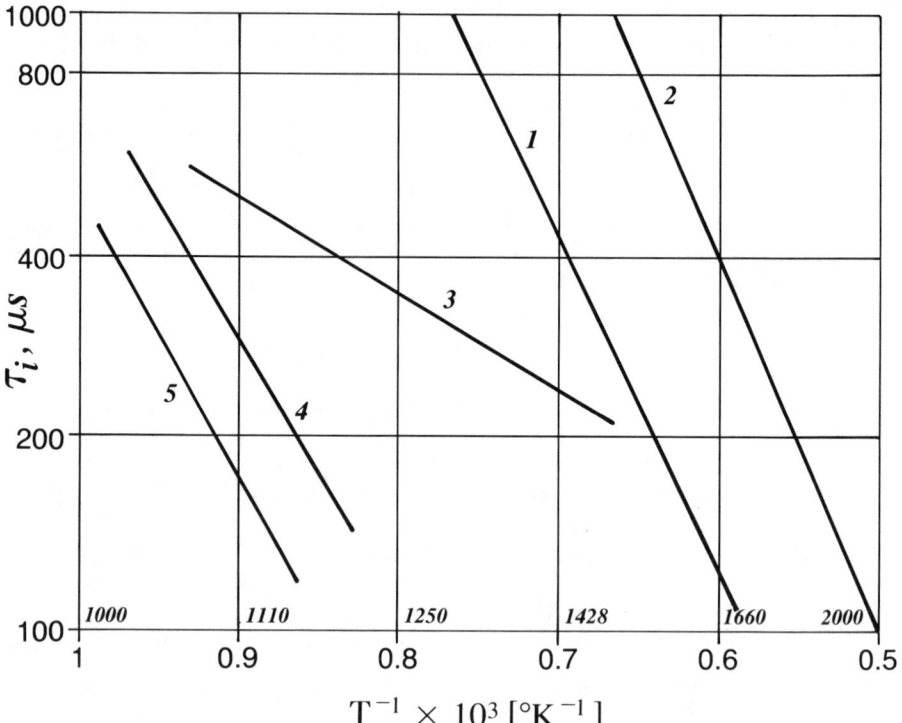

Fig. 3 Ignition delays in kerosene droplets in oxygen and air (see Table 1 for definition of curves).

Table 1 Ignition, pressure and oxygen concentration for Kerosene and ignition

Curve	P_2(MPa)	Gas
1	4.0	Air
2	1.0	
3	10.0	
4	1.0	Oxygen
5	4.0	

(3), and heptane (4) in air at P_2 = 4-5 MPa were reported by Borisov and Kogarko (1960). The τ_i values for solid hydrocarbons (Chang and Schultz-Grunow 1972) are greater than those for drops.

It is quite clear that the ignition delay of a spray should be greater than that of stoichiometric gaseous mixtures. This seems quite obvious if one considers the

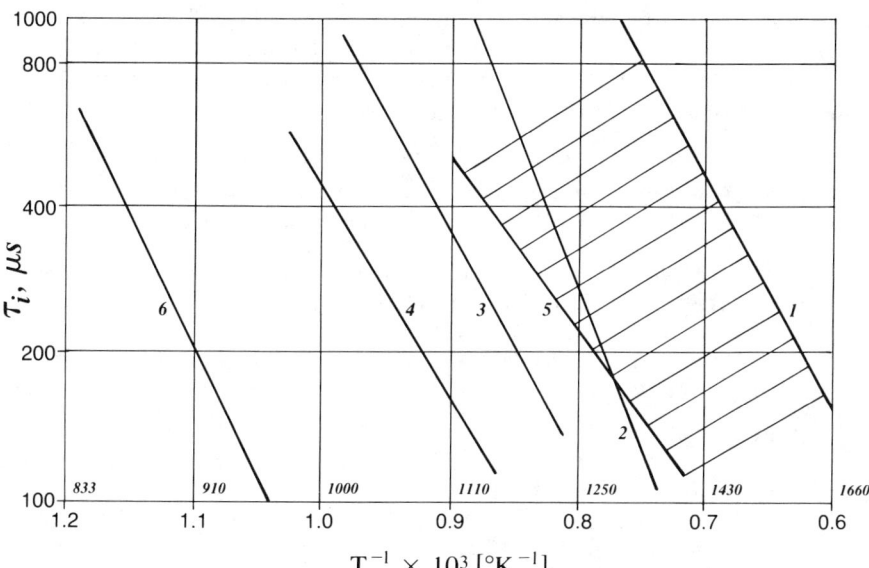

Fig. 4 Ignition delays for gaseous and liquid fuels in air (see Table 2 for definition of curves).

Table 2 Gaseous and liquid fuels used in this study

Curve	Fuel	P_2(MPa)
1	kerosene	4
2	methane	4-5
3	butane	4-5
4	heptane	4-5
5	decalene	4
6	hydrogen	4

processes of atomization, evaporation, and mixing of atomized fuel with the surrounding gas. The finite times required for intensive breakup of drops in shock waves and for mixing of atomized fuel with the gaseous oxidant coupled with the Arrhenius temperature dependence of chemical reactions results in the ignition delay approaching the breakup time of the drops at a certain temperature T_2 and in a weak dependence of the ignition delay on temperature. The fairly strong dependence of the ingition delay on temperature obtained in this work for some hydrocarbons was not observed by Kauffman and Nicholls (1971), Kauffman et al. (1971), Wierzba et al. (1974), and Miyasaka and Mizutani (1975a, 1975b). Most of the data reported by those authors were obtained for high temperatures ($\tau_i \leq 100$ μs), at which the droplet breakup time and the induction period for chemical reactions were comparable. Analysis of the data by Kauffman and Nicholls (1971) and Kauffman et al. (1971) and of the results of the study presented here leads to the conclusion that the rates of chemical reactions are the rate limiting steps for droplet ignition in an oxygen atmosphere at T < 900 K and in air mixtures at T ≤ 1500 K.

It is extremely difficult to estimate the effect of chemical factors in the measurement of self-ignition delays of sprays within $10^{-5} < \tau_i < 10^{-4}$ s because of rather pronounced dynamic response of the mixture formation process in this region. This is the probable explanation of the observations that the effect of chemico-kinetic factors on the self-ignition process was not observed by Lu and Slagg (1972, 1974) and that some physical properties (for instance, the drop diameter) affect the ignition.

Effects of Mixture Composition

The relationship between the characteristic times of chemical and physical processes in the submillisecond region of self-ignition delays of pure fuels may be used to understand the role of chemical kinetics in the modification of ignition phenomenom by additives. The effects of additives on ignition delay were observed to fall into one of three general types (Gelfand et al. 1978; Borisov et al. 1979), as illustrated in Fig. 5 for self-ignition of hydrocarbon fuels in air. Figure 5a represents the general type for which the ignition delay τ_i for the mixture at the low-temperature limit for ignition temperature and activation energy is between the ignition delays and activation energies for the pure components A and B. An example of such a system is the mixture of alkyl nitrates and hydrocarbons, say, isopropyl nitrate + kerosene in air. The second general type shown in Fig. 5b includes those mixtures whose ignition delays are larger than the delays of their pure components. Methyl-cyclopentadenyl-tricarbonyl-manganese + kerosene mixture was found in this investigation to be an example of this type of system. In the third general type, Fig. 5b, the mixture has ignition delays which are shorter than those of the components in the temperature range under investigation. Dodecatriene + 1,4-butandiol dinitrate is an example of such a mixture. It should be noted that there is a certain

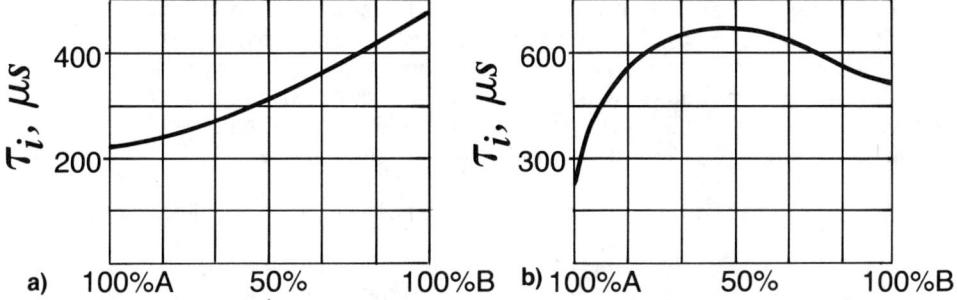

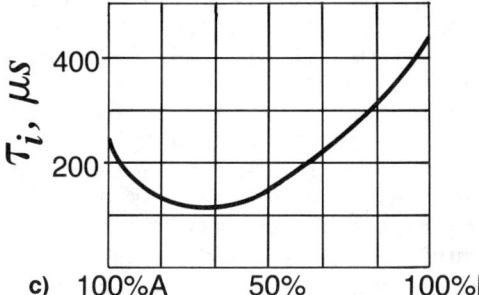

Fig. 5 The effect of additives of ignition delays of hydrocarbon fuels. a) isopropyl nitrate with kerosene; b) pentadenyl-tricarbonyl-manganese with kerosene; c) dodecatriene with 1.4-butandiol dinitrate.

selectivity in the effect of additives on hydrocarbons of different classes. If an additive to saturated hydrocarbons slightly influences the effective activation energy and T*, the addition of alkyl nitrates to compounds like kerosene causes appreciable changes in both the effective activation energy and T*. This conclusion can easily be drawn from the analysis of the results of this investigation and the data of Lu and Slagg (1972, 1974) for standard grade heptane and kerosene with isopropyl nitrate.

The effect of physical factors on ignition appeared to be insignificant in the submillisecond region in this experiment. For instance, the variation of particle dimensions by a factor of 3 and cooling of the liquid to -30°C had no effect on the ignition delay. Variation of gaseous heat-transfer properties through the substitution of 0.8 He-0.2 O_2 mixture for air did not produce a significant change in the self-ignition delay. It can easily be shown that the replacement of air by the 0.8 He + 0.2 O_2 mixture does not affect liquid atomization parameters but increases heat fluxes to the droplets. The effect of various additives on the self-ignition of combustible mixtures, similar to that reported above for heterogeneous systems, had already been observed, in part, in experiments with gaseous mixtures (Grillo and Slack 1976; Chang and Schultz-Grunow 1972; Kumar and Bhaskaran 1976).

Rationalization of Single- and Multiple-Droplet Results

Measurements of ignition delays for single drops (Kawada et al. 1973) and for a system of drops of the same fuel (Miyasaka and Mizutani 1975) do not agree with each other. In addition, the latter results do not support either the conclusions on the impossibility of drop ignition at a temperature $T < T^*$ or those on a significant dependence of ignition delays upon temperature. These discrepancies may be explained as follows. Kauffman et al. (1971), Lu and Slagg (1972, 1974), Wierzba et al.(1974), and, especially, Miyasaka and Mizutani (1975) have not accurately estimated the temperature of the ignition process. They postulated that the ignition process occurred at the temperature behind the reflected wave in pure gas. A thorough analysis of the flow pattern near the injected system of drops indicates that the actual temperature during ignition should be higher. The flow pattern due to the shock wave interaction with a column of drops perpendicular to the shock tube axis is illustrated in the above works. A normal shock is formed at the upstream side of the drop column during drop deformation transversely to the flow owing to close spacing between the drops. Because of the inertia of the drops, the parameters behind the incident shock are close to those behind the reflected wave. In addition, oblique shocks, formed at the side surfaces of the drops, interact with one another. Thus the temperature and pressure distribution in the flow is complex and the temperature and pressure at some point exceed those for the parameters behind the reflected wave. As a consequence, ignition delays, measured behind the incident wave for droplet columns, appear to be, in many cases, less than delays measured for single droplets behind reflected waves. In addition to the difference in real temperatures the differences in shattering pattern for the single- and multiple-droplet experiments also affect the ignition delay. When the droplet column is positioned closer to the end plate, the reflected wave diminishes the effect of the gas flow behind the incident wave. The excess fuel for the multiple-droplet experiments can be estimated from the relation:

$$\alpha = 0.166\pi \, \rho_f / (\Delta l/d)^3 \rho$$

where α is the relative fuel concentration; Δl, the distance between adjacent fuel droplets; d, the droplet diameter; ρ, the total gas density; and ρ_f the density of the liquid. The ignition delays reported by Kauffman et al. (1971) and by Lu and Slagg (1972, 1974), were observed in mixtures with $\alpha \approx$ 10-40. Those reported by Miyasaka and Mizutani (1975) were observed in mixtures with α in the range of 27 < α < 310. For stoichiometric mixtures in oxygen and air, α is about 0.3 and 0.06, respectively. In terms of the two-phase media dynamics, the incident shock wave encounters a dense medium with a liquid to gas concentration ratio of more than 10. Thus shock wave reflection occurs when the incident wave reaches the droplet column and again when the incident wave is reflected from the end plate waves. It is obvious that for systems with such high equivalence ratios, the ignition delay times may not be comparable to the results obtained in experiments with single droplets. Because of low droplet density, the interaction of shock waves and the formation of normal shock at the front boundary of the drop column, typical of the experiments by Kauffman and Nicholls (1971) and Miyasaka and Mizutani (1975), will not occur in most cases of physical significance. As a consequence, experiments with single drops are of greater practical interest. As an example, the results reported by Miyasaka and Mizutani (1975) cannot explain the observations of Laurent et al. (1973) on the absence of self-ignition of atomized fuel in air or the inherent instability of detonations in mixtures of atomized fuel with oxidizing gas. The results of studies of self-ignition of single droplets are more useful for explaining both cases cited.

Conclusions

The self-ignition of sprays in the submillisecond time region depends primarily on chemico-kinetic factors and may be controlled within a sufficiently wide range by fuel additives. Ignition of pure fuel sprays in gaseous flows is not possible below a critical gas temperature, but ignition at lower temperatures can be achieved with introduction of additives into the fuel (or by selection of another fuel). Ignition delays depend considerably on the rates of liquid atomization and mixing for a time less than 10^{-4} s, i.e., when the rate of chemico-kinetic processes exceeds the rate of physical processes.

References

Borisov, A. A. and Kogarko, S. M. (1960) On ignition delay measurements at high temperatures. Izv. Akad. Nauk SSR (8), 1348-1353.

Borisov, A. A., Gelfand, B. E., Eremenko, L. T., Timofeev, E. I., and Tsyganov, S. A. (1979) Features of the ignition of liquid fuel mixtures, Dokl. Akad. Nauk SSR 247, 1176-1179.

Chang, C. M. and Schultz-Grunow, F. (1972) Theoretical and experimental study of the gas phase ignition in a stagnant hot oxidizing gas. AIAA J. 8, 107-114.

Gelfand, B. E., Kalinin, V. N., Petrunin, A. B., Stepanov, V. V., Timofeev, E. I., Tsyganov, S. A., Zakharkin, A. I., and Zhigach, A. F. (1978) Ignition of drops of boronorganic coompounds in air behind shock waves. Dokl. Akad. Nauk SSR 240, 627-629.

Grillo, A. and Slack, M. W. (1976) Shock tube study of ignition delay times in CH_4-O_2-Ar mixtures. Combust. Flame 27, 377-381.

Kadota, T., Hiroyasu, M., and Oya, N. (1976) Spontaneous delay of a fuel drop in high pressure and high temperature gaseous environments. Bull. JSME. 19, 437-445.

Kadota, T., Hyroyasu, H., and Igura, S. (1970) Spontaneous ignition delay of fuel sprays in high pressure gaseous environments. Trans. JSME 41, 1559-1567.

Kauffman, G. W. and Nicholls, J. A. (1971) Shock wave ignition of liquid fuel drops. AIAA J. 9, 880-885.

Kauffman, C. W., Nicholls, J. A. and Olzman, K. A. (1971) The interaction of an incident shock wave with liquid fuel drops. Combust. Sci. Technol. 3, 165-178.

Kawada, H., Yoshizawa, Y., and Kurokawa, K. (1973) Ignition of fuel droplet in the stagnant gas region produced by colliding shock waves. Acta Astron. 1, 753-760.

Kumar, R. R. and Bhaskaran, K. A. (1976) Shock tube study of the effect of unsymmetric dimethyl hydrazine on the ignition delay of CH_4-O_2-Ar mixtures. Combust. Sci. Technol. 8, 107-112.

Laurent, F., Bellet, J. C., Soustre, J., and Manson, N. (1973) Shock induced combustion of kerosene with use of isopropyl nitrate additive. Combus. Inst. Eur. Symp., Academic Press, London-New York, pp. 230-245.

Lu, P. L. and Slagg, N. (1974) Chemical aspects in shock ignition of fuel drops. Astron. Acta 1, 1219-1226.

Lu, P. L. and Slagg, N. (1972) Chemical aspects in the shock initiation of fuel drops. Astron Acta. 17, 693-702.

Miyasaka, K. and Mizutani, Y. (1975a) Ignition of sprays behind a reflected shock. Mod. Dev. Shock Tube Res. pp. 429-436. Shock Tube Research Society, Kyota, Japan.

Miyasaka, K. and Mizutani, Y. (1975b) Ignition of sprays by an incident shock. Combust. Flame 25, 177-186.

Mullins, B. P. (1953) Studies on the spontaneous ignition of fuels injected into a hot air stream. Fuel 32, 451-492.

Wierzba, A. S., Kauffman, C. W., Nicholls, J. A. (1974) Ignition of partially shattered liquid fuel drops in a reflected shock environment. Combust. Sci. Technol. 9, 233-245.

Hydrocarbon Induced Acceleration of Methane-Air Ignition

R. Zellner* and K.J. Niemitz†
Institut für Physikalische Chemie der Universität Göttingen, W. Germany
and
J. Warnatz‡
Institut für Physikalishce Chemie der Technischen Hochschule Darmstadt, W. Germany
and
W.C. Gardiner Jr.,§ C.S. Eubank,π and J.M. Simmie#
University of Texas, Austin, Texas

Abstract

Ignition delay times in air for lean CH_4, CH_4/C_2H_6, CH_4/C_2H_8, and $CH_4/\underline{i}\text{-}C_4H_{10}$ mixtures were determined by time resolved i.r. (C-H stretch) and uv($^2\pi - ^2\Sigma$) absorption measurements following reflected shock heating. Total densities were 2.5×10^{-5} mole/cm^3, corresponding to a pressure of 3.3 atm at 1600 K. In agreement with previous work, it was found that the induction time of a 1% CH_4 mixture is markedly shortened by addition of 0.1% of higher hydrocarbon. The systematic effect of additives can be characterized quantitatively with sufficient resolution to discriminate between the effect of different ones. A computer modeling study of induction times was performed using a recent flame chemistry model. It was found that the ignition delay times of CH_4 alone and their acceleration caused by hydrocarbon additives can be quantitatively accounted for if 10% decay of CH_4 is used as the induction time criterion. However, the shapes of the computed fuel profiles differ from the experimental ones, indicating an incompleteness in the combustion chemistry model for very lean mixtures.

Presented at the 8th ICOGER, Minsk, USSR, Aug. 23-26, 1981. Copyright © 1982 by J. Warnatz by the American Institute of Aeronautics and Astronautics with permission.
*Privatdocent.
†Research Assistant.
‡Privatdocent. Permanent address: Department of Applied Physical Chemistry, University of Heidelberg, FRG.
§Professor of Chemistry, Department of Chemistry.
πResearch Assistant, Department of Chemistry. Permanent address: Environmental Research Laboratory, NOAA, Boulder, Colo.
#Visiting Scientist, Department of Chemistry. Permanent address: Chemistry Department of University College, Galway, Ireland.

Introduction

An effort to understand the factors controlling the detonability of gas mixtures is required by the need to assess hazards related to liquefied natural gas (LNG) spills. Theoretical approaches are possible using a two-front or a combined induction-reaction zone model (Korobeinikov et al. 1972) to describe the interaction between the kinetics of the combustion reactions and the fluid dynamics of propagation (Ficket and Davis 1979). One of the principal quantities of interest in these models is the ignition delay time (τ_i) of a gas mixture of assumed composition as a function of temperature and total density. It can be obtained in principle from chemical models (Oran et al. 1981; Atkinson 1982) or, more reliably, from experiments.

There is a wealth of quantitative information about the ignition delays of methane in shock-heated O_2-Ar mixtures (Lifshitz et al. 1971; Tsuboi and Wagner 1974) and limited information about the ignition of CH_4 in air, as well as about the ignition of other hydrocarbons in air. Also, it is known from previous experiments that methane ignition is accelerated by the presence of relatively small amounts of larger hydrocarbons (Lifshitz et al. 1971; Crossley et al. 1972; Eubank et al. 1981). This acceleration is of practical importance and also provides an opportunity to test our understanding of the ignition chemistry of hydrocarbon/air mixtures.

We recently reported a shock-tube study of the ignition of simulated natural gas mixtures in air (Eubank et al. 1981). It was found in that work that the ignition delay of 1% CH_4 in air is shortened considerably by the addition of 0.1-0.2% C_2H_6, C_3H_8, and $\underline{n}$-C_4H_{10}. The present paper reports an extension of our previous work in two respects: 1) By further experiments we characterized the effects of single additives in order to discriminate between different ones; and 2) a computer modeling study was performed. The common goal of these investigations was to derive quantitative accounting of the observed accelerations on the basis of a set of elementary reactions and to develop truncated chemical models and correlation formulas of ignition delays useful for incorporation into theories of detonability.

Experimental

The shock-tube apparatus has been described in detail elsewhere (Olson et al. 1979). Induction times were determined by monitoring time-dependent absorption behind reflected shock waves. For most of the experiments we used fuel concentration measurements using 3.39-μm He-Ne line absorption (Olson et al. 1978). These experiments were supplemented by 1) OH ($X^2\Pi - A^2\Sigma^+$) absorption measurements around 306.5 nm employing a conventional resonance light source (Ernst et al. 1977); and 2) i.r. emission studies between 3.1 and 3.8 μm (CH_4) and around 4.5 μm (CO_2). Shock properties were computed from the incident shock velocity using standard methods (Gardiner et al. 1981) and JANAF (JANAF Thermochemical Tables) or other standard (Chao et al. 1973; Chen et al. 1975) thermochemical data sources. Gas mixtures were prepared manometrically and allowed to stand at least 24 h before use. Total densities behind reflected shock waves were kept constant areound 2.5×10^{-5} mole/cm^3, corresponding to a pressure of 3.3 atm around 1600 K. The following test gas mixtures were used:

1) 2% CH_4 in 20% O_2 and 78% Ar
2) 2% CH_4 in air
3) 1% CH_4 in air
4) 1% CH_4 + 0.1% C_2H_6 in air
5) 1% CH_4 + 0.1% C_3H_8 in air
6) 1% CH_4 + 0.1% $\underline{i}$-C_4H_{10} in air

The gases were Ar (99.999%, Matheson); O_2 (99.9%, Big Three Industries); CH_4 (99.54%, Phillips 66); C_2H_6 (99.96%, Phillips 66); C_3H_8 (99.6%, Matheson); $\underline{i}$-C_4H_{10} (99.5%, Matheson); and air (raw laboratory air for mixture 2, Linde hydrocarbon-free grade air for mixtures 3-6).

Computer Modeling

Induction times of CH_4 and CH_4/additive ignition were modeled by computing time-dependent species concentrations for adiabatic, constant density conditions, neglecting flow and transport. Given a reaction mechanism containing S species and R elementary reactions

$$\sum_{n=1}^{S} \nu_{in}^{(r)} A_n \rightarrow \sum_{n=1}^{S} \nu_{in}^{(p)} A_n \quad i=1, \ldots, R$$

the differential equations for the concentrations c_j of species j and gas temperature can be written (Warnatz 1978, 1981)

$$\frac{dc_j}{dt} = \sum_{i=1}^{R} k_i \left(\nu_{ij}^{(p)} - \nu_{ij}^{(r)} \right) \cdot \prod_{n=1}^{S} c_n^{\nu_{in}^{(r)}} \quad (1)$$

IGNITION IN A METHANE-AIR MIXTURE

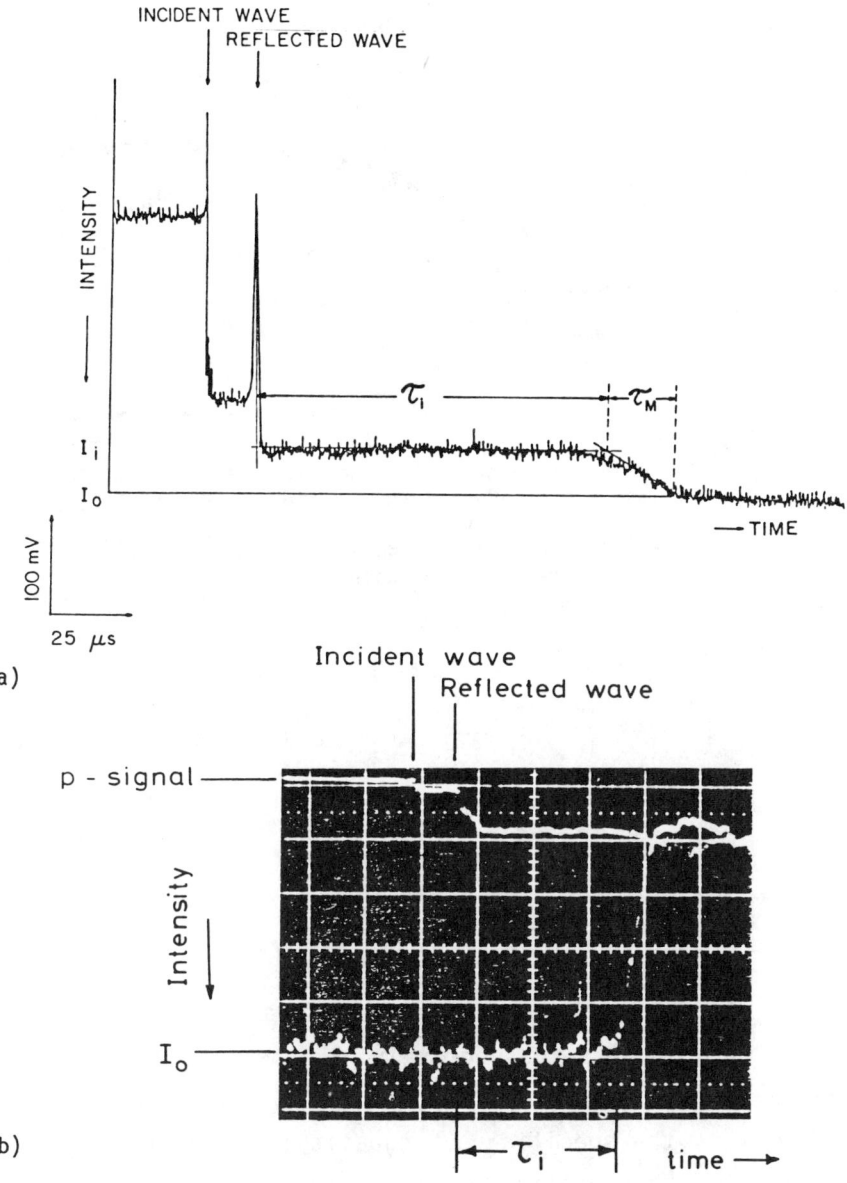

Fig. 1 Experimental records of induction time measurements.
a) Transmitted intensity of 3.39- μm He-Ne laser beam showing its attenuation due to absorption of CH_4 (see text); 1% CH_4 in air, T = 1470 K. b) OH (A-X) absorption profile; 1% CH_4 in air, T = 1470 K.
c) Infrared emission intensities between 3.1 and 3.8 μm (C-H stretch) and around 4.5 μm (CO_2 ν_3); 1% CH_4 in air, T = 1508 K; full sweep 1 ms.

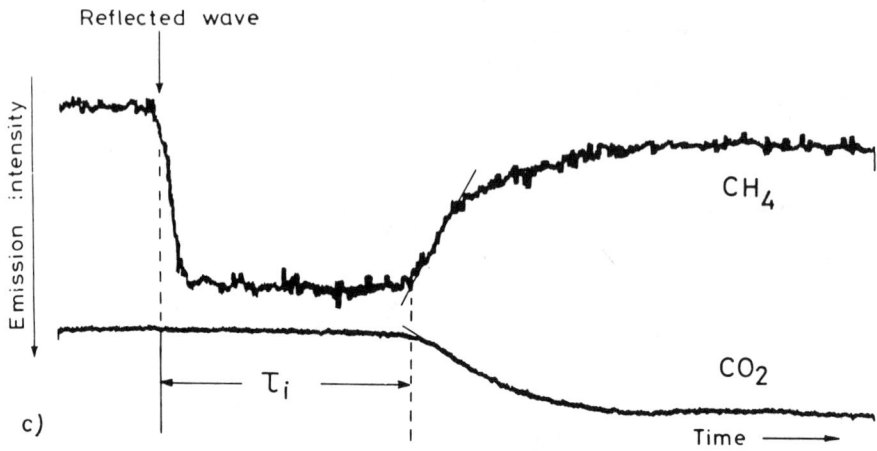

Fig. 1 (cont.) Experimental records of induction time measurements. a) Transmitted intensity of 3.39-μm He-Ne laser beam showing its attenuation due to absorption of CH_4 (see text); 1% CH_4 in air, T = 1470 K. b) OH (A-X) absorption profile; 1% CH_4 in air, T = 1470 K. c) Infrared emission intensities between 3.1 and 3.8 μm (C-H stretch) and around 4.5 μm (CO_2 ν_3); 1% CH_4 in air, T = 1508 K; full sweep 1 ms.

and

$$\frac{dT}{dt} = - \frac{1}{\delta \cdot \bar{c}_p} \cdot \sum_j \hat{H}_j r_j \tag{2}$$

where

$$\bar{c}_p = \sum_j w_j \hat{c}_{p,j}$$

$$r_j = M_j \frac{dc_j}{dt}$$

where $\nu_{in}^{(r)}$, $\nu_{in}^{(p)}$ are the stoichiometric coefficients for reagents and products, respectively; k_i the rate constants in the form $A_i T^{B_i} \exp(-E_{Ai}/RT)$, $i = 1, \ldots, R$; δ the density; $c_{p,j}$ the specific heat capacity; w_j the mass fraction; H_j the specific enthalpy; and M_j the molecular weight of species j. Those reactions that proceed in both forward and reverse directions were included separately using rate constant expressions compatible with the corresponding equilibrium constants. This system of ordinary differential equations of first order was solved using the methods of Gear (1971).

The reaction mechanism used is presented in Table 1. It is based on a recent evaluation of experimental data (Warnatz 1982). Non-Arrhenius behavior of bimolecular reaction rate constants, as well as the density dependence of unimolecular reaction rate constants were incorporated wherever such behavior has been experimentally verified. Earlier versions of the mechanism outlined in this table have been tested in modeling studies of flame velocity and concentration profiles of laminar flames (Warnatz 1978, 1981). It was applied here without any adjustments to rate constants or product channels. It was therefore a severe test to see how this model would perform under conditions of an induction phase, where, contrary to the situation in a flame, branching reactions with O_2 are crucial.

Although the temperature change due to chemical conversion during the induction phase is small, the temperature behind reflected shocks in air does change significantly. It is initially much higher than its vibrational equilibrium value owing to the slow vibrational relaxation of N_2, which occurs on a time scale longer than, or perhaps comparable to, the induction time. Unfortunately, the rate of vibrational relaxation of nitrogen by hydrocarbons has not been reported. For purposes of discovering the magnitude of the effect, we carried out additional computations for the frozen vibrational temperature.

Results and Discussion

Representative experimental records are shown in Fig. 1. Fig. 1a shows a 3.39-μm absorption profile monitoring the decay of aliphatic hydrocarbon. Owing to the strong decrease of the 3.39-μm absorptivity by CH_4 with increasing temperature (Olson et al. 1978), the signal decreases in two stages upon passage of the incident and reflected shock waves (Eubank et al. 1981). The induction time τ_i is defined by the transition to rapid decrease of CH_4 at the onset of the main reaction zone. Figure 1b shows an OH 306.5-nm absorption profile. Incident and reflected shock front arrival are determined in this case by means of a pressure pulse obtained by a transducer in the plane of the optical observation point. Owing to the sensitivity of the OH absorption technique (Ernst et al. 1977), a rise of the absorption intensity is observable even at small chemical conversions. The concentration of OH remains high throughout the main reaction zone and decays slowly during the postreaction period. The induction time is defined here by the transition to rapid rise of the OH absorption signal.

Table 1 Reaction mechanism and rate constants
$$k = AT^B \exp(-E_A/RT) \text{ cm}^3/\text{mole-s}$$

Reaction						$A/\text{cm}^3\text{mole}^{-1}\text{s}^{-1}$	B	E_A/kcal
(1) H	+	O_2	→	OH	+ O	2.20(+14)	0.0	16.80
(2) O	+	OH	→	H	+ O_2	1.00(+13)	0.0	0.0
(3) O	+	H_2	→	OH	+ H	1.50(+07)	2.00	7.55
(4) H	+	OH	→	O	+ H_2	6.70(+06)	2.00	5.56
(5) OH	+	H_2	→	H_2O	+ H	1.00(+08)	1.60	3.30
(6) H	+	H_2O	→	OH	+ H_2	4.60(+08)	1.60	18.55
(7) OH	+	OH	→	H_2O	+ O	1.50(+09)	1.14	0.0
(8) H_2O	+	O	→	OH	+ OH	1.50(+10)	1.14	17.24
(9) H	+	O_2 + M	→	HO_2	+ M	8.00(+17)[a]	-0.80	0.0
(10) H	+	HO_2	→	OH	+ OH	1.50(+14)	0.0	1.00
(11) H	+	HO_2	→	H_2	+ O_2	2.50(+13)	0.0	0.70
(12) OH	+	HO_2	→	H_2O	+ O_2	1.50(+13)	0.0	0.0
(13) O	+	HO_2	→	OH	+ O_2	2.00(+13)	0.0	0.0
(14) CO	+	OH	→	CO_2	+ H	4.40(+06)	1.50	- 0.75
(15) CO_2	+	H	→	CO	+ OH	1.50(+14)	0.0	26.40
(16) H	+	CH_4	→	H_2	+ CH_3	2.20(+04)	3.00	8.75
(17) O	+	CH_4	→	OH	+ CH_3	1.20(+07)	2.10	14.00
(18) OH	+	CH_4	→	H_2O	+ CH_3	1.60(+06)	2.10	2.45
(19) H	+	CH_3	→	CH_4		1.07(+35)	-6.50	9.10
(20) CH_4			→	CH_3	+ H	1.35(+33)[b]	-5.50	108.00
(21) O	+	CH_3	→	CH_2O	+ H	7.00(+13)	0.0	0.0
(22) CH_3	+	O_2	→	CH_2O	+ H+O	7.00(+12)	0.0	25.60
(23) CH_3	+	CH_3	→	C_2H_6		1.26(+49)	-10.75	17.77
(24) C_2H_6			→	CH_3	+ CH_3	8.91(+50)[b]	-10.75	105.40
(25) CH_3	+	CH_3	→	C_2H_5	+ H	8.00(+14)	0.0	26.60
(26) H	+	C_2H_5	→	CH_3	+ CH_3	3.00(+13)	0.0	0.0
(27) CH_3	+	CH_3	→	C_2H_4	+ H_2	1.00(+16)	0.0	32.00
(28) CH_3	+	H	→	CH_2	+ H_2	1.80(+14)	0.0	15.00
(29) OH	+	CH_2O	→	CHO	+ H_2O	8.00(+13)	0.0	1.50
(30) H	+	CH_2O	→	CHO	+ H_2	2.50(+13)	0.0	4.10
(31) O	+	CH_2O	→	CHO	+ OH	3.00(+13)	0.0	3.40

Table continued on next page

Table 1 (cont.) Reaction mechanism and rate constants
$$k = AT^B \exp(-E_A/RT) \text{ cm}^3/\text{mole-s}$$

Reaction						$A/\text{cm}^3\text{mole}^{-1}\text{s}^{-1}$	B	E_A/kcal
(32) O_2	+	CHO	→	CO	+ H_2O	1.20(+13)	0.0	0.90
(33) H	+	CHO	→	CO	+ H_2	3.00(+14)	0.0	0.0
(34) CHO	+	M	→	CO	+ H + M	1.50(+14)	0.0	14.70
(35) H	+	CH_2	→	CH	+ H_2	5.00(+13)	0.0	0.0
(36) O	+	CH_2	→	CO	+ H + H	5.00(+13)	0.0	0.0
(37) O_2	+	CH_2	→	CO	+ OH + H	6.00(+13)	0.0	2.40
(38) O	+	CH	→	CO	+ H	4.00(+13)	0.0	0.0
(39) O_2	+	CH	→	CO	+ OH	2.00(+13)	0.0	0.0
(40) H	+	C_2H_6	→	H_2	+ C_2H_5	5.40(+02)	3.50	5.20
(41) O	+	C2H6	→	OH	+ C2H5	3.00(+07)	2.00	5.10
(42) OH	+	C_2H_6	→	H_2O	+ C_2H_5	6.30(+06)	2.00	0.65
(43) CH_3	+	C_2H_6	→	CH_4	+ C_2H_5	5.50(-01)	4.00	8.29
(44) O	+	C_2H_5	→	ACAL[c]	+ H	5.00(+13)	0.0	0.0
(45) O_2	+	C_2H_5	→	HO_2	+ C_2H_4	2.00(+12)	0.0	5.00
(46) C_2H_5			→	C_2H_4	+ H	2.19(+42)[b]	-8.75	53.59
(47) H	+	C_2H_4	→	C_2H_5		1.10(+42)	-8.75	15.49
(48) C_2H_5	+	C_2H_5	→	C_2H_4	+ C_2H_6	1.10(+13)	0.0	0.0
(49) O	+	C_2H_4	→	CHO	+ CH_3	2.00(+13)	0.0	2.30
(50) OH	+	C_2H_4	→	CH_2O	+ CH_3	2.63(+40)	-8.50	10.61
(51) H	+	C_2H_4	→	H_2	+ C_2H_3	1.50(+14)	0.0	10.20
(52) H	+	C_2H_3	→	H_2	+ C_2H_2	8.00(+12)	0.0	0.0
(53) O	+	C_2H_3	→	KETE[d]	+ H	1.00(+13)	0.0	0.0
(54) C_2H_3			→	C_2H_2	+ H	9.56(+27)[b]	-4.75	47.45
(55) H	+	C_2H_2	→	C_2H_3		5.25(+28)	-4.75	10.85
(56) O	+	C_2H_2	→	CH_2	+ CO	4.10(+08)	1.50	1.70
(57) OH	+	C_2H_2	→	KETE	+ H	1.55(+33)	6.25	10.98
(58) H	+	ACAL	→	CH_3	+ CO + H_2	4.00(+13)	0.0	4.20
(59) O	+	ACAL	→	CH_3	+ CO + OH	5.00(+12)	0.0	1.80
(60) OH	+	ACAL	→	CH_3	+ CO + H_2O	1.00(+13)	0.0	0.0

Table continued on next page

Table 1 (cont.) Reaction mechanism and rate constants

$$k = AT^B \exp(-E_A/RT) \text{ cm}^3/\text{mole-s}$$

Reaction						$A/\text{cm}^3\text{mole}^{-1}\text{s}^{-1}$	B	E_A/kcal
(61) OH	+	KETE	→	CH_2O	+ CHO	2.80(+13)	0.0	0.0
(62) O	+	KETE	→	CHO	+ CHO	2.00(+13)	0.0	2.3
(63) H	+	KETE	→	CH_3	+ CO	7.00(+12)	0.0	3.0
(64) C_2H_5	+	CH_3	→	C_3H_8		1.00(+13)	0.0	0.0
(65) H	+	C_3H_8	→	H_2	+ C_2H_7	1.30(+14)	0.0	9.7
(66) H	+	C_3H_8	→	H_2	+ IPRO	1.00(+14)	0.0	8.3
(67) O	+	C_3H_8	→	OH	+ C_3H_7	3.00(+13)	0.0	5.7
(68) O	+	C_3H_8	→	OH	+ IPRO	2.60(+13)	0.0	4.4
(69) OH	+	C_3H_8	→	H_2O	+ C_3H_7	3.70(+12)	0.0	1.6
(70) OH	+	C_3H_8	→	H_2O	+ IPRO	2.80(+12)	0.0	0.8
(71) CH_3	+	C_3H_8	→	CH_4	+ IPRO	1.00(+11)	0.0	9.6
(72) CH_3	+	C_3H_8	→	CH_4	+ C_3H_7	1.00(+11)	0.0	9.6
(73) C_3H_8			→	CH_3	+ C_2H_5	8.00(+16)[b]	0.0	85.0
(74) H	+	IPRO[e]	→	C_3H_8		2.00(+13)	0.0	0.0
(75) H	+	C_3H_7	→	C_3H_8		2.00(+13)	0.0	0.0
(76) O_2	+	C_3H_7	→	HO_2	+ C_3H_6	1.00(+12)	0.0	5.0
(77) O_2	+	IPRO	→	HO_2	+ C_3H_6	3.00(+12)	0.0	5.0
(78) C_3H_7			→	C_2H_4	+ CH_3	3.00(+14)[b]	0.0	33.0
(79) C_3H_7			→	C_3H_6	+ H	1.00(+14)[b]	0.0	37.3
(80) H	+	C_3H_6	→	C_3H_7		4.50(+12)	0.0	2.7
(81) IPRO			→	C_3H_6	+ H	2.00(+14)[b]	0.0	38.7
(82) H	+	C_3H_6	→	IPRO		6.00(+12)	0.0	1.2
(83) O	+	C_3H_6	→	CH_3	+ CH_3 + CO	5.00(+12)	0.0	0.6
(84) OH	+	C_3H_6	→	ACAL	+ CH_3	7.00(+12)	0.0	0.0
(85) $N_{2(v=0)}$			⇌	N_2 (vibrationally equilibrated)		4.50(+06)[b]	0.0	25.9

a Units are $\text{cm}^6/\text{mole}^2\cdot\text{s}$. b Units are s^{-1}. c ACAL = acetaldehyde, CH
d KETE = ketene, CH_2CO. e IPRO = iso-Propyl radical, $\underline{i}\text{-}C_3H_7$.

Results obtained this way appear to be consistently larger, by some 20%, than τ_i measurements based on CH_4 consumption. Figure 1c is an experimental record of thermal emission intensities of CH_4 and CO_2. As can be seen, combustion progress as evidenced by CH_4 disappearance is mirrored by CO_2 appearance.

Figure 2 shows a set of computed profiles. Compared to the experimental observations, distinct differences can be noted:

1) The decrease of fuel concentration occurs much more gradually than observed in the experiments.

2) A large decrease of CH_4 is predicted long before any perceptible appearance of CO_2. Methane is almost fully converted to intermediates, primarily CH_3, CH_2O, and CO, prior to the appearance of final products.

3) The increase of OH through the observed concentration range (essentially from 10^{-9} to 10^{-8} mole/cm^3) is predicted to be much slower than is observed experimentally.

In view of the different appearances of the computed and observed profiles, we chose to define the computed τ_i as the time at which 10% of the total methane is consumed. It can be seen from Fig. 2 that if other criteria (e.g., onset of rapid conversion or rise of CO_2) were chosen, the computed τ_i values would roughly double.

CH_4 Ignition in O_2/Ar and in Air

We had noted in previous experiments (Eubank et al. 1981) that the induction time of a 2% CH_4 mixture is shortened by a factor of roughly 1.6 upon replacing "synthetic" air (O_2/Ar = 20/80) by raw laboratory air (Fig. 3). It was shown that the apparent acceleration was not due to chemical effects caused by the moisture content or reactive trace impurities in laboratory air (Eubank et al. 1981). Also, it is unlikely that an acceleration of this magnitude could be caused by the enhanced third body collision efficiency of N_2 compared to Ar in unimolecular reactions, since the overall induction time is insensitive to the unimolecular reaction rate constants. We conclude that it is caused by vibrational nonequilibrium of N_2, which causes the translational temperature of the gas to be higher than its equilibrium value.

We therefore added vibrational relaxation of N_2

$$N_2 \text{ (v=0)} + M \rightleftarrows N_2 \text{ (vibrationally equilibrated)} + M$$

to the kinetic model, using for its rate constant an Arrhenius expression fitted to the pure-N_2 Landau-Teller parameters given by Millikan and White (1963). A typical example of a time-dependent temperature profile--computed

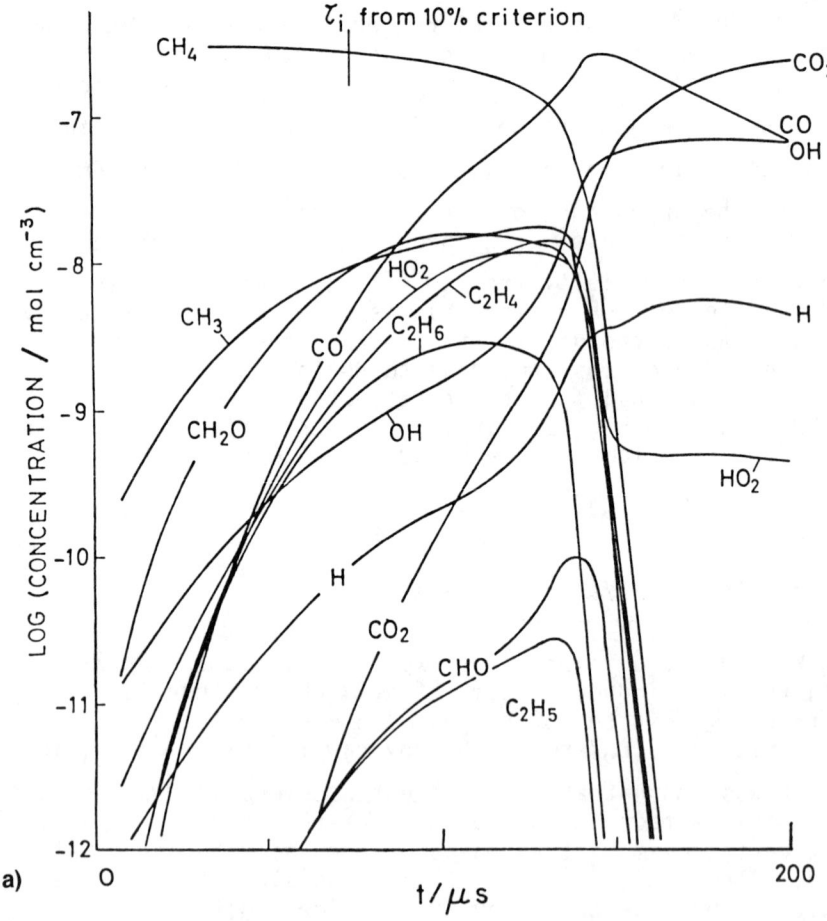

Fig. 2 Concentration profiles computed using the mechanism of Table 1; 1% CH_4 in air, initial temperature 1690 K. The induction time derived from such profiles by the 10% fuel decay criterion is indicated by the vertical bar. The discrepancy between this definition of τ_i and the onset of the main reaction zone is discussed in the text.
a) Logarithmic plot. b) Linear plot.

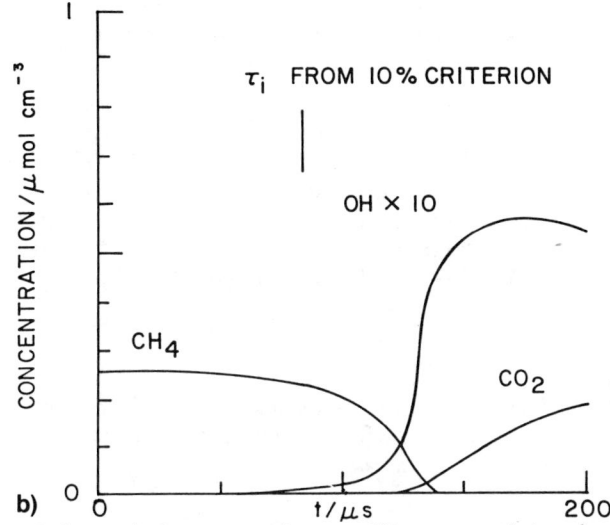

Fig. 2 (cont.) Concentration profiles computed using the mechanism of Table 1; 1% CH$_4$ in air, initial temperature 1690 K. The induction time derived from such profiles by the 10% fuel decay criterion is indicated by the vertical bar. The discrepancy between this definition of τ_i and the onset of the main reaction zone is discussed in the text. a) Logarithmic plot. b) Linear plot.

according to Eq. (2)--is shown in Fig. 4. The computed temperature decreases slowly from an initial value T_i and rises again at the end of the induction time as combustion begins. The most important point is that the temperature during the entire induction time exceeds the vibrational equilibrium temperature, which is never reached prior to ignition, even at the highest temperatures considered. The differences between the two are computed to be 60, 96, and 139 K for vibrational equilibrium temperatures of 1300, 1600, and 1900 K. No such effect is caused by O_2, since its much faster vibrational relaxation (Millikan and White 1963) is essentially immediate on the time scale of ignition experiments.

Figure 3 shows the comparison between experimental and computed induction times for 2% CH$_4$ in O_2/Ar and in air. The agreement is satisfactory in the O_2/Ar case, while the induction time shortening upon replacing Ar by N_2 is over estimated at higher temperatures by the vibrational nonequilibrium assumption.**

**The rate constant given by Millikan and White (1963) is for relaxation in N_2 only. It is likely that the process is faster in the presence of hydrocarbons. Our computations must therefore be considered to yield only the maximum downward displacement of the induction time curves.

While the experimental induction times are well represented in Arrhenius form $[\tau_i/s = 1.7 \pm 0.8)\, 10^{-11}$ $\exp(216\, \text{kJmole}^{-1}/RT)$ for the 2% CH_4 mixture in O_2/Ar (Eubank et al. 1981)], in agreement with previous observations (Lifshitz et al. 1971; Tsuboi and Wagner 1974; and Crossley et al. 1972), the modeling results suggest some non-Arrhenius behavior (Fig. 3). Similar effects have been observed in simulations of ignition delays of H_2/O_2 mixtures (Oran et al. 1981; Meyer and Oppenheim 1970).

CH_4/Additive Ignition in Air

Experimentally determined ignition delays for four mixtures of 1% CH_4 and 1% CH_4/0.1% additive (C_2H_6, C_3H_8, i-C_4H_{10}) are shown in Fig. 5. Each additive shortens the

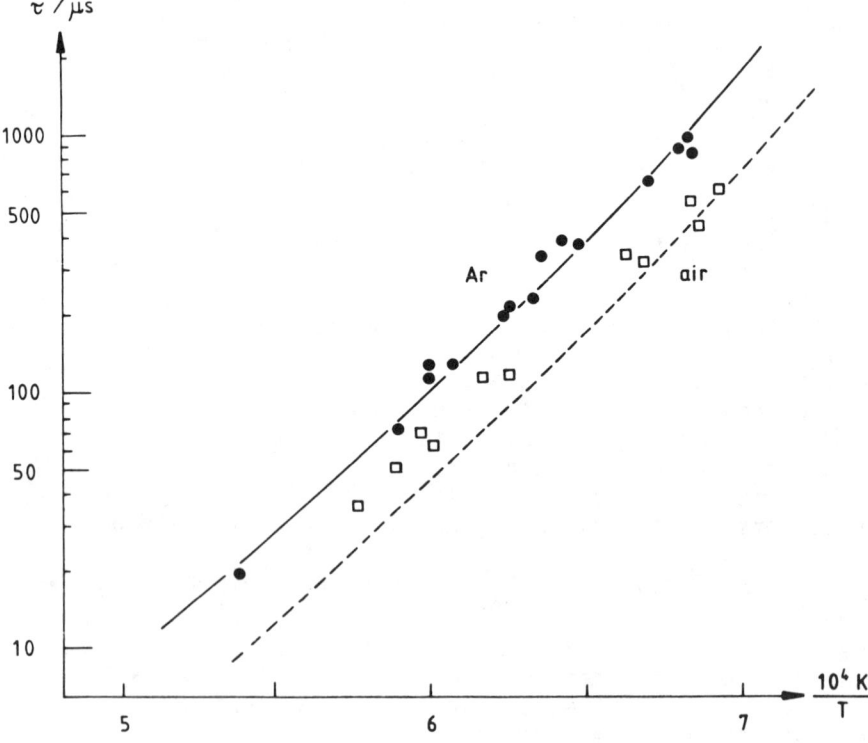

Fig. 3 Arrhenius representation of induction times of a mixture of 2% CH_4 in Ar/O_2 = 80/20 (●) and in air (□). The temperature scale is defined for both mixtures by the reflected shock temperature computed from the incident shock speed, assuming full vibrational equilibration in both waves. The lines are the corresponding results from computations (see text).

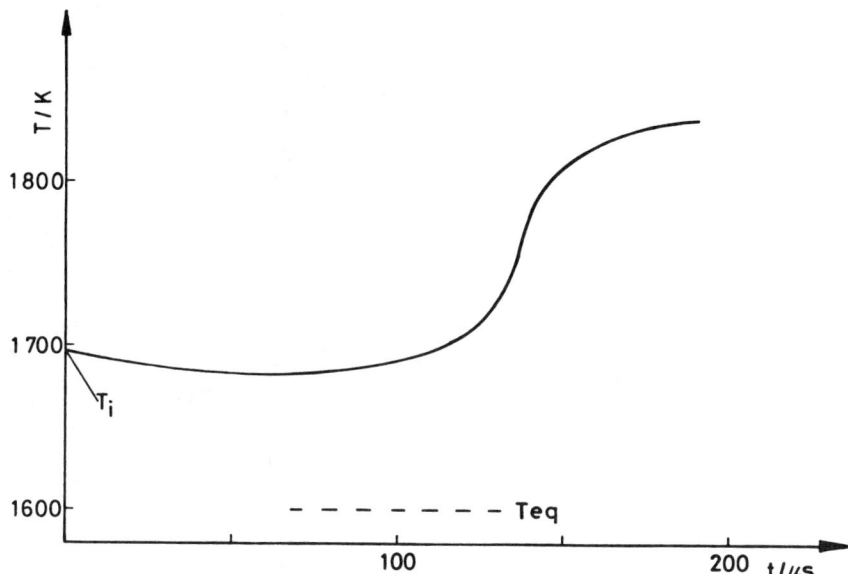

Fig. 4 Computed temperature profile for ignition of 1% CH_4 in air, including explicit account of N_2 vibrational relaxation. Corresponding species concentration profiles are those of Fig. 2. T_i = 1690 K is the computed post-reflected-shock temperature for vibrationally cold N_2; T_{eq} = 1600 K is the corresponding temperature for fully equilibrated N_2.

induction time by roughly the same amount, but distinct differences between the effects of different additives can be perceived. The most prominent effect is the increase in temperature dependence upon replacing C_2H_6 by C_3H_8 and $\underline{i}$-C_4H_{10}. This effect had not been noted in our previous experiments (Eubank et al. 1981), where only mixtures of additives were used. In contrast, ignition delays of C_2H_6, C_3H_8, and C_4H_{10} in alkane/argon-simulated-air mixtures (Burcat et al. 1971) have temperature dependences corresponding to activation energies of roughly 165 kJ, appreciably less than the 216 kJ value for ignition of CH_4. On the other hand, the induction times observed here for 1% CH_4/0.1% additive mixtures are considerably longer than one would calculate, based on the data of Burcat et al. (1971), for the ignition of the additive alone. This shows that induction times of mixed fuels are not determined by the ignition behavior of the most easily ignited component (Eubank et al. 1981).

Figure 6 shows comparisons between experiment and simulation for the ignitions of 1% CH_4, 1% CH_4/0.1% C_2H_6, and 1%

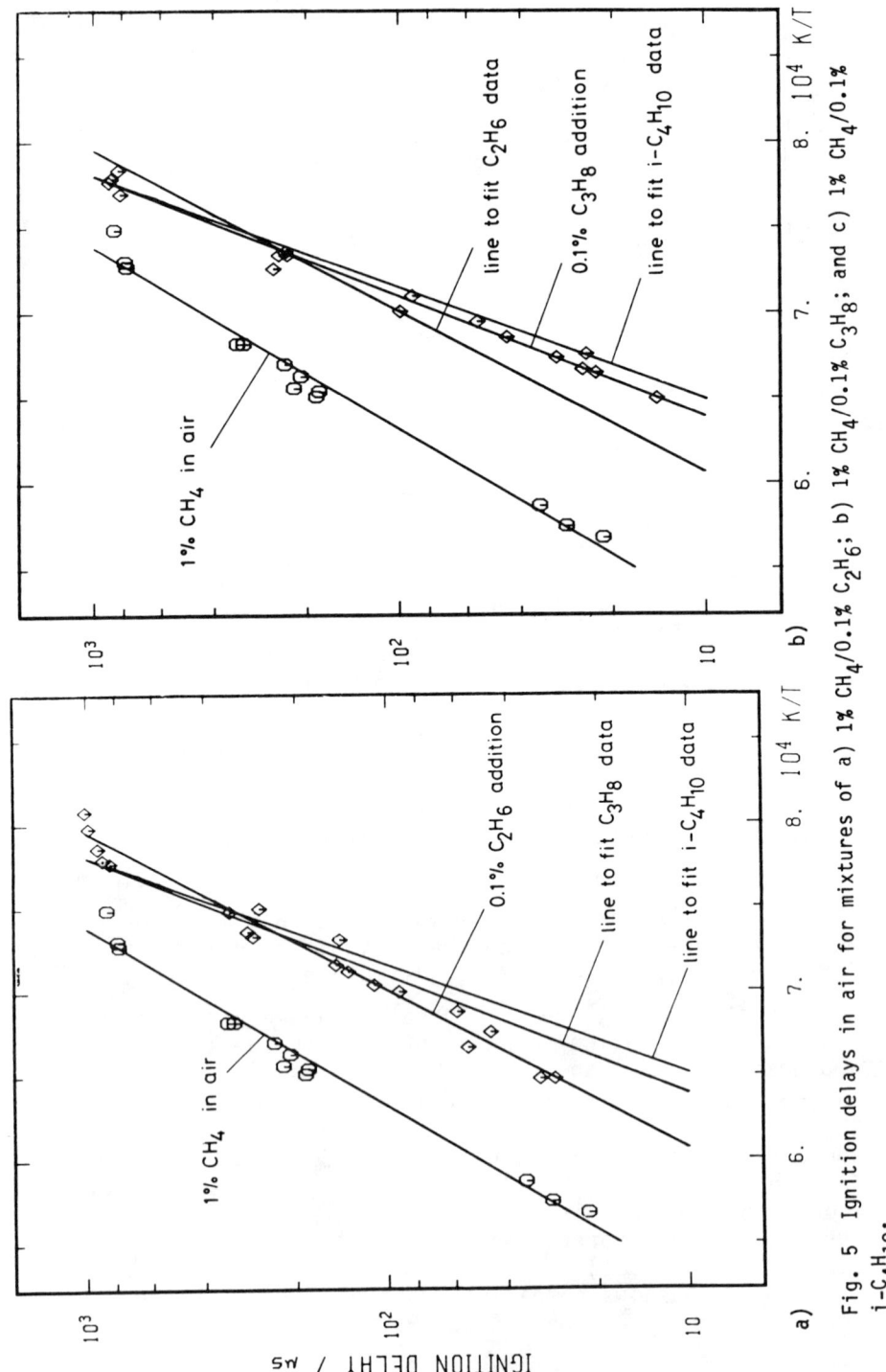

Fig. 5 Ignition delays in air for mixtures of a) 1% CH_4/0.1% C_2H_6; b) 1% CH_4/0.1% C_3H_8; and c) 1% CH_4/0.1% $i-C_4H_{10}$.

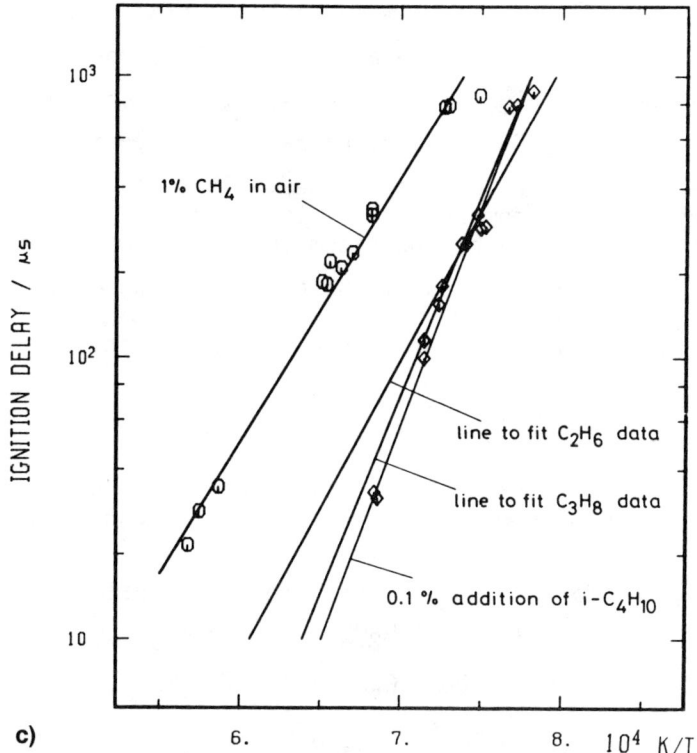

Fig. 5 (cont.) Ignition delays in air for mixtures of a) 1% CH_4/0.1% C_2H_6; b) 1% CH_4/0.1% C_3H_8; and c) 1% CH_4/0.1% $\underline{i}$-C_4H_{10}.

CH_4/0.1% C_3H_8 mixtures in air.[††] Considering the shortcomings of the computed profiles noted above, the agreement is satisfactory, particularly since the mechanism in Table 1 had previously been applied only to flame chemistry (Warnatz 1978, 1981), and its extension to induction processes had not been tested prior to this study.

The modeling results, despite their disagreements with the observed forms of the CH_4, OH, and CO_2 profiles, still provide information about the source of the observed accelerations of CH_4 ignition by additives. There are two a priori possibilities.

1) Initiations: The rates of thermal decomposition of higher alkanes are considerably faster than that of CH_4. At

---

[††]No simulation has yet been made for the ignition of a 1% CH_4/0.1% $\underline{i}$-C_4H_{10} mixture owing to a lack of high-temperature data on the thermal decomposition rate of $\underline{i}$-C_4H_{10}.

1600 K, for example, the primary fragmentation of C_2H_6 into two CH_3 radicals occurs several hundred times more rapidly than the decomposition of CH_4. However, this strong enhancement of the CH_3 production rate is only mildly reflected in shorter induction times. This is in part due to the relatively (compared to H) low reactivity of CH_3 and the inhibition caused by CH_3 recombination. The primary effect is therefore caused by the coupling with chain branching reactions.

2) Branching: The primary branching reaction in all mixtures is

$$(1) \quad H + O_2 \rightarrow OH + O$$

Since this reaction is not rate limiting in the presence of 20% O_2, enhancement of the branching rate is connected with the production rate of H atoms. In the case of C_2H_6 addition, the following reactions are additional H atom sources:

$$(41) \quad O + C_2H_6 \rightarrow OH + C_2H_5$$

$$(42) \quad OH + C_2H_6 \rightarrow H_2O + C_2H_5$$

$$(43) \quad CH_3 + C_2H_6 \rightarrow CH_4 + C_2H_5$$

The C_2H_5 radical is thermally unstable and rapidly produces H atoms:

$$(46) \quad C_2H_5 \rightarrow C_2H_4 + H$$

Reactions (41-43) are therefore effectively chain branching. No such routes exist for CH_4 as reagent, where the CH_3 radical formed by the analogs of (41-43) is thermally stable and has a branching reaction with O_2:

$$(22) \quad CH_3 + O_2 \rightarrow CH_2O + O + H$$

that is slow (Bhaskaran and Frank 1980) compared to reaction (1).

As mentioned before, there is disagreement between experimental and computed profiles. Whereas the experiment shows sharp, simultaneous transitions at the onset of the main reaction zone, our model calculation predicts more gradual reaction. In other words, the model predicts slow combustion of CH_4 to intermediates and hence a deceleration of a branched chain explosion. The origin of this in the

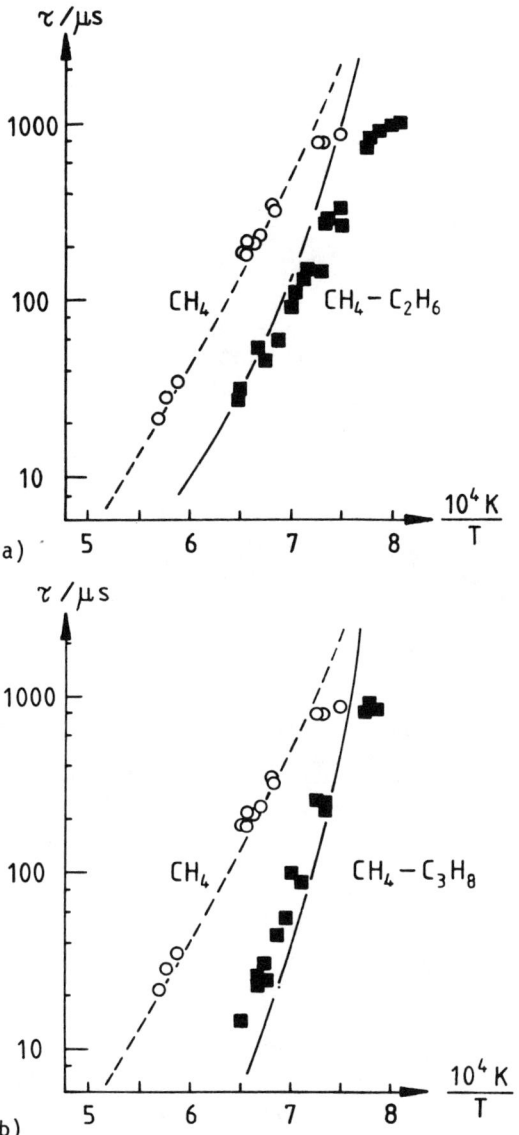

Fig. 6 Comparison between experimental (points) and computed (lines) ignition delays in air for mixtures of a) 1% CH_4/0.1% C_2H_6 and b) 1% CH_4/0.1% C_3H_8.

model may be identified by inspection of Fig. 7, where the main flow of the CH_4 combustion chemistry in excess O_2 is shown schematically. There are two points where inhibition of the branched chain reaction occurs.

1) Recombination of CH_3 to C_2H_6: Owing to the relatively slow rate of reaction between CH_3 and O_2, CH_3 tends

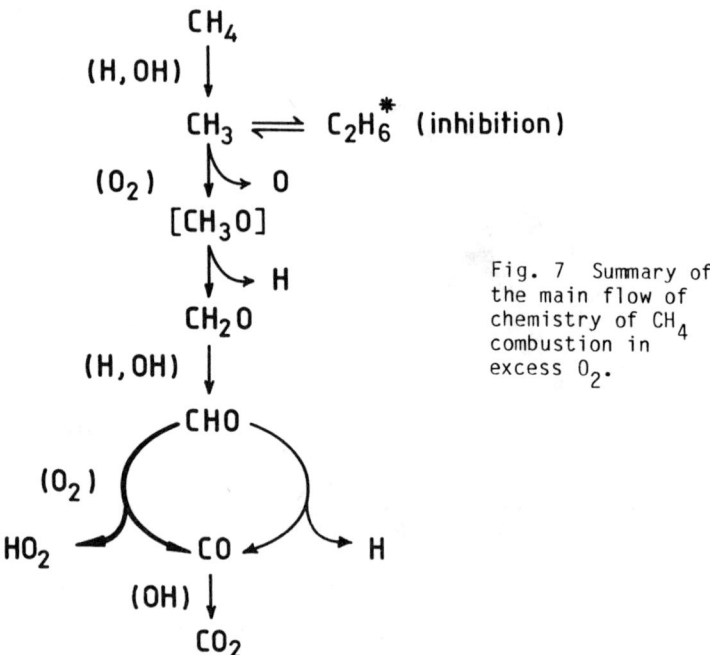

Fig. 7 Summary of the main flow of chemistry of CH_4 combustion in excess O_2.

to accumulate as C_2H_6. The occurrence of this reaction explains why addition of "instantaneous" CH_3 sources such as C_2H_6 or di-t-butylperoxide has a relatively small effect on the induction times of CH_4/O_2 mixtures (Tsuboi and Wagner 1974).

2) Reactions of CH_2O: Following attack on CH_2O by H and OH, the subsequent reaction of CHO is computed to be mainly with O_2,

(32) $CHO + O_2 \rightarrow CO + HO_2$

accounting for 60% of its total removal at 1600 K. Unlike H atoms, the HO_2 radical does not participate in chain branching. Therefore CH_2O is computed to act as a chain termination agent.

The deceleration effect imposed by the assumed CH_2O chemistry could be removed by suitable adjustment of rate constants, for example, by lowering the rate constant for reaction (32). The magnitude of change in k_{32} required, however, is probably beyond the error limits that may be accepted for values of this rate constant obtained from direct rate constant measurements. Hence we are left with incomplete knowledge of the detailed combustion chemistry. This incompleteness may be apparent only in lean mixtures, whose chemistry therefore should be further investigated.

Acknowledgments

Acknowledgment is made to the donors of the Petroleum Research Fund, administered by the American Chemical Society, for partial support of this research. This research was also supported by the Robert A. Welch Foundation, the U.S. Army Research Office, the Fonds der Chemischen Industrie and the Deutsche Forschungsgemeinschaft. One of the authors (R. Z.) gratefully acknowledges a Heisenberg Stipendium from the Deutsche Forschungsgemeinschaft.

References

Atkinson, R. and Bull, D.C. (1983) Kinetic modelling of ethane/air detonability. Shock Waves, Explosions, and Detonations: AIAA Progress in Astronautics and Aeronautics (edited by Bowen, Manson, Oppenheim, and Soloukhin) Vol. 87, pp. 318-332. AIAA, New York.

Bhaskaran, K. A., Frank, P., and Just, T. (1980) Proceedings of the 12th International Symposium on Shock Tubes and Waves (edited by A. Lifshitz and J. Rom), p. 503. Magnes Press, Jerusalem.

Burcat, A., Scheller, K., and Lifshitz, A. (1971) Shock tube investigation of comparative ignition delay times for C_1-C_5 alkanes. Combust. Flame 16, 29.

Chao, J., Wilhoit, R. C., and Zwolinski, B. J. (1973) Ideal gas thermodynamic properties of ethane and propane. J. Phys. Chem. Ref. Data 2, 427.

Chen, S. S., Wilhoit, R. C., and Zwolinski, B. J. (1975) Ideal gas thermodynamic properties and isomerization of n-butane and isobutane. J. Phys. Chem. Ref. Data 4, 859.

Crossley, R. W., Dorko, E. A., Scheller, K., and Burcat, A. (1972) The effect of higher alkanes on the ignition of methane-oxygen-argon mixtures in shock waves. Combust. Flame 19, 373.

Ernst, J., Wagner, H. G., and Zellner, R. (1977) Direct rate measurements for $OH + OH \rightarrow H_2 + O$ in the range 1200-1800 K, Ber. Bunsenges. Phys. Chem. 81, 1270.

Eubank, C. S., Rabinowitz, M. J., Gardiner, W. C. Jr., and
 Zellner, R. (1981) 18th Symposium (International) on
 Combustion, p. 1767. The Combustion Institute, Pittsburgh, Pa.

Ficket, W. and Davis, W. C. (1979) Detonation. University of
 California Press, Berkeley, Calif.

Gardiner, W. C. Jr., Walker, B. F., and Wakefield, C. B. (1981)
 Mathematical methods for modeling chemical reactions in shock
 waves. Shock Waves in Chemistry and Chemical Technology
 (edited by A. Lifshitz) Chap. 7. Marcel Dekker, New York.

Gear, C. W. (1971) Numerical Initial Value Problems in Ordinary
 Differential Equations. Prentice Hall, Englewood Cliffs, N. J.

JANAF Thermochemical Tables (1971) (2nd ed.) NSRDS-NBS 37.

Korobeinikov, V. P., Levine, V. A., Markov, V., and Chernyi, G. G.
 (1972) Astronaut. Acta 17, 529.

Lifshitz, A., Scheller, K., Burcat, A., and Skinner, G. (1972)
 Multistage ignition of hydrocarbon combustion. Combust.
 Flame 19, 311.

Meyer, J. W. and Oppenheim, A. K. (1970) 13th Symposium
 (International) on Combustion, p. 1153. The Combustion
 Institute, Pittsburgh, Pa.

Millikan, R. C. and White, D. R. (1963) Systematics of
 vibrational relaxation. J. Chem. Phys. 39, 3209.

Olson, D. B., Mallard, W. G., and Gardiner, W. C. Jr. (1978)
 High temperature absorption of the 3.39-μm He-Ne laser line
 by small hydrocarbons, Appl. Spectrosc. 32, 489.

Olson, D. B., Tanzawa, T., and Gardiner, W. C., Jr. (1979)
 Thermal decomposition of ethane, Int. J. Chem. Kinet. 11, 23.

Oran, E. S., Boris, J. P., Young, T., Flanigan, M., Burks, T. and
 Picone,, M., (1981) 18th Symposium (International) on
 Combustion, p. 1641. The Combustion Institute, Pittsburgh, Pa.

Tsuboi, T. and Wagner, H. G. (1974) 15th Symposium (International) on Combustion, p. 883. The Combustion Institute,
 Pittsburgh, Pa.

Warnatz, J. (1978) Calculation of the structurue of laminar flat
 flames I: Flame velocity of freely propagating ozone
 decomposition flames. Ber. Bunsenges. Phys. Chem. 82, 193.

Warnatz, J. (1981) 18th Symposium (International) on Combustion,
 p. 369. The Combustion Institute, Pittsburgh, Pa.

Warnatz, J. (1982) Survey of Elementary reaction rate constants
 in the C/H/O system. Chemistry of Combustion Reactions
 (edited by W. C. Gardiner Jr.). Springer-Verlag, New York.

Extinction of In-Flight Engine Fuel-Leak Fires with Dry Chemicals

Robert L. Altman*
NASA Ames Research Center, Moffett Field, Calif.

Abstract

When fuels leak onto surfaces of an operating engine they can ignite when engine case temperatures exceed 550°C. As aircraft flight speeds are increased, engine case temperatures, bleed air temperatures, maximum air velocities, and fire extinguishant storage temperature requirements also increase, making the task of extinguishing fuel-leak fires in flight even more difficult. We have undertaken to find new fire extinguishants that are more effective than the CF_3Br, CF_2Br_2, and CF_2ClBr now in use.

Besides testing commercially available dry chemicals, such as $NaHCO_3$, $KHCO_3$, KCl, and $KC_2N_2H_3O_3$ (ICI Monnex), we have tried to develop and test new dry-powder fire extinguishants. Specifically, our interest has been in developing new dry-power extinguishants that, when discharged into a jet engine fuel-leak fire, would stick to the hot surfaces. Ater putting out the initial fire, the extinguishant residue would act chemically as an antireignition barrier, even when the jet fuel, JP-4, continued to leak onto the heated surface.

Introduction

Previous fire extinguishment tests with Halons like CH_3Br, CH_2ClBr, CF_3Br, and CF_2ClBr have shown that the minimum Halon concentration in the gas phase above a liquid

---

Presented at the 8th ICOGER, Minsk, USSR, Aug. 23-26, 1981. This paper is declared a work of the U.S. Government and therefore is in the public domain.
*Research Scientist.

pool or spray fire required to extinguish the fire at first increases with increased airflow at low flow rates but then decreases at still higher flow rates. In both situations, however, the total weight of extinguishant required to put out the fire increased with increasing airflow (Hirst et al. 1976, 1977). To counter this, an increase in the Halon discharge rate by increasing the stored nitrogen pressure or by increasing the extinguishant discharge temperature will decrease the total weight of Halon required for complete extinguishment (Dyer et al. 1977a, 1977b). The reduced weight effectiveness of Halons with increased airflow has induced Graviner, the manufacturer of the Concorde nacelle fire extinguishing system, to install a pair of airflow-reducing flaps upstream of the compressor; the flaps reduce the nacelle airflow to a minimum before the Halon is discharged after an engine fuel-leak fire has been detected (Davis 1971).

The longer fuel-leak fires burn before extinguishment is begun, the harder they are to extinguish; also, long-burning fires can start up again once the extinguishant is exhausted because the surroundings are then hot enough to reignite the fuel (Fenwal Inc. 1964; Klueg and Demaree 1969; Sommers 1970). Therefore we have devised another experimental procedure for rating the effectiveness of fire extinguishants in controlling fuel-leak fires. In this technique, the effectiveness of an extinguishant is measured in terms of the time delay between initial extinguishment and reignition; throughout the test, fuel continues to drip on the heated surface until reignition occurs.

The fuel used in this study was a wide-cut jet fuel, JP-4 or Jet-B, whose distillation temperature range and flash point are significantly lower than that of the kerosene-base jet fuels, JP-5, used by the U.S. Navy, and JP-8 or Jet-A, used in commercial transport aircraft. The JP-4 has a lower average molecular weight than either JP-5 or JP-8 (Ott 1970), so both the spontaneous ignition temperature and the hot-surface ignition temperature of JP-4 are greater than those of either of the kerosene-base jet fuels (Gardner and Whyte 1971; Whyte and Gardner 1975).

Test Procedures -- Static Test

Klueg and Demaree (1969) reported that the hot-surface ignition of jet fuel occurred more readily when the fuel was released in a continuous stream rather than in a spray discharge and that delayed ignition of a spray discharge often produced violent explosions. In line with these observations, the initial experimental setup devised by

EXTINCTION OF FUEL-LEAK FIRES WITH DRY CHEMICALS

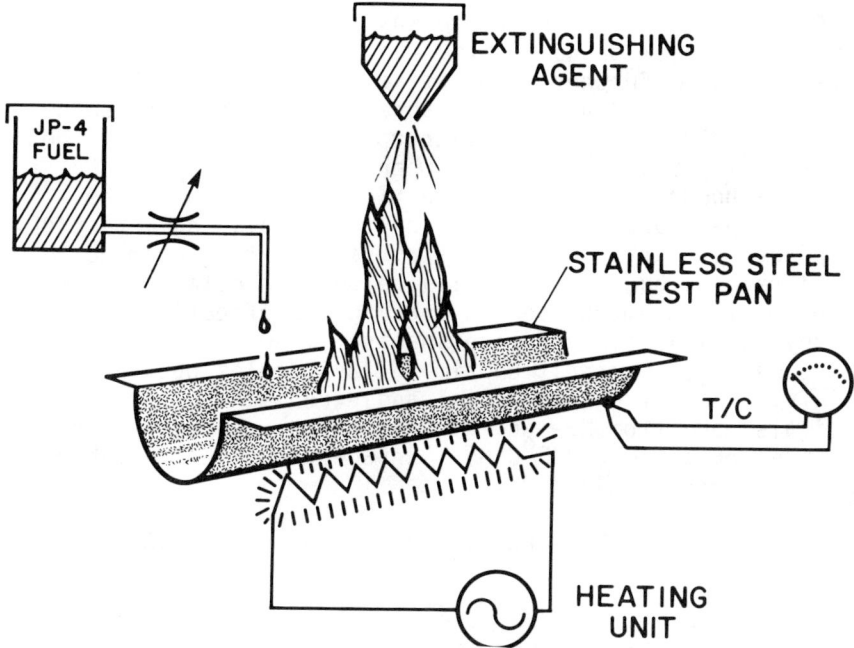

Fig. 1 Initial static nacelle fire test facility (schematic).

Altman et al. (1983) is shown schematically in Fig. 1. The jet fuel, JP-4, fed by gravity from an overhead container, leaked down a 0.1-cm-i.d. tube onto a heated semicylindrical stainless steel surface so shaped to fit into a Lindberg heating unit of the same configuration. The metal surface was heated to a temperature between 700 and 900°C by the nichrome heating element embedded in the Lindberg heating unit located directly underneath the curved metal surface. One of the two parameters used to rank the effectiveness of fire extinguishants was the hot-surface temperature, which was determined by a thermocouple welded to the semicylindrical surface at a point close to that where the fuel drop made initial contact with the hot metal surface.

To further describe the experimental procedure, suppose that the steady-state temperature of the hot surface was a nominal 700°C, as determined from the recorded emf output of the thermocouple. The dripping of the jet fuel was then started, and very soon after the first fuel drop hit the hot surface, the temperature of the surface dropped to, say, 650°C (because of fuel evaporative cooling). Shortly thereafter, the thermocouple temperature began to rise because the fuel drops had burst into flame, and then a given weight of dry chemical fire extinguishant was discharged onto the plate in the same area where the drops had first landed. If the flame was extinguished, the continuous stream of nonburning fuel drops striking the hot surface induced further evaporative cooling and the temperature dropped again. Because the fire extinguishing powder now blanketed the thermocouple-to-surface weld, the nominal temperature rose again when the drops burst into flame, although to a temperature higher than $700°C$; the higher temperature was a result of the insulating effect of the powder blanket. The time from first extinguishment to second reignition, the so-called reignition delay time, was the prime measurement used in ranking the effectiveness of the dry chemical fire extinguishants.

Some of the reignition time delay results obtained from an initial survey of commercial and experimental dry chemicals made from Na- and K-bicarbonate, carbonate, and chloride, etc., on a $775°C$ hot surface are given in Table 1. The rank order obtained,

$$K_2CO_3 > KHCO_3 > KCl > NaHCO_3 > Na_2CO_3 > NaCl$$

Table 1 Average time for reignition of jet fuel drip on $775°C$ hot surface

Dry chemical fire extinguishant powder	Sample size, g		
	2.5	5	7.5
	Time to reignite, s		
$KHCO_3$ (Ansul Purple K[a])	30	100	220
KCl (Pyrochem Super K[a])	3	17	70
$NaHCO_3$ (Ansul Dry Powder[a])	3	4	8
$NH_4H_2PO_4$ (Ansul Foray[a])	3	2	2
$KC_2N_2H_3O_3$ (ICI Monnex[a])	2	2	2
Na_2CO_3 (Ansul NaX[a])	1	1	1
$NaCl$ (Ansul Met-L-X[a])	0	0	2
$NaAl(OH)_2CO_3$ (Kaiser Dawsonite)	40	125	340
K_2CO_3 (Reagent Powder)	20	300	No Ignition
K_2CO_3 (Reagent Granular)	12	38	260
$KHCO_3$ (Reagent Powder)	40	60	200
$KHCO_3$ (Reagent Granular)	15	40	150
Al_2O_3 (Reagent Alumina)	2	2	15
H_2SiO_3 (Reagent Silicic Acid)	2	2	12

[a] These are commercial extinguishants.

is about the same as that reported by other investigators (Dolan 1957; Lafitte and Bouchet 1959; Lafitte et al. 1965; Friedrich 1960; Lee and Robertson 1960). Our experiments with two different particle size distributions of $KHCO_3$ and K_2CO_3, all made from reagent chemicals, demonstrated that reduction in particle size increased the reignition time delay just as it increased fire extinguishant weight effectiveness. In agreement with these earlier investigations, the data in Table 1 also show that given the effectiveness of any sodium salt, the analogous potassium salt always seemed to be still more effective. We were, however, surprised to observe the special effectiveness of K_2CO_3 and $NaAl(OH)_2CO_3$; the first being an experimental fire extinguishant prepared in our laboratories from the reagent chemical. As a result, we arranged for the Ansul Company to undertake the preparation of an improved experimental dry chemical fire extinguishant (Riley and Stauffer 1976) from K_2CO_3 with additives to improve the powder flow and to reduce hydration and also to develop in-house an alternate method of preparing Na- and $KAl(OH)_2CO_3$ by heating dry powder mixtures of $Al(OH)_3$ and Na- or $KHCO_3$ in a CO_2 atmosphere (Altman et al. 1982).

Van Tiggelen et al. (1963) and DeWitte et al. (1964) separate fire extinguishants into two classes: 1) those that interfere with the flame chemical reactions, and 2) those that cool the flame, that is, change the mechanism of flame propagation rather than merely reduce the overall rate of chemical reaction. Since, as mentioned in the description of our experimental procedure, the presence of fire extinguishing powder altered the hot-surface heat-transfer properties, we also tried alumina, Al_2O_3, and silicic acid, H_2SiO_3, as reignition delay "baseline" test materials; alumina because no chemical change would be expected, and silicic acid because it could decompose to yield only water and SiO_2. At least in amounts of 7 gm or less, these possible flame temperature reducers were equally effective to the poorest chemical flame reaction interference agents, $NaHCO_3$, Na_2CO_3, $NaCl$, $(NH_4)H_2PO_4$, or $KC_2N_2H_3O_3$ (ICI Monnex).

The initial fire test apparatus shown in Fig. 1 was then modified in order to better control the fuel-air ratio and make the reignition delay time determinations more reproducible. This apparatus, designed by Ling and Mayer, (Altman et al. 1983) is shown schematically in Fig. 2, and a description of some of these modifications follows.

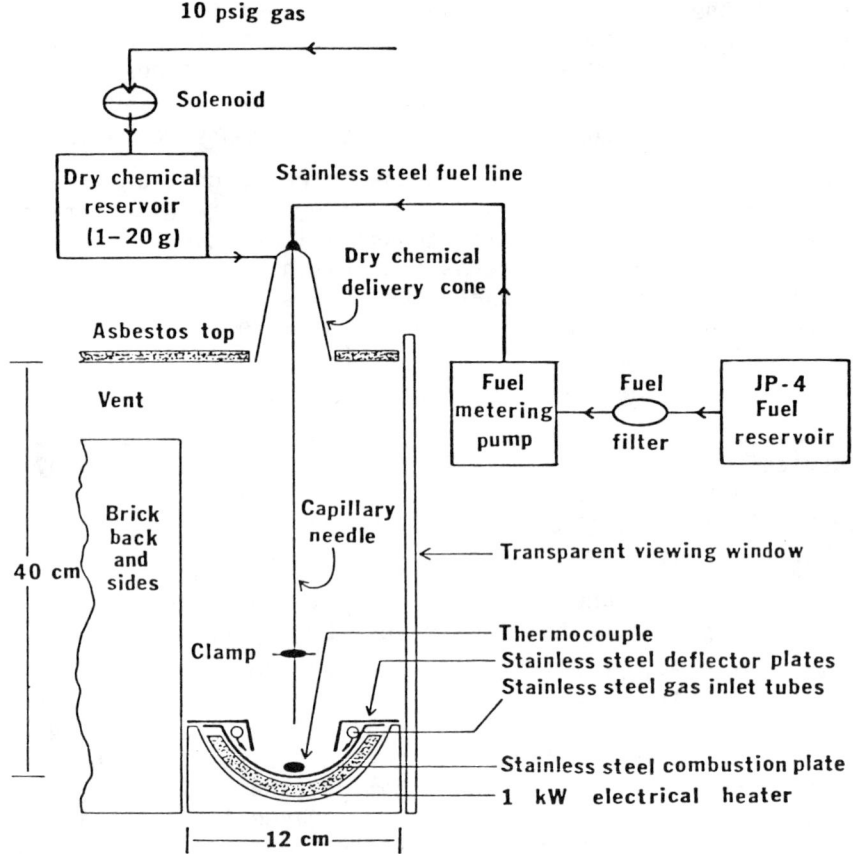

Fig. 2 Improved static nacelle fire test facility (schematic).

A metering pump replaced the gravity feed technique used for fuel delivery in the initial setup, and in order to more precisely direct the liquid flow to the thermocouple hot-plate contact point, a long hypodermic capillary needle of ∿ 0.01 cm i.d. led the output of the pump down to a point 5 cm above the hot surface. The jet fuel left the capillary at this point and fell to the hot surface, where the heat induced the formation of droplets which bounced around until they burst into flame. A gas inlet tube was also installed along both of the upper edges of the semicylindrical hot surface to increase the air replenishment rate, previously controlled only by downward diffusion of air through the vaporized fuel and gaseous combustion products. In each of the 0.6-cm-i.d. air inlet tubes were drilled 11 downward-

Table 2 Reignition delay time for jet fuel drip on hot-surface with 10 g of commercial and experimental dry chemicals

Dry chemicals	Delay time, s	
	700 C	900 C
K_2CO_3 (Ansul Prep)	150 ± 80	...
$KHCO_3$ (Ansul Purple-K)	69 ± 20	7 ± 4
$KC_2N_2H_3O_3$ (ICI Monnex)	55 ± 30	7 ± 2
KCl (Pyrochem Super-K)	33 ± 4	...
NaCl (Pyrochem BCD)	33 ± 15	5 ± 2
$(NH_4)H_2PO_4$ (Ansul Foray)	12 ± 2	...
$(NH_4)H_2PO_4$ (Pyrochem TUW-156)	8 ± 3	...
SnI_2 (68% I)	380 ± 80	2 ± 2
KI (76% I)	> 900	2 ± 2
NaI (85% I)	600 ± 60	3 ± 2
CI_4 (98% I)	None	...
$SnCl_4 \cdot 2H_2O$	26 ± 3	...
SnO	15 ± 5	...
$Na_2WO_4 \cdot 2H_2O$	17 ± 2	...
Na_2WO_4	8 ± 2	...

directed holes of 0.16 cm diam, spaced equidistantly over a 10 cm length, in order to make the airflow interact more directly with the fuel drops formed by the collision of the fuel stream with the heated surface.

A deflector plate was installed on top of the gas inlet tube in order to reduce upward air movement, and a metal gauze strip, omitted from Fig. 2, was attached directly to the lower edge of the deflector plate to prevent fuel drops formed by interaction of the fuel stream with the heated surface from ricochetting up toward the air inlet tube and bursting into flame under the deflector plate. A fuel flow rate of from 3 to 4 ml/min and an airflow rate of 2 liters (STP)/min was used in most of the testing done in this apparatus, and some of the results obtained are given in Table 2.

We were, of course, pleased to see that the Ansul K_2CO_3 preparation was the best of the lot of industry-produced dry chemicals. Because early experimental work (Muench and Klein 1949) had demonstrated that the most weight-effective Halon was one containing iodine, that is, CH_3I, we tried to develop a dry chemical iodide. Of such iodides tested -- SnI_2, KI, NaI, and CI_4 -- CI_4 was ineffective in delaying reignition even at $700°C$; all the other iodides turned out to be less effective at $900°C$ than the commercial dry chemicals. The other tin salts listed in Table 2 were tried to see whether the increased effectiveness of SnI_2 over CI_4 had something to do with the tin. Sodium tungstate, Na_2WO_4, with and without water, was tested because Lewis and Von Elbe (1961) cited some experimental data on the greater effectiveness of Na_2WO_4 over that of KCl as a surface coating on glass in removing H-free radicals. We have observed that the hydrate of Na_2WO_4 loses all of its water below $150°C$, and the dehydrated salt is reported to melt at $\sim700°C$ (Okada et al. 1978) and to vaporize by decomposition to gaseous Na_2WO_4, Na, and polymers of WO_3 in the 1100-1200°C range (Spitzin et al. 1975; Spoliti et al. 1980). From a mass spectrometric study of H_2-O_2 flames ($\sim1700°C$) to which tungsten and potassium halides had been added, Farber and Srivastava (1973) showed that a gas phase reaction like

$$K_2WO_4(g) + H(g) \rightarrow K(g) + KHWO_4(g)$$

was thermodynamically favorable under these temperature conditions. If such an analogous reaction were important in the Na_2WO_4-reignition delay process, then the effectiveness of the hydrate should be the same as that of the dehydrated salt. But the experimental data of Table 2 show that the hydrate is about twice as effective, indicating that the effect of water release from the hydrate in delaying hot-surface reignition is as great as the chemical effectiveness of Na_2WO_4 in reducing H-free radicals, if this does indeed occur with hot-surface temperatures below $1000°C$.

Table 3 shows some of the results obtained with other experimental dry chemicals. Obviously, Na- and $KAl(OH)_2CO_3$, sodium and potassium dawsonite, are superior to Na- and K-bicarbonate. Since either of the dawsonites can be considered to be an addition produce of boemite, AlOOH, and the appropriate alkali metal bicarbonate, the effectiveness of K-dawsonite, KD, might then be expected to be some mole-fraction weighted sum of the effectiveness of $KHCO_3$ and AlOOH. But the ignition delay data for $KHCO_3$ in Table 2 and for AlOOH and KD in Table 3 show that the effectiveness of

Table 3 Reignition delay time for jet fuel drip on a hot-surface with 10 g of experimental dry chemicals

Dry chemicals	Delay time, s	
	750°C	900°C
$NaAl(OH)_2CO_3$ [a]	296 ± 50	6 ± 3
$KAl(OH)_2CO_3$ [a], (KD)	153 ± 15	10 ± 4
$Al(OH)_3$	100 ± 30	3 ± 2
$AlOOH$ [a]	48 ± 35	None
Al_2O_3	28 ± 12	None
B_2O_3	5 ± 3	2 ± 1
H_3BO_3	10	...
$KD + B_2O_3$ (10%) [b]	62 ± 28	6 ± 2
K_3AlF_6	14 ± 5	...
$Al(OH)_3 + SnI_2$ (7% I) [b]	204 ± 20	8 ± 1
$Al(OH)_3 + KI$ (8% I) [b]	233 ± 56	...
$Al(OH)_3 \cdot KI$ (7% I) [c]	72 ± 3	8 ± 1
$AlOOH \cdot KI$ (7% I) [c]	131 ± 7	15 ± 4
$Al_2O_3 \cdot KI$ (7% I) [c]	> 900	50 ± 12
$KAl(OH)_2CO_3 + SnI_2$ (6% I) [b]	520 ± 52	51 ± 3
$KAl(OH)_2CO_3 \cdot SnI_2$ (6% I) [c]	419 ± 61	50 ± 2
$KAl(OH)_2CO_3 + KI$ (7% I) [b]	500 ± 90	13 ± 4
$KAl(OH)_2CO_3 \cdot KI$ (7% I) [c]	> 900	50 ± 14

[a] Prepared at arc from reagent chemicals. [b] Mechanical mixture. [c] Mechanical mixture preheated before test (weight percent iodine in mixture).

KD is clearly greater than the effectiveness obtained by averaging the effectiveness of an equimolar combination of $KHCO_3$ and $AlOOH$. Boron trioxide, B_2O_3, was tested because it melts at 450°C to form a glass; however, it seems to be no more effective than Al_2O_3, and a mixture of KD and B_2O_3 is even less effective than pure KD. We conclude this section by noting that the effectiveness of all mixtures of KI with either Al_2O_3 or KD seems superior to that of either aluminum-containing constituent, especially at the higher temperatures, but further discussion is deferred until the presentation of the effect of airflow rate on extinguishment effectiveness.

Dynamic Test

A schematic of the dynamic fire test facility designed by Myronuk and Fish of the Ames Research Center is shown in Fig. 3 (Altman et al. 1983). The test section downstream of the blower contains a stainless steel flame holding surface heated to 800-900°C. Ambient air flowed over the surface at rates from 6 to 36 m/s; and jet fuel was leaked onto the surface at a rate of 250 cm^3/min. As in the static testing, once a steady hot-surface temperature was obtained at a given airflow rate, the fuel leak was initiated and made to ignite on the hot surface within 1 s. After ignition, a specific mass of extinguishant was discharged within 1 s onto the heated surface, and, with the fuel leak uninterrupted, the time between initial extinguishment and reignition was recorded as the prime parameter of extinguishment effectiveness. The other variables were airflow rate and hot-surface temperature.

Because of their potential toxicity we could not obtain static reignition delay time data with Halon extinguishants. Therefore our first objective was to obtain dynamic results with these materials in order to establish a baseline; some of the results are given in Table 4. To explain the data tabulation, all four of the Halons tested extinguished the fuel-leak fire at an airflow rate of 6 m/s and did prevent its reignition for 2 s; however, 39 g of CH_2ClBr were required, but only 21 g of CF_2Br_2. In all these tests the fire reignited after 2 s because the Halon was being continuously diluted by the airflow when the Halon discharge ended after the first second. When the airflow was increased to 36 m/s, even more Halon extinguishant was required to keep the fire from reigniting for as long as 20 s: 40 g of either CF_3Br or CF_2ClBr and 60 g of either CF_2Br or CH_2ClBr. The reported upper limit of the reignition delay time is 20 s because when the fire was kept from reigniting for a longer time the hot-surface temperature declined significantly as a result of fuel evaporative cooling. From data such as these we rank these extinguishants as follows:

$$CF_2Br_2 \approx CF_2ClBr \geq CF_3Br \approx CH_2ClBr$$

CF_2Br_2 and CH_2ClBr are no longer under consideration as engine nacelle fire extinguishants because they have much lower safe exposure limits than CF_2ClBr and CF_3Br (Botteri

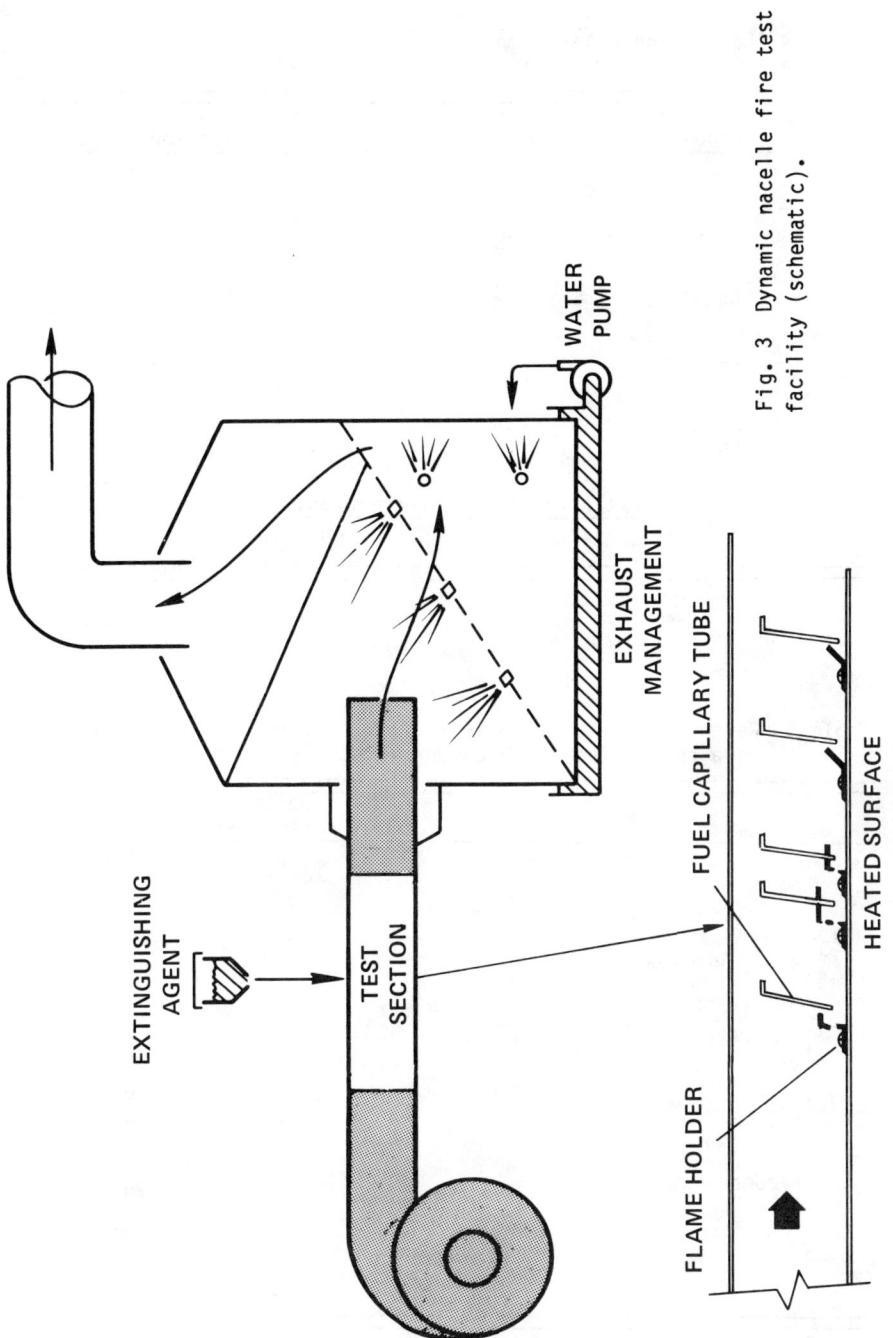

Fig. 3 Dynamic nacelle fire test facility (schematic).

Table 4 Reignition delay time vs airflow rate for jet fuel drip at 800 C hot-surface for various Halon extinguishants

Halon number	Chemical composition	g	Delay time, s at various airflows, m/s	
			6	36
1202	CF_2Br_2	21	2	...
		40	...	20
1211	CF_2ClBr	24	2	...
		40	...	20
1301	CF_3Br	35	2	...
		60	...	20
1011	CH_2ClBr	39	2	...
		60	...	20

Extinguishment effectiveness
$CF_2Br_2 \approx CF_2ClBr \geq CF_3Br \approx CH_2ClBr$

Table 5 Reignition delay time vs airflow rate for jet fuel drip on a hot-surface with commercial dry chemicals

Dry chemicals		g	Delay time, s at various airflows, m/s	
			6	36
$KHCO_3$[a]	(Ansul PKP)	30	20	2
		50	...	20
	(Ansul X)	20	20	< 1
		30	...	20
$NaHCO_3$[a] (Ansul + 50 C)		30	20	0
		50	...	< 1
$KC_2N_2H_3O_3$[a] (ICI Monnex)		10	2	0
		20	20	20
KCl[a] (Pyrochem Super-K)		30	< 1	0
		50	...	20
K_2CO_3[a] (Ansul Prep)		8	0	20
		20	2	20
		30	20	...

[a] 800°C, extinguishment effectivness, $K_2CO_3 > KC_2N_2H_3O_3 > KHCO_3 > NaHCO_3 > KCl$.

et al. 1972). However, all four of these halons were tested by us to see whether our method of ranking fire extinguishant effectiveness did give relative ratings similar to that obtained, for example, by Klueg and Demaree (1969) who found CF_3Br slightly better than CF_2Br_2 and both better than CF_2ClBr but all three of these better than CH_2ClBr. Much of the other earlier work has been reprinted by the National Fire Protection Association (1954) and is also discussed in Gann (1975).

A similar procedure was then carried out with commercial dry chemicals; these results are given in Table 5. As shown in Table 5, the extinguishment effectiveness of these chemicals was ranked as follows:

$$K_2CO_3 > KC_2N_2H_3O_3 > KHCO_3 > NaHCO_3 > KCl$$

Other workers report potassium carbonate, K_2CO_3, to be more than three times as weight effective as Na_2CO_3 or KCl, with KCl slightly more weight effective than Na_2CO_3 (DeWitte et al. 1964). Na_2CO_3 is reported to be from 2 to 3 times as effective as Li_2CO_3 or $NaHCO_3$ (Friedrich 1960), and $KHCO_3$ has been reported as effective as Na_2CO_3 but more effective than $NaHCO_3$ insofar as the reduction in burning velocity is a useful measure of fire extinguishment potential (Rosser et al. 1963). As for $KC_2N_2H_3O_3$, Underwriters' Laboratories (1974) rates it much more weight effective than $KHCO_3$, which is as effective as KCl in extinguishing heptane pan fires burning in near-stagnant air.

Since the best of the commercial lots reported in Table 5 was Ansul's K_2CO_3, Li- and Na-carbonate dry powders prepared from reagent chemicals were also tested as an exercise in varying the alkali metal element in the carbonate; the results are shown in Table 6. Tables 5 and 6 also show that while the effectiveness of either 30 g of pure KD and KCl is nil at an airflow rate of 36 m/s, a 2:1 mixture of KD and KCl by weight increased the weight effectiveness of KCl from three to five times. The results were similar for a 9:1 mixture of KD and KI when compared with the results for pure KI. Since both KCl and KI are volatile at these temperatures, the increase in KI and KCl effectiveness in these mixtures with KD could be due to the creation of a diffusion barrier for the gaseous alkali halide molecules by the KD or to the creation of some new chemical compound between the alkali metal halide and KD,

Table 6 Reignition delay time vs airflow rate for jet fuel drip on a hot-surface with experimental dry chemicals

Dry chemicals	g	Delay time, s at various airflows, m/s	
		6	36
Li_2CO_3 [a]	30-40	<1	0
Na_2CO_3 [a]	30-40	<1	0
$KAl(OH)_2CO_3$ [b] "KD"	30	20	0
KD + KCl (32%) [c]	10-20	2	20
KI [a]	40	<1	0
KD + KI (10%) [c]	20	3	20
Al_2O_3 + KI (9%) [c]	40	0	0
KD + B_2O_3 (10%) [c]	20	20	0

[a] 900°C. [b] 800°C. [c] 800°C, powder mixture.

thus reducing the volatility of both the alkali metal halide and decomposition products of KD.

To shed some light on these alternatives, a mixture of alumina and KI having approximately the same KI content as the mixtures of KI with KD was tested. Since no reignition delay resulted over the entire airflow range with even twice the total mass as the KD plus KI mixture, the diffusion barrier idea seems incorrect. Still another way of testing this idea is to try to increase the stickiness of KD, for if KD makes either the KCl or KI stick to the hot surface longer, thereby increasing the effectiveness, then increasing the stickiness of KD should increase its own effectiveness to something like its static effectiveness, shown in Table 3. For this test a mixture of KD with 10% B_2O_3 was prepared, but as shown in Table 6, it was no more effective than pure KD, even though some evidence of glass formation was apparent because the steel surface was very difficult to clean.

As for the possibility of some new chemical compound between KCl or KI and KD being produced by the hot surface when mechanical mixtures of alkali metal halides and KD are tested, an intimate mixture of KI and the starting materials for making KD were heated as if to make KD. Since no chemical reaction is expected between KI and either of the $KHCO_3$ or $Al(OH)_3$ materials for making KD, the preheated mixture should show similar fire extinguishing properties to those of the mechanical mixture of KI and KD given in Table 6. Various such mixtures of the precursors of KD with from

Table 7 Reignition delay time vs airflow rate for jet fuel drip on an 800°C hot-surface with experimental dry chemicals

Dry chemicals[a]	g	Delay time, s at various airflows, m/s	
		6	36
KD·KI (5%)	20	1	<1
	25	20	...
KD·KI (9%)	15	<1	20
KD·KI (18%)	20	<1	20
Al_2O_3·KI (9%)	40	0	0
KD·SnI_2 (5%)	15	20	20
KD·SnI_2 (10%)	15	20	20
KD·SnI_2 (20%)	10	20	20

[a]Mechanical mixture preheated before use.

5 to 18% KI were heated as if to make KD; their reignition delay time properties, which are tabulated in Table 7, do not seem to be significantly different from those of a mechanical mixture of KI and KD.

In a separate study, we have shown that heating SnI_2 with the precursors of KD, that is, $KHCO_3$ and $Al(OH)_3$, yields KI, AlOOH, and SnO_2 by the following gross reaction:

$$SnI_2 + 2KAl(OH)_2CO_3 \rightarrow SnO_2 + 2KI + 2AlOOH + H_2\uparrow + 2CO_2\uparrow$$

An infrared investigation of the solid state after it was heated demonstrated the presence of SnO_2, KI, and AlOOH, and the entire disappearance of SnI_2 and KD when the starting mole ratios of SnI_2 to KD precursors were 1:1 or 1:2. As expected, KD remained when the SnI_2 to KD precursors were 1:4. For all three samples of the KD SnI_2 end product reported in Table 7, the KD is in great enough excess to give a product containing SnO_2, AlOOH, KD, and KI. Further experimental work is necessary to determine why the fire extinguishment effectiveness of this material is better than that of the KD KI preparations.

Conclusion

Certain dry chemicals developed and tested in our laboratories seem to have greater weight effectiveness than the Halons in current use for controlling fuel-leak fires, particularly in the presence of high airflow rates. Ansul's

K_2CO_3 and $KHCO_3$, as well as ICI's Monnex, $KC_2N_2H_3O_3$, are better than CF_2ClBr and CF_3Br in delaying the hot-surface reignition of fuel-leak fires after initial extinguishment. Our experimental dry chemical formulations of potassium dawsonite, $KAl(OH)_2CO_3$, and KCl or KI seem to be even more weight effective than the commercial dry chemicals already mentioned, but the applicability and effectiveness of dry chemicals in controlling engine nacelle fires has yet to be demonstrated in test aircraft. However, we plan to participate further in such a test activity in the not too distant future and these results will be reported in another paper.

Acknowledgment

Most of the experimental work was done in the facilities of the Chemical Research Projects Office of Ames Research Center with the assistance of Professors A. C. Ling, L. A. Mayer, and D. J. Myronuk, and their students, all from San Jose State University, San Jose, Calif.

References

Altman, R. L., Ling, A. C., Mayer, L. A., and Myronuk, D.J. (1983) Development and testing of dry chemicals in advanced extinguishing systems for jet engine nacelle fires. Final Report 82-T-002 for Joint Technical Coordinating Group on Aircraft Survivability, Washington, D.C.

Altman, R. L. (1982) Synthesis of Dawsonites, U.S. Patent 4,356,157.

Botteri, B. P., Cretcher, R. E., and Kane, R. W. (1972) An appraisal of halogenated fire extinguishing agents. Aircraft Applications of Halogenated Fire Extinguishing Agents, National Academy of Sciences Symposium Proceedings, Washington, D.C., pp. 215-238.

Davis, R. A. (1971) Concorde power plant fire protection system. Aircr. Eng, 43, 26.

Dewitte, M., Vrebosch, J., and van Tiggelen, A. (1964) Inhibition and extinction of premixed flames by dust particles, Combust. Flame 8, 257.

Dolan, J. E. (1957) The suppression of methane/air ignitions by fine powders. Sixth Symposium (International) on Combustion, p. 787. Reinhold Publishing Corp., New York.

Dyer, J. H., Majoram, M. J., and Simmons, R. F. (1977a) The extinction of fires in aircraft jet engines - Part III, Extinction of fires at low airflows. Fire Technol. 13, 126.

Dyer, J. H., Majoram, M. J., and Simmons, R. F. (1977b) The extinction of fires in aircraft jet engines - Part IV, Extinction of fires by sprays of bromochlorodifluoromethane. Fire Technol. 13, 223.

Farber, M. and Srivastava, R. D. (1973) A mass spectrometric investigation of reactions involving tungsten and molybdenum with potassium-seeded H_2/O_2 flames. Combust. Flame 20, 33-42.

Fenwal, Inc., Ashland, Mass. (1964) Simulated flight test investigation of the effectivenss of a lightweight, aircraft, fixed, fire-extinguishing system. Contract N600(19)59572, Navy Bureau of Aeronautics.

Friedrich, M. (1960) Mode of action of dry fire extinguishing agents. U.S. Naval Research Laboratory Translation No. 804, Washington, D.C.

Gann, R. C. (ed.) (1975) Halogenated fire suppressants. American Chemical Society Symposium Series 16, Washington, D.C.

Gardner, L. (1971) Jet fuel specifications. AGARD Conference Proceedings No. 84 on Aircraft Fuels, Lubricants, and Fire Safety, Paper No. 1, Hague, Netherlands.

Hirst, R., Farenden, P. J., and Simmons, R. F.(1976) The extinction of fires in aircraft jet engines - Part I, Small-scale simulation of fires. Fire Technol 12, 266.

Hirst, R., Farenden, P. J., and Simmons, R. F.(1977) The extinction of fires in aircraft jet engines - Part II, Full-scale fire tests. Fire Technol. 13, 59.

Klueg, E. P. and Demaree, J. E. (1969) An investigation of in-flight fire protection with a turbo fan powerplant installation. NA-69-26, Federal Aviation Administration, Washington, D.C.

Lafitte, P. and Bouchet, R. (1959) Suppression of explosion waves in gaseous mixtures of means of fine powders. Seventh Symposium (International) on Combustion, p. 504. Butterworths Scientific Publications, London.

Lafitte, P., Delbourgo, C. R., Combourieu, J., and Dumont, J. C.(1965) The influence of particle diameter on the specificity of fine powders in the extinction of flames. Combust. Flame 9, 357.

Lee, T. G. and Robertson, A. F. (1960) Extinguishment effectiveness of some powdered materials on hydrocarbon fires. Fire Res. Abstr. Rev. 2, 13.

Lewis, B. and Von Elbe, G. (1961) Combustion, Flames and Explosions of Gases, 2nd ed., p. 27. Academic Press, New York.

Muench, N. P. and Klein, H. A. (1949) Fire protection of jet engine aircraft. MCREXE-664-466-K, Wright-Patterson Air Force Base, Ohio.

National Fire Protection Association Q48-8 (1954) The halogenated extinguishing agents. Boston, Mass.

Okada, K., Miyake, M., Iwai, S., Ohno, H., and Furukawa, K. (1978) Structural analysis of molten Na_2WO_4, J. Chem. Soc, Faraday Trans., 74, 1141.

Ott, E. E. (1970) Effects of fuel slosh and vibration on the flammability hazards of hydrocarbon turbine fuels within aircraft fuel tanks. AFAPL-TR-70-65, Wright-Patterson Air Force Base, Ohio.

Riley, J. and Stauffer, E. E. (1976) Ansul Project No. 33511-05102 -- potassium carbonate fire extinguishing agent.

Rosser, W. A. Jr., Inami, S.H., and Wise, H. (1963) The effect of metal salts on premixed hydrocarbon-air flames. Combust. Flame 7, 107.

Sommers, D. E. (1970) Fire protection tests in a small fuselage-mounted turbojet engine and nacelle installation. RD-70-57, Federal Aviation Administration, Washington, D.C.

Spitzin, V. I., Nifantyeva, R. M., Glazunov, M. P. and Drobasheva, T. I. (1975) Mass spectrometric study of the vaporization of sodium-tungsten bronzes of various composition. Dokl. Akad. Nauk (USSR) 224, 1356.

Spoliti, M., Piacente, V., Bencivenni, L., Ferro, D., and Cesaro, S. N. (1980) Infrared and mass spectrometric analysis of sodium molybdate and tungstate. High Temp. Sci., 12, 215.

Underwriter's Laboratories Fire Protection Equipment List (1974), Chicago, Illinois.

van Tiggelen, A., Vrebosch, J., Dewitte, M., De Geest, J., and Remmerie, P. (1963) Inhibition of flame reactions. RTD-TDR-63-4011, Wright-Patterson Air Force Base, Ohio.

Whyte, R. B. and Gardner, L. (1975) Wide-cut vs kerosene fuels, fire safety and other operational aspects. AGARD Conference Proceedings No. 166 on Aircraft Fire Safety, Paper No. 3, Rome, Italy.

Chapter V. Nonequilibrium Systems

Linearized Kinetic Models for Polyatomic Gases and Mixtures of Gases: Application to Vibrationally Relaxing Flows

R. Brun,* G. Duran,† P.C. Philippi,‡ M.F. Dourieu,† and R. Tosello†
Université de Provence, Marseille, France

Abstract

The application of the Gross-Jackson process allows the construction of kinetic models for the linearized collisional terms of the Boltzmann equation relative to the mixtures of polyatomic gases. One of these models is presented and adapted here, which correctly describes the energetic relaxation effects related to the internal modes. This model includes a reduced number of collision frequencies, and it is also shown here that these frequencies can be connected to phenomenological and specific relaxation times, so that computing nonequilibrium flows represents a relatively easy task. However the nonequilibrium must remain "weak" owing to the linearization. Two applications of the relaxation equations so obtained are performed. The first one concerns the steady nozzle flow of a binary mixture $CO-N_2$ with near-equilibrium conditions. The energetic departure from equilibrium is compared with the one deriving from the use of relaxation equations, and differences appear about the level of vibrational freezing. A disagreement is also evidenced in the second example, in which a computational analysis of the thermal boundary layer developing on a plane wall behind a reflected shock is done: It is shown that the present linearized method correctly predicts the return to equilibrium of the boundary layer,

---
Presented at the 8th ICOGER, Minsk, USSR, Aug. 23-25, 1981. Copyright © American Institute of Aeronautics and Astronautics, Inc., 1982. All rights reserved.
*Maitre de Recherche, Laboratoire de Dynamique et Thermophysique des Fluides.
†Graduate Student.
‡Assistant Professor, Universite de Santa Catarina, Bresil.

contrary to the usual method, which is invalid in
dissipative near-equilibrium flows.

Introduction

It is well known that situations in which rotational or vibrational relaxations may exist are not always correctly described by the usual relaxation equations, irrespective of the physical model used (Kogan et al. 1979). There are two principal, and also well-known, examples characteristic of this disagreement: First, the steady expansion flow in supersonic nozzles for which - at least for some gases - the vibrational freezing level is not the same in experimental and theoretical cases (Hurle et al. 1964; Teare et al. 1970; and Von Rosenberg et al. 1971). Second, the phase of coming back to equilibrium in unsteady dissipative flows (Brun et al. 1980). These examples will be examined hereafter, but it is obvious that relaxation models particularly adapted to near-equilibrium situations are needed. Thus linearized models for polyatomic gases and mixtures of gases have been constructed (Philippi and Brun 1981) that seem to give the correct behavior of the near-equilibrium flows encountered in many experiments.

One of the aims of the present paper is, first, to use one of these models which contains only a reduced number of collision frequencies and to show that these frequencies can be connected to phenomenological relaxation times, so that it is possible, without too long calculations, to compute nonequilibrium flows using experimental values for these relaxation times or, at least, values deduced from experiments and phenomenological models. The second purpose is the application of this model to the description of the vibrational relaxation in near-equilibrium situations.

Thus two examples are examined: 1) determination of the energetic vibrational nonequilibrium in an axisymmetric nozzle with a weakly expanding flow in the divergent section, and 2) calculation of the vibrational nonequilibrium in the unsteady boundary layer developing at an end wall behind a reflected shock wave.

In both examples, the present model seems to be more adapted than the usual relaxation equations for describing the near-equilibrium zones of these flows.

Analysis of the Model

The present model derives from the Gross-Jackson procedure (1959), which consists in modeling the collisional terms in the Boltzmann equation written for a mixture of polyatomic gases, the model being a fortiori valid for a pure gas. The Gross-Jackson-type models are linearized models available for problems where nonequilibrium effects are small, many details of the flow being described, however (Philippi and Brun 1981), and the models are more or less accurate since the method is based on a series expansion.

The Boltzmann equation giving the distribution function f_{ip} for the p component in a quantum state i representing both a rotational and a vibrational quantum number is written symbolically (binary mixture p, q):

$$\frac{df_{ip}}{dt} = J_{pp} + J_{pq} \qquad (1)$$

where J_{pp} and J_{pq} are respectively the direct and cross-collision terms.

If, classically, ϕ_{ip} is the nondimensional perturbation of the zeroth-order distribution function f_{ip}^o obtained with $J_{pp}^o + J_{pq}^o = 0$, we have

$$f_{ip} = f_{ip}^o (1 + \phi_{ip})$$

For our purpose, f_{ip}^o is chosen here as a Maxwell-Boltzmann distribution function defined with molecular density n_p, mean temperature T, and velocity $\underset{\sim}{u}$ of the mixture. Other choices are of course possible (Brun and Zappoli 1977).

After having decomposed ϕ_{ip} on orthogonal bases [Maxwell eigenfunctions (ME) and Wang-Chang and Uhlenbeck polynomials (WCU)], J_{pp} and J_{pq} are written:

$$J^{(N)}(\phi_{ip}) = - \gamma(N) \phi_{ip} + K^{(N)} \phi_{ip}$$

where N is the order of the expansion; $\gamma(N)$ is a positive constant; and $K^{(N)}$ is the projection of the operator on ME and WCU bases (order N) in spaces p and q.

The complete expressions of the models N=1, N=2, and N=3 have been written by Philippi and Brun (1981). Every model has an H theorem, but the accuracy increases with order N: For example, from hydrodynamic points of view, only the model N=3 gives the correct behavior (thermal diffusion, heat transfer, ...). However, for many problems, the model N=2 may be sufficient. This model satisfies the relaxation of moments of the first and second order, such as velocity, momentum, and temperatures, so it is adapted for the type of problem of interest here, that is, the relaxation of internal energy modes.

Two types of relaxation terms can be distinguished in the model N=2: first, those which are associated with the nonequilibria δn_p, $\delta \underline{u}(=\underline{u}_p-\underline{u})$, and δT_p^{tr} (translational temperature nonequilibrium of the p component), and second, δT_p^{in} (internal temperature nonequilibrium of the p component). These terms tend to give the mixture an equilibrium state characterized by n_p, $\underline{u}$, and T. The other relaxation terms tend to equalize all velocities and temperatures of the gases.

Thus we use the model N=2. It is not written here, but the principal point in which we are interested is that it possesses only five collision frequencies in the cross-term J_{pq} and only two in the direct term J_{pp}. These collision frequencies are associated with the different relaxation processes, and we must first examine the physical meaning of each of them for a subsequent connection with phenomenological relaxation times, particularly those which are associated with the exchanges of internal energy: These last collision frequencies are only three. Thus if $\Delta\varepsilon_p$ is the nondimensional change of internal energy for a molecule p in a binary collision, and $\Delta\varepsilon_{pp}$ (or $\Delta\varepsilon_{pq}$) the total balance of this energy in the collision pp (or pq), we can easily show the following results for the three types of collision frequencies.

1) The collision frequency noted $\langle\Delta\varepsilon_p\Delta\varepsilon_q\rangle_{pq}$ is associated with collisions pq in which there is a mutual exchange of internal energy. For the case of vibrational exchanges, the corresponding frequency $\langle\Delta\varepsilon_{vp}\Delta\varepsilon_{vq}\rangle_{pq}$ concerns the so-called VV collisions between the components p and q.

2) The collision frequency noted $\langle(\Delta\varepsilon_{pp})^2\rangle_{pp}$ is associated with collisions with interchange of translational and internal energies in collisions pp, so that for vibrational exchanges they concern the TV collisions between p molecules.**

**In fact, VV collisions for the p component are also included in these collisions, but their global energetic effect is negligible.

3) The collision frequency noted $\langle \Delta\varepsilon_p \Delta\varepsilon_{pq} \rangle_{pq}$ also concerns the collisions with interchange of translational and internal energies, but in collisions pq (TV collisions also).

Connection with Phenomenology

From a practical point of view, the previous frequencies are difficult to calculate, so it is interesting to connect them to phenomenological (finally experimental) relaxation times.

To do this, we adopt the following method: Starting with the Boltzmann equation (1) and multiplying by an internal energy of a molecule p, E_{ip} (vibrational or rotational), summing on the quantum levels, and integrating in the velocity space, one obtains a formal relaxation equation for the mean (macroscopic) internal energy:

$$n_p \frac{dE_p^{int}}{dt} + \frac{\partial \cdot q_p^{int}}{\partial \underset{\sim}{r}} = \sum_i E_{ip} \int (J_{pp} + J_{pq}) \, d\underset{\sim}{V} \quad (2)$$

where $n = \sum_p n_p$ and q_p^{int} are the fluxes fo the internal energy considered.

Now, for calculating the collisional term, we assume a Maxwell-Boltzmann distribution function defined with the two temperatures T_p^{tr} and T_p^{in} (the same for the component q). Neglecting the translational relaxation ($T_p^{tr} = T_q^{tr} = T^{tr}$) and assuming that T^{tr}, T_p^{in}, and T_q^{in} are not very different (Brun and Chickaoui 1979), simplifications can be easily done, and after tedious but not difficult calculations, we obtain

$$\frac{dE_p^{int}}{dt} + \frac{1}{M_p} \frac{\partial q_p^{int}}{\partial \underset{\sim}{r}} = (x_p \langle \Delta\varepsilon_{pp}^2 \rangle_{pp} + x_q \langle \Delta\varepsilon_{pq} \Delta\varepsilon_p \rangle_{pq}$$

$$- x_q \langle \Delta\varepsilon_p \Delta\varepsilon_q \rangle_{pq}) (\overline{E_p^{int}} - E_p^{int})$$

$$+ x_q \langle \Delta\varepsilon_p \Delta\varepsilon_q \rangle_{pq}) (\overline{E_q^{int}} - E_q^{int})$$

where $x_p = n_p/n$ and $\overline{E}^{int}$ are the equilibria of the internal energy defined at T^{tr}.

Thus the collision frequencies, previously defined, have a clear physical meaning (Herzfeld and Litovitz 1965) and obvious assimilations can be done. For example, in the case of vibrational relaxation, τ_{TV} and τ_{VV} being the respective phenomenological TV relaxation times, we have

$$<(\Delta\varepsilon_{vpp})^2>_{pp} = (1/\tau_{TV})_{pp}$$

$$<\Delta\varepsilon_{vpq}\Delta\varepsilon_{vp}>_{pq} = (1/\tau_{TV})_{pq}$$

$$<\Delta\varepsilon_{vp}\Delta\varepsilon_{vq}>_{pq} = (1/\tau_{VV})_{pq}$$

Thus the collision frequencies associated with the relaxation processes of the internal energy in our linearized model N=2 can be replaced by relaxation times for which the experimental values are generally known (Lambert 1977).

Then, taking into account the previous model, using the first-order Boltzmann equation (with N=2),

$$\frac{df_{ip}^{o}}{dt} = J^{(N)}(\phi_{ip})$$

and multiplying this equation by E_{ivp}, summing, and integrating, we obtain finally the linearized nonequilibrium for the mean vibrational energy of the p component (per molecule), that is,

$$\frac{E_{vp} - \bar{E}_{vp}(T)}{kT} = \frac{C_{vp}C_{TR}}{C_{TRV}^2} \frac{\partial \underline{u}}{\partial \underline{r}} \tau_p$$

$$\times (1 + x_q \frac{\tau_q - \tau_p}{(C_{vp}/k)(\tau_{VV})_{pq} + x_p(C_{vp}/C_{vq})\tau_q + x_q\tau_p}) \quad (3)$$

where $\bar{E}_{vp}(T)$ is the equilibrium vibrational energy defined at T; C_{vp}, C_{TR}, and C_{TRV} are respectively the vibrational specific heat of the p component, the translational and rotational specific heat of the mixture, and the translational rotational and vibrational specific heat of the mixture. Furthermore,

$$\tau_p^{-1} = x_p/(\tau_{TV})_{pp} + x_q/(\tau_{TV})_{pq}$$

So, we have found an expression, Eq. (3), which is now easily available for applications in specific flows. However, the validity of the method must be controlled in each case, since it is based on an expansion as a function of a small parameter τ/θ where τ is a characteristic relaxation time (in fact, the longest one of the flow under consideration) and θ is a characteristic flow time. In Eq. (3), which gives the nonequilibium, this ratio appears in the form $\tau_p(\partial.\underline{u}/\partial\underline{r})$, which gives the order of magnitude of the expression and which must remain much smaller than 1 everywhere.

Examples of Flow Calculations

From a practical point of view, in the present method, the flow was first calculated as if it were in equilibrium with the Navier-Stokes (or boundary-layer) equations written with mean quantities u, T, and n_p. Then the nonequilibrium of the internal energy is calculated with Eq. (3). A comparison is made with the energetic nonequilibrium obtained by the usual method derived from the use of relaxation equations coupled with Euler or Navier-Stokes equations.

Example 1

The first example deals with the steady supersonic expansion of the mixture $CO-N_2$ in a nozzle (vibrational relaxation). An axisymmetric nozzle is chosen for calculating the vibrational energy evolution. The convergent part is 5 cm long. The reservoir pressure p_R and temperature T_R are respectively 120 atm and 2500 K. The asymptotic semiangle of the divergent part is only 0.6 deg, so the present method can be considered as valid. First, the classical method using relaxation equations of the Landau-Teller type (harmonic oscillator model) and one-dimensional Euler equations is used; an unsteady computation method (MacCormack type) has been chosen, which avoids tedious iterations for determining the critical section; and the equations are written under a conservative form and solved with a finite-difference scheme of a predictor-corrector type which is an explicit second-order scheme (MacCormack and Baldwin 1975).

Second, the present method is applied; i.e., as previously stated, the calculation is made with equilibrium equations and then, the vibrational energy E_v for CO (and

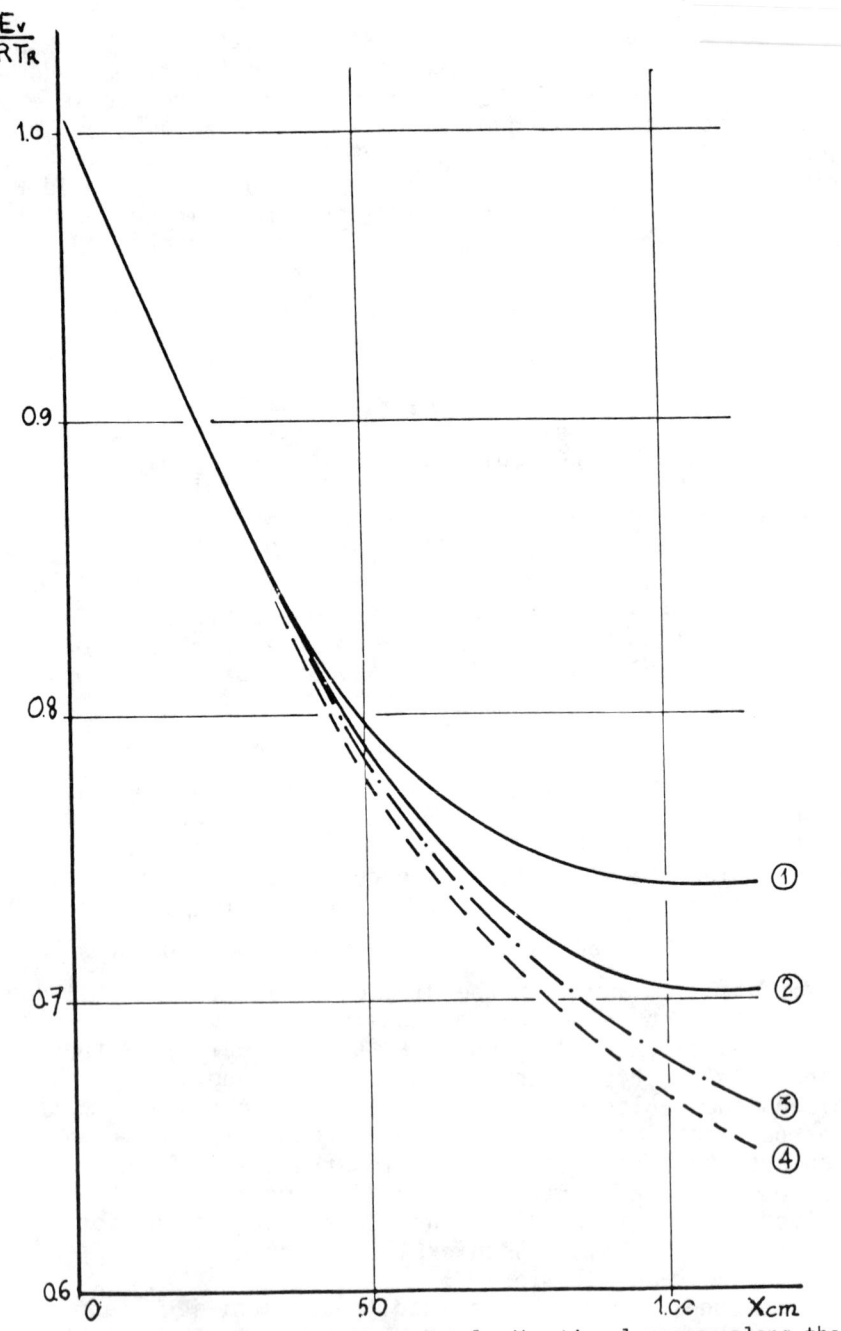

Fig. 1 Evolution of the nondimensional vibrational energy along the nozzle (TR: reservoir temperature) 1) Usual method, 2) present method, 3) Equilibrium energy defined at T (present method), 4) Equilibrium energy defined at T_{TR} (usual method) (CO, $x_p = x_q = 0.05$).

N_2) is calculated with Eq. (3), in which the values of relaxation times given by Lambert (1977) are used. The comparison between both methods is shown in Fig. 1, where the evolution of the vibrational energy along the nozzle is drawn for both cases. The principal aspect of this comparison lies in the level of freezing, which is different ($\sim 7\%$) in either method, the level given by the present method being the smallest one. This fact is qualitatively in agreement with experimental results, but no quantitative comparison has been done because no experimental results are available for nozzles with slowly divergent sections. On the other hand, the present calculation is not applicable to rapidly expanding flows, for which numerous experiments have been made. However, it is well known that experimental values for E_v are nearer to the equilibrium than those obtained from the usual method. The same behavior had been observed for pure gases (Brun et. al. 1979).

Example 2

The second example deals with the unsteady boundary layer at an end wall behind a reflected shock (vibrational

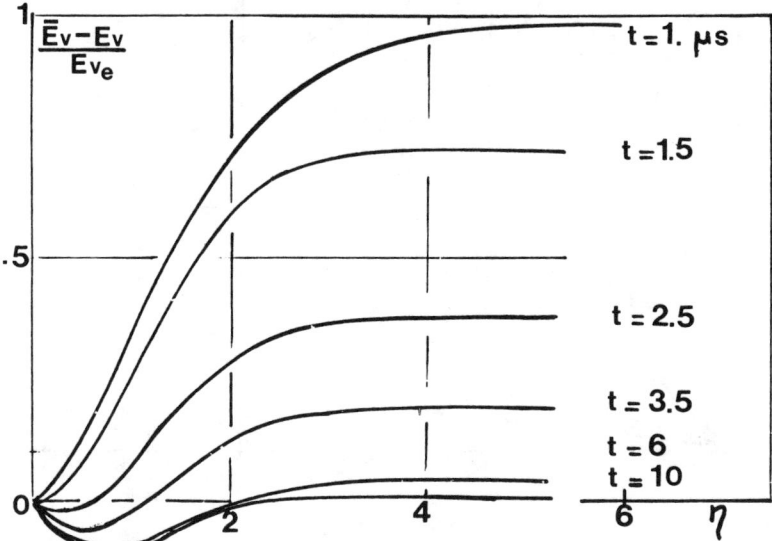

Fig. 2 Example of the time evolution of the nondimensional energetic departure from equilibrium. Gas: CO_2; incident shock Mach number: 3, 5; initial pressure: 4-mm Hg; usual method (η: nondimensional distance from the wall, E_{ve}: inviscid vibrational energy).

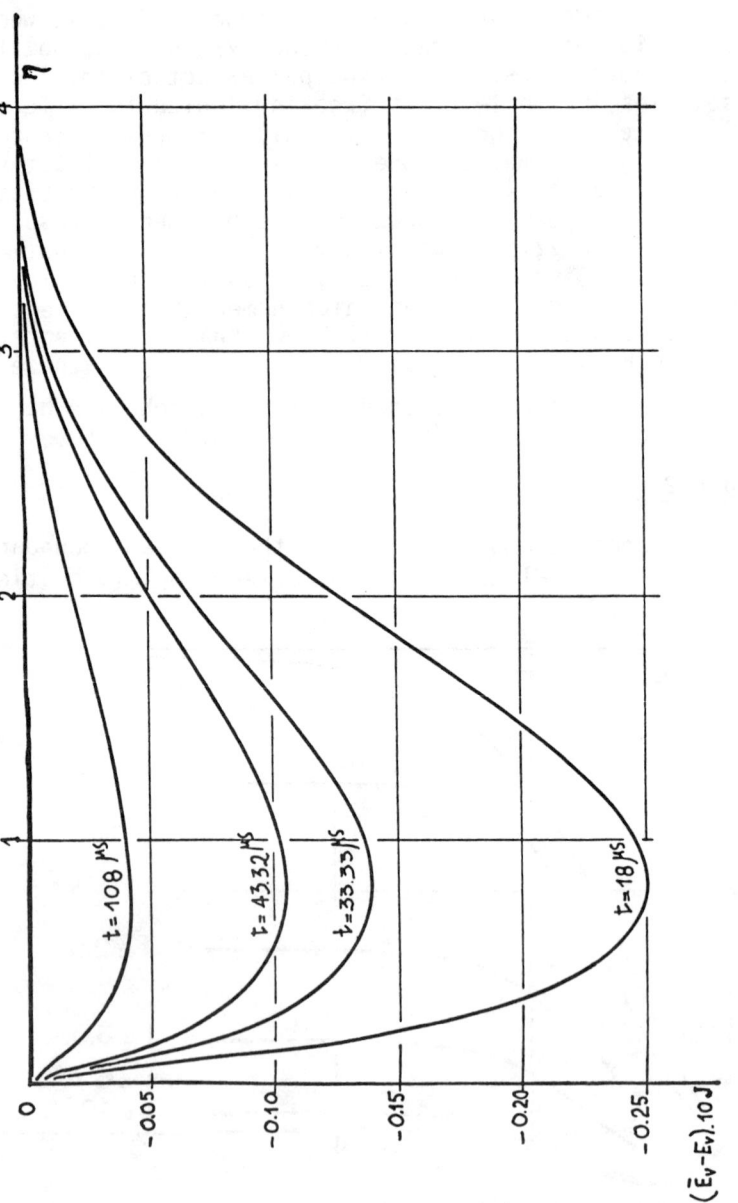

Fig. 3 Final phase of the return to equilibrium. Same conditions as in F (present method).

relaxation). The inviscid flows behind the incident shock and behind the reflected shock are assumed to be in vibrational nonequilibrium. These flowfields are calculated by a classical method of characteristics and the spatiotemporal evolution of the different quantities is determined, including the unsteady behavior of the reflected shock.

For the end-wall boundary layer the equations are first transformed by a method of Von Mises type, then the computation is performed with an implicit Crank-Nicholson finite-difference scheme. As in the previous example, the computation is first made with the usual nonequilibrium equations, including a relaxation equation (pure gas: CO_2 and harmonic oscillator model). An example of the results for the time evolution of the departure from equilibrium $\overline{E}_v - E_v$ is plotted as a function of a reduced distance from the wall η in Fig. 2. Taking into account the hypothesis of the equilibrium at the wall ($\eta=0$), the initial values of $\overline{E}_v - E_v$ just after the reflection of the shock increase in a monotonous way from zero to their inviscid values. With time, these values tend to zero (equilibrium), but a frozen zone ($\overline{E}_v - E_v < 0$) appears near the wall and, when $t \to \infty$, a steady profile for $\overline{E}_v - E_v$ is obtained. This zone results from the competition between the diffusion of vibrational energy from hot regions to the wall and the relaxation which cannot bring back the system to equilibrium during the same time scale owing to a long relaxation time next to the wall.

However, the asymptotic steady profile obtained by this usual method does not seem to be plausible, since, as time passes, the whole system must come back to equilibrium and a nonequilibrium zone must not persist. So the present method is applied for the near-equilibrium situation when the time is sufficient to ensure that the method is valid; and, in this case, the frozen zone also appears (Fig. 3) as in the strong nonequilibrium case, but when $t \to \infty$, this zone disappears and the whole flowfield tends to an equilibrium state, which seems physically correct.

Conclusion

For strong nonequilibrium flows, the usual methods (with relaxation equations) can and must be used, but they fail for near-equilibrium situations, at least for dissipative flows. Thus a universal method is needed to cover all situations. This method is now in progress (Kogan et al. 1978), but applications are still to be done.

References

Brun, R. and Chickaoui, A. (1979) First-order terms in the vibrational relaxation equations. Rarefied Gas Dynamics. Proceedings of the 11th International Symposium on Rarefied Gas Dynamics, (edited by R. Campargue), pp. 743-750.

Brun, R. and Zappoli, B. (1977) Model equations for a vibrationally relaxing gas. Phys. Fluids 20, 1441-1447.

Brun, R., Zappoli, B., and Zeitoun, D. (1979) Comparison of two computation methods for vibrational nonequilibrium flows. Phys. Fluids 22, 786-787.

Brun, R., Zeitoun, D., and Dourieu, M. F. (1980) Vibrationally relaxing boundary layers. Euromech Colloquium 126: Aerodynamic Problems Related to Hypersonic Flight, Berlin.

Gross, E. P. and Jackson, E. A. (1959) Kinetic models and the linearized Boltzmann equation. Phys. Fluids 2, 432-441.

Herzfeld, K. F. and Litovitz, T. A. (1965) Absorption and Dispersion of Ultrasonic Waves (2nd ed.). Academic Press, New York.

Hurle, I. R., Russo, A. L., and Hall, J. G. (1964) Spectroscopic studies of vibrational nonequilibrium in supersonic nozzle flow. J. Chem. Phys. 40, 2076-2089.

Kogan, M. N., Galkin, V. S., and Makashev, N. K. (1979) Generalized Chapman-Enskog method: derivation of the nonequilibrium gasdynamics equations. Rarefied Gas Dynamics. Proceedings of the 11th International Symposium on Rarefied Gas Dynamics. (edited by R. Campargue), pp. 693-734.

Lambert, J. D. (1977) Vibrational and Rotational Relaxation in Gases. Clarendon Press, Oxford.

MacCormack, R. W. and Baldwin, B. S. (1975) A numerical method for solving the Navier-Stokes equations. Paper presented at the AIAA 13th Aerospace Sciences Meeting, Pasadena, Calif.

Philippi, P. C. and Brun, R. (1981a) Kinetic modeling of polyatomic gas mixtures. Physica 105 A, 147-168.

Philippi, P. C. and Brun, R. (1981b) Kinetic models and relaxation effects in the Knudsen layer of gas mixtures. Rarefied Gas Dynamics: AIAA Progress in Astronautics and Aeronautics (edited by S. S. Fisher), Vol. 74, pp. 428-443. AIAA, New York.

Teare, J. D., Taylor, R. L. and Von Rosenberg, C. W. Jr. (1970) Observations of vibration-vibration energy pumping between diatomic molecules, Nature 225, 240-243.

Von Rosenberg, C. W. Jr., Taylor, R. L., and Teare, J. D. (1971) Vibrational relaxation of CO in nonequilibrium nozzle flow. J. Chem. Phys. 54, 1974-1977.

Theoretical Model for Sound Output from a Pulsating Arc Discharge

N.I. Kidin* and V.B. Librovich†
Academy of Sciences, Moscow, USSR
and
M.L. Vuillermoz‡ and J.P. Roberts§
Polytechnic of the South Bank, London, England

Abstract

A theoretical model, based on hydrodynamic relationships, for sound emission from a pulsating arc discharge in a combustion zone is described. A first-order perturbation solution is given which includes a relaxation time constant to account for chemical and thermal radiation effects. Comparison of predictions from this model with previously published experimental results shows that this relaxation time must be less than 1.5 ms, indicating that chemical and thermal radiation effects play an insignificant role in this sound generating process for frequencies above 100 Hz.

Nomenclature

A = activation energy, J/kg
C_v = specific heat at constant volume, J/kgK
C_p = specific heat at constant pressure, J/kgK
c = local velocity of sound, m/s
F = electrical source term defined in text, W/m^3

Presented at the 8th ICOGER, Minsk, USSR, Aug. 23-26, 1981. Copyright © American Institute of Aeronautics and Astronautics, Inc.,1983. All rights reserved.
*Senior Research Worker, Institute for Problems in mechanics.
†Deputy Director, Institute for Problems in Mechanics.
‡Senior Lecturer.
§Principal Lecturer.

k_o = rate constant, $m^{-3}K^{-n}$
m^o = radial mass flow rate per unit area, kg/m^2s
P_o = ambient pressure, Pa
p^o = acoustic pressure, Pa
R = gas constant per unit mass, J/kgK
S = surface area of source, m^2
t = time, s
T = temperature, K
V = volume of source, m^3
v = particle velocity, m/s
W_a = sound power output when $\omega\tau \gg 1$, W
W_b = sound power output when $\omega\tau \ll 1$, W
γ = principal specific heat ratio
ΔU = heat of reaction, J
ρ = local mean density, kg/m^3
σ = Stephan constant, $Jm^{-2}K^{-4}$
τ = relaxation time defined in text, s
ϕ = power input per unit volume, W/m^3
ψ = input electrical power, W

Subscripts and Superscripts

1 = time-independent quantities
' = time-dependent perturbations

Introduction

A pulsating arc discharge placed in the flame region of a working combustion chamber provides a convenient means of exciting the various acoustic modes of the system. With an externally controlled arc and internal sound pressure monitor, the results of interactions between the acoustic field in the chamber and the combustion process can be observed directly.

An experimental investigation of the sound emission from a modulated discharge within a combustion zone has been carried out and the results previously reported (Medvedev et al. 1982). The present paper is concerned with the development of a theoretical model which attempts to account for the experimental observations.

The experimental arrangement described above consists of two adjustable stainless steel electrodes placed one above the other in the luminous region of a simple Meker burner. The energizing potential across the electrodes was a fixed dc component with a superimposed audio-frequency signal of variable amplitude and frequency. Various fuel gases were used: methane, butane, and pentane, as well as a range of primary air/fuel gas ratios.

The sound output from the discharge was found to be a simple function of the electrical input and to be largely independent of the nature of the flame and the electrode separation over the frequency range investigated. The following model provides an explanation for these findings.

Theoretical Model

The arc discharge is assumed to heat a volume of gas surrounding it; we can thus treat this acoustic source as a homogeneous chemical reactor with electrical discharge and thermal radiation. The corresponding equations are

$$\frac{\partial \rho}{\partial t} + \mathrm{div}(\rho \vec{v}) = 0 \tag{1}$$

$$\rho \frac{\partial \vec{v}}{\partial t} + \rho(\vec{v} \nabla)\vec{v} = -\mathrm{grad}\, P_0 \tag{2}$$

$$\frac{\partial}{\partial t}\left(\rho \frac{v^2}{2} + \rho C_v T\right) = -\mathrm{div}\left[\rho \vec{v}\left(\frac{v^2}{2} + \frac{P_0}{\rho} + C_v T\right)\right] + \phi \tag{3}$$

Gas velocities are considerably less than the local velocity of sound so these equations can be treated as though the body of the gas were at rest. With the help of the Gauss-Ostrogradsky theorem these equations can be integrated over the surface of the source to give equations in terms of spatial averages:

$$\frac{d\overline{\rho}}{dt} = -(\rho v)_r \frac{S}{V} \tag{4}$$

$$\frac{d}{dt}(\overline{\rho v}) = -(\rho v^2 + P_0)_r \frac{S}{V} \tag{5}$$

$$\frac{d}{dt}\overline{\frac{\rho v^2}{2}} + \overline{\left(\rho C_v T\right)} = -\left[\rho v \left(\frac{v^2}{2} + C_p T\right)\right]_r \frac{S}{V} + \overline{\phi} \tag{6}$$

where the overbar sign denotes average values over the volume of the source, for example,

$$\bar{\rho} = (1/V) \int_{vol} \rho \, dV$$

and the subscript r denotes the value of a parameter at a distance r from the center of the sphere.

For convenience, we introduce a new source variable

$$\vec{m} = (\rho \vec{v})$$

and rewrite the equations as

$$\frac{d\bar{\rho}}{dt} = -m_r \frac{S}{V} \tag{7}$$

$$\frac{d\bar{m}}{dt} = -\left(\frac{m^2}{\rho} + \rho RT\right)_r \frac{S}{V} \tag{8}$$

$$\frac{d}{dt}\left(\overline{\frac{m^2}{2\rho}} + \overline{\rho C_v T}\right) = -\left(\frac{m^3}{2\rho^2} + mC_p T\right)_r \frac{S}{V} + \bar{\phi} \tag{9}$$

If the equations are linearized using first-order perturbation methods (e.g., $\bar{\rho} = \rho_1 + \rho'$), we obtain

$$\frac{d\rho'}{dt} = -m' \frac{S}{V} \tag{10}$$

$$\frac{dm'}{dt} = -R(\rho_1 T' + \rho' T_1)_r \frac{S}{V} \tag{11}$$

$$\rho_1 C_v \frac{dT'}{dt} + C_v T_1 \frac{d\rho'}{dt} = -(C_p T_1 m')_r \frac{S}{V} + \phi' \tag{12}$$

The variables ρ' and m' can be eliminated from these equations by assuming the source to be small compared to the wavelength, so that the pressure can be treated as spatially uniform throughout its volume. Thus

$$\frac{d\rho'}{dt} = -m' \frac{S}{V} \tag{13}$$

$$\frac{dm'}{dt} = -R(\rho_1 T' + \rho' T_1) \frac{S}{V} \tag{14}$$

$$\rho_1 C_v \frac{dT'}{dt} = -RT_1 m' \frac{S}{V} + \phi' \tag{15}$$

From Eqs. (13) and (14),

$$\frac{d^2 m'}{dt^2} = -\frac{RS}{V}\left(\rho_1 \frac{dT'}{dt} - \frac{ST_1 m'}{V}\right) \tag{16}$$

and from Eq. (15)

$$\frac{d^2 m'}{dt^2} = \frac{V}{SRT_1}\left(\frac{d^2\phi'}{dt^2} - \rho_1 C_v \frac{d^3 T'}{dt^3}\right) \tag{17}$$

By equating these expressions and making use of $C_p - C_v = R$ and $\gamma RT = c^2$, we obtain

$$\left(\frac{V}{Sc}\right)^2 \frac{d^3 T'}{dt^3} - \frac{dT'}{dt} = \frac{1}{\rho_1 C_v}\left[\left(\frac{V}{Sc}\right)^2 \frac{d^2\phi'}{dt^2} - \frac{\phi'}{\gamma}\right] \tag{18}$$

This equation is solvable, but can be simplified further by noting that $(V/Sc)^2 < 10^{-10}$ s^2 (physically this is the assumption that the source is small compared to the wavelength) and hence for the frequency range of interest both $d^3 T'/dt^3$ and $d^2 \phi'/dt^2$ can be neglected. We thus obtain

$$\frac{dT'}{dt} = \frac{1}{\rho_1 C_p} \phi' \tag{19}$$

and hence, from Eq. (15),

$$m' = \frac{V}{RT_1 S} \phi' \frac{\gamma - 1}{\gamma} \tag{20}$$

Source Term

The net power per unit volume is assumed to be the sum of the electrical and chemical inputs and thermal radiation losses:

$$\phi = W_{el} + W_{ch} + W_{rad} \quad (21)$$

where W_{el} is the instantaneous electrical power per unit volume;

$$W_{ch} = -\Delta U k_o T^n \exp(-A/RT) \quad (22)$$

is the chemical energy release rate per unit volume (ΔU is assumed negative for exothermic reactions); and

$$W_{rad} = -\sigma (T^4 - T_o^4) S/V \quad (23)$$

is the rate of radiant energy loss per unit volume.

From Eq. (4) and assuming as before small time-dependent perturbations in temperature, we obtain

$$\rho_1 C_p \frac{dT'}{dt} = W'_{el} + W_{1_{ch}} \left(n + \frac{A}{RT_1}\right) \frac{T'}{T_1} - 4 \frac{S}{V} \sigma T_1^3 T' \quad (24)$$

Expressing this in terms of a relaxation time, τ, we have

$$\frac{dT'}{dt} = -\frac{T'}{\tau} + \frac{W'_{el}}{\rho_1 C_p} \quad (25)$$

where

$$\frac{1}{\tau} = \left[-W_{ch}\left(n + \frac{A}{RT_1}\right)\frac{1}{T_1} + \frac{4\sigma T_1^3 S}{V}\right] \frac{1}{\rho_1 C_p} \quad (26)$$

If the time-dependent electrical input per unit volume is sinusoidal of the form $F\cos(\omega t)$ the total time-dependent source function can then be expressed as

$$\phi' = -\rho_1 C_p \frac{T'}{\tau} + F\cos(\omega t) \quad (27)$$

The steady-state solution of Eqs. (6) and (7) is

$$T' = \frac{F\tau}{\rho_1 C_p} \frac{\cos(\omega t) + \omega\tau \sin(\omega t)}{1 + \omega^2 \tau^2} \qquad (28)$$

Output Sound Power

Sound emission from a pulsating source can be given in terms of the source strength alone (the product of mean velocity amplitude at the surface and the surface area) if the source is small compared to the wavelength. For the problem in hand this quantity may be obtained from Eqs. (15), (20), and (28). For simplicity, and in order to examine the essential dependence of acoustic emission on electrical input as against chemical and thermal radiation effects, expressions for the sound power radiated are obtained for the two extreme conditions $\omega\tau \gg 1$ and $\omega\tau \ll 1$, the former ignoring all effects other than the electrical power input, the latter condition where chemical and thermal effects dominate.

For $\omega\tau_r \gg 1$ from Eq. (28),

$$T' = \frac{F\tau_r}{\rho_1 C_p} \frac{\sin(\omega t)}{\omega \tau_r} = \frac{F \sin(\omega t)}{\omega \rho_1 C_p} \qquad (29)$$

and

$$\phi' = \rho_1 C_p \frac{dT'}{dt} = F\cos(\omega t) \qquad (30)$$

and from Eq. (20),

$$m' = \rho_1 v' = \frac{V}{RT_1 S} \frac{\gamma - 1}{\gamma} F\cos(\omega t) \qquad (31)$$

The acoustic pressure at a distance r from a small source is given by (see, for example, Morse and Ingard 1968)

$$p = \frac{\rho_1}{4\pi r} \frac{d}{dt} (Sv') \qquad (32)$$

Letting $\psi' = FV$, the total time-dependent electrical power amplitude, then

$$p = -\frac{\rho_1}{4\pi r} \frac{1}{P_0} \frac{\gamma-1}{\gamma} \psi' \omega \sin(\omega t) \qquad (33)$$

The corresponding total sound power output will be

$$W_a = 4\pi r^2 \frac{p^2}{\rho_1 c} = \frac{4\pi r^2}{\rho_1 c} \left[\frac{\rho_1 \psi'}{4\pi r P_0} \frac{\gamma-1}{\gamma} \omega \sin(\omega t) \right]^2 \qquad (34)$$

or

$$\overline{W}_a = \frac{1}{2} \frac{\rho_1 \omega^2}{4\pi c P_0^2} \left(\frac{\gamma-1}{\gamma}\right)^2 \qquad (35)$$

Similarly it may be shown that for $\omega\tau \ll 1$,

$$\overline{W}_b = \frac{1}{2} \frac{\rho_1 \omega^4 \tau^2 \psi'^2}{4\pi c P_0^2} \left(\frac{\gamma-1}{\gamma}\right)^2 \qquad (36)$$

We thus see that sound output power is a function of the frequency of the input electrical signal, both directly through ω and indirectly through the relaxation time τ.

Comparison with Experiment

Sound pressure level measurements reported by Medvedev et al. (1982) show unmistakably that under a variety of conditions the sound output power is proportional to the square of the input electrical source term, and for constant input electrical power it is proportional to the square of the frequency of the applied signal in the frequency range from 100 Hz to 10 kHz. These results indicate that Eq. (10) above fully describes the sound generating process over the frequency range investigated, and that the combined thermal and chemical relaxation time must be greater than about 1.5 ms. The dominant sound producing mechanism is therefore Joulean heating above at least 100 Hz.

For additional confirmation of the theory, a set of measurements was taken with a square wave electrical input signal and a comparison made between the resulting sound output pressure and the first differential of the input power.

From Eq. (32) the acoustic pressure is seen to be proportional to the first differential of the velocity v'. As $v' = m'/\rho_1$ from Eq. (20), it therefore follows that the acoustical pressure should be

$$p \propto \frac{d}{dt}\phi' \propto \frac{d}{dt}\psi' \qquad (37)$$

Positive and negative spikes were indeed observed corresponding to the leading and trailing edges of the square wave input, with the height of the pressure spikes proportional to the first differential of the input power, (Fig. 1).

Comparison with Other Models

Other models have been proposed for the sound emission from a pulsating electrical discharge.

Sodha et al. (1978) and Burchard (1967) postulate Joulean heating as the sound producing mechanism, while McQueen (1981) assumes constant current density though the arc and hence variations in the volume with varying current.

Sodha et al. (1978) derived the acoustical pressure generated solely from consideration of the pressure disturbances within the flame rather than source strength. Such an approach predicts that the sound output should be inversely proportional to the frequency of the applied signal, which is difficult to reconcile with the experimental observations reported.

Burchard (1967) reports a spectrum with a slope of nearly 6 dB/octave, but his accompanying analysis predicts a slope of only 3 dB/octave. The difference arises from the assumptions made in solving the wave equation; namely that the discharge behaves as a cylindrical rather than a spherical source over the frequency range investigated.

In the McQueen (1981) model it follows that the radius of the arc discharge is proportional to the square root of the absolute value of the current through it. In this case the acoustical pressure from the arc should be proportional to the second derivative of the electrical input. The absence of such pressure waves in our measurements (we observe only the first derivative) would

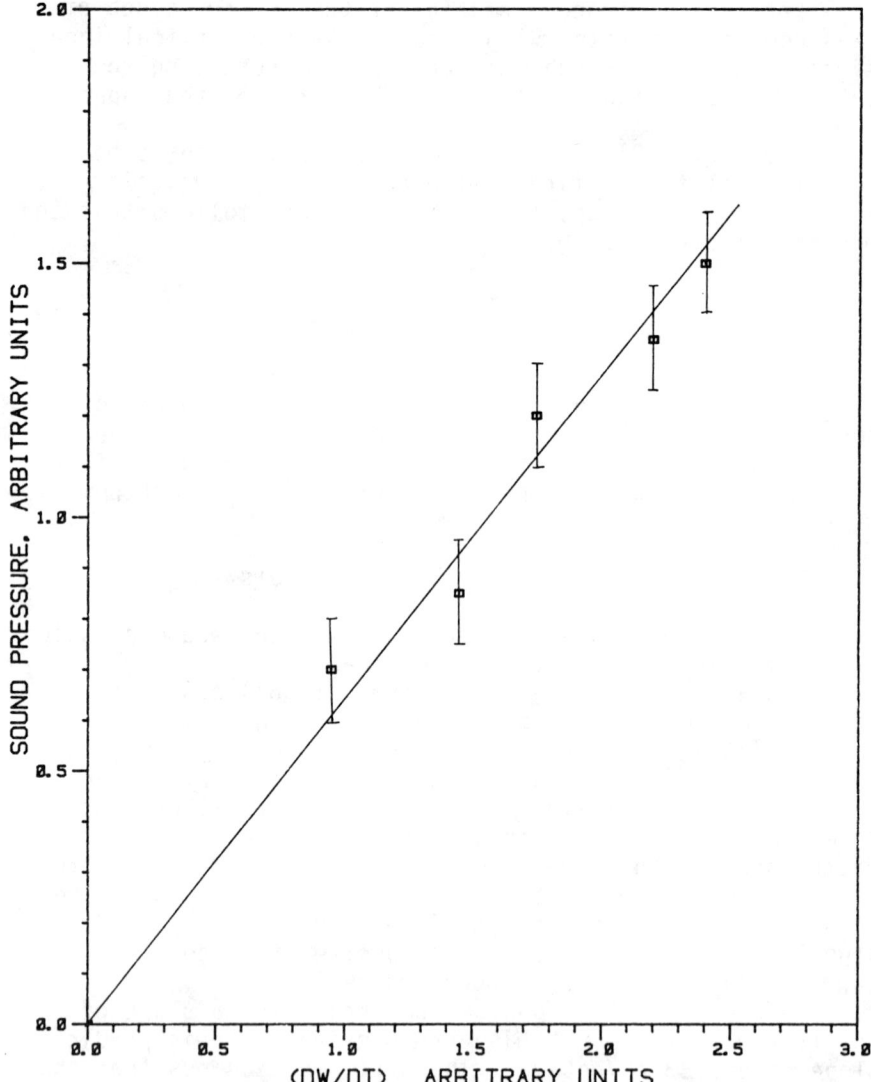

Fig. 1 Sound pressure against first differential of electrical power for a square wave input.

indicate that this mechanism is of little significance in the arrangement under study.

Discussion

An analysis of previous experimental results in the context of the simple model proposed has shown that a

modulated arc discharge can provide a useful sound source
for investigating the acoustic characteristics of a working
combustion system. The strength of such a source can be
obtained from known constants and the input electrical power
only, and, under the conditions of interest, chemical,
thermal radiation, and internal plasma effects play no
Further work to investigate the characteristics of the arc
source operating in a combustion system is now being
undertaken.

Acknowledgments

The project of which this work forms a part is
partially funded by the Science Research Council. We also
wish to thank the Royal Society and the British Council for
their support.

References

Burchard, J. (1967) Preliminary investigation of the
electrothermal loud speaker. Combust. Flame 13, 82-86.

McQueen, D. (1981) The electric arc furnace as a noise source.
Noise Control Eng. 15, 89-95.

Medvedev, N., Vuillermoz, M., and Roberts, J. (1982) An
experimental study of sound generation by a modulated arc
discharge within a flame. Combust. Flame (in press).

Morse, P. M. and Ingard, K. U. (1968) Theoretical Acoustics, p.
310. McGraw-Hill, New York.

Sodha, M. S., Tripathi, V. K., and Sharma, J. K. (1978) Flame
loudspeakers. Acustica 40, 68-69.

Chapter VI. Lasers

Gain Coefficient and Vibrational Temperature Measurements in Shock Tube Driven CO_2-GDL, $CO_2 + N_2 + He$, and $CO_2 + CO(N_2) + H_2$ Mixtures

S.S. Novikov,* V.M. Doroschenko,† and N.N. Kudryavtsev†
Academy of Sciences, Moscow, USSR

Abstract

A method for determining CO_2 upper and lower laser level temperatures (populations) and hence obtaining comprehensive information on population inversion generation in CO_2 lasers has been developed. The method is based on the simultaneous measurement of intensities in corresponding vibrational rotational bands of CO_2-laser active medium emitting molecules. The measurements of radiation intensities in 4.3- and 2.7-μm CO_2 bands were performed in CO_2 gasdynamic laser (GDL) on CO_2 and N_2 + He mixtures. The vibrational temperatures of 00^01 and 10^00 vibrational levels of carbon dioxide, and small-signal gain coefficient on 00^01 10^00 laser transition K_0 were calculated. These results revealed a faster relaxation of the CO_2 upper laser level in comparison with the theoretical calculation carried out in this work. The method based on the spontaneous radiation intensity measurements in a 4.3-μm band of CO_2 and 4.7-μm band of CO and the gain coefficient K_0 determined by the probing-laser technique were used for performance optimization of a CO_2 GDL on $CO_2 + N_2 + H_2$ mixture with a high hydrogen content (50%). At stagnation temperature 1100-1200 K and pressure 1.5 + 20 atm, the maximum gain

---

Presented at the 8th ICOGER, Minsk, USSR, Aug. 23-26, 1981.
Copyright © 1983 by S. S. Novikov, V. M. Doroshenko, and N. N. Kudryavtsev. Published by the American Institute of Aeronautics and Astronautics with permission.
*Professor, Moscow Institute of Chemical Physics.
†Senior Scientist, Moscow Institute of Chemical Physics.

coefficient of about 0.5 m^{-1} sufficient for generation was attained.

Introduction

Comprehensive information on population inversion generation in molecular lasers can be obtained by the measurements of vibrational temperatures of upper and lower laser levels (Kudryavtsev et al. 1979a). The gain coefficient on the 00^01-10^00 transition may be determined from vibrational temperatures of the asymmetric and combined (symmetric and bending) CO_2 modes. Kudryavtsev et al. (1979b) have developed a method for deducing these temperatures from spontaneous radiation intensity measurements in the 4.3-μm band. In this work the vibrational temperatures of 00^01 and 10^00 CO_2 laser levels and the gain coefficient on 00^01-10^00 transition in CO_2 gasdynamic lasers (GDL) were determined from simultaneous radiation intensity measurements in the resonator cavity of 4.3- and 2.7-μm bands.

The relaxation rate of the CO_2 lower laser level is increased by molecular hydrogen (Röm and Stricker 1974). Experimental (Kudryavtsev et al. 1976) and theoretical (Makarov and Losev 1975) investigations have shown that maximum gain coefficient K_0 is obtained with 5-10% hydrogen in the active medium. In this work, the influence of high hydrogen content (40-50%) on CO_2-GDL performance was studied. The radiation intensity of the 4.3-μm CO_2 and 4.7-μm CO bands (CO resulting from chemical reactions in CO_2 + H_2 mixture) and gain coefficient on CO_2 00^01-10^00 transition were measured simultaneously. Application of vibrational temperatures determination methods (Kudryavtsev et al. 1979 b; Kudryavtsev and Novikov 1981) permitted the determination of vibrational temperatures of CO-T_5, asymmetric-T_3, and combined-T_2 CO_2 modes. Laser performance optimization was carried out and a gain coefficient of 0.4-0.5 m^{-1} was attained for mixtures containing up to 50% of molecular hydrogen.

Determination of CO_2 and CO Vibrational Temperatures

Vibrational temperature determination of asymmetric-T_3 and combined-T_2 CO_2 modes from i.r. radiation intensity

Table 1 Integrated radiation intensity of CO_2 and CO bands changing (in percent) at 1% increase of each vibrationally nonequilibrated gas parameter: T_3 or T_5 = 1500 K, T_2 = 500 K, T = 300 K, P = 5×10^{-2} atm, X = P_{CO_2}, D = 10^{-2} atm/cm, D = length of radiating volume, P_{CO_2} = partial CO_2 pressure.

Parameters of vibrationally nonequilibrated gas	Bands, μm		
	4.3	2.7	4.7
T	0.023	0.0071	0.074
P	0.300	0.0287	0.275
T_3 or T_5	2.850	2.6	2.870
T_2	0.785	4.48	...

measurements in 4.3- and 2.7-μm bands was carried out by numerical solution of the set of equations:

$$L_{4.3} = \int C_{4.3}(\omega) I_{4.3}(\omega, T_3, T_2) d\omega \quad (1)$$
$$L_{2.7} = \int C_{2.7}(\omega) I_{2.7}(\omega, T_3, T_2) d\omega$$

where $L_{4.3}$ and $L_{2.7}$ are experimentally measured values of radiation intensity of CO_2 4.3- and 2.7-μm bands, respectively; $C_{4.3}(\omega)$ and $C_{2.7}(\omega)$ are the spectral transmittance of corresponding optical systems. The spectral radiation intensity of a 4.3-μm band, $I_{4.3}(\omega, T_3, T_2)$, and a 2.7-μm band, $I_{2.7}(\omega, T_3, T_2)$, in vibrationally nonequilibrated conditions was calculated according to Kudryavtsev and Novikov (1981). Small-signal gain coefficient K_0 for the rotational line P 20 of 00^01-10^00 CO_2 vibrational transition was determined by vibrational temperatures T_3 and T_2 obtained from the solution of Eqs. (1). From the data represented in Table 1 it follows that the dependences of vibrationally nonequilibrated CO_2 radiation intensity in a 4.3-μm band on vibrational temperature T_3 and in a 2.7-μm band on vibrational temperatures T_3 and T_2 are the most critical (Kudryavtsev and Novikov 1981). This provides for high accuracy of vibrational temperature determination based on solution of Eqs. (1). In this investigation, errors in vibrational temperatures T_3 and T_2 determination were 5-8%, in gain coefficient K_0, 20-30%, while the usual error in i.r. radiation intensity measurements constituted 10-15%. The described method was used for investigations of CO_2 GDL on $CO_2 + N_2$ + He (H_2O) mixtures.

Appreciable concentrations of carbon monoxide may be formed in $CO_2 + N_2 + H_2$ laser mixtures due to $CO_2 + H_2$ chemical reactions. CO molecules significantly influence CO_2-GDL population inversion generation (Anderson 1976; Losev 1977). In this investigation of mixtures containing hydrogen along with CO_2 vibrational temperatures measurements, CO vibrational temperature T_5 was determined by using measurements of carbon monoxide radiation in a 4.7-μm band. Separation of 4.3-μm CO_2 and 4.7-μm CO bands radiation, overlapping in vibrationally nonequilibrated conditions in a 4.4-4.6-μm spectral region, while using wideband transmittance optical systems $C(\omega)$, is impossible. That is why the total 4.3- and 4.7-μm bands radiation intensity was registered in the optical system with λ = 4.75 + 0.6-μm band transmittance. In another system with λ = 4.36 + 0.1-μm band transmittance, CO_2 4.3-μm band radiation intensity was registered mainly. The CO_2 small-signal gain coefficient K_0 was measured simultaneously by the probing-laser technique (Anderson 1976; Losev 1977). In this method CO_2 - T_3, T_2 and CO - T_5 vibrational temperatures were calculated from a numerical solution of the following equations:

$$L_{4.75} = \int C_{4.75}(\omega)[I_{4.3}(\omega, T_3, T_2) + I_{4.7}(\omega, T_5)]d\omega$$

$$L_{4.36} = \int C_{4.36}(\omega)[I_{4.3}(\omega, T_3, T_2) + I_{4.7}(\omega, T_5)]d\omega$$

$$K_0 = K_0(T_3, T_2) \tag{2}$$

T_3, T_2, and T_5 vibrational temperature calculation accuracies are predetermined by a strong dependence of the 4.3-μm band and radiation intensity on T_3 and that of the 4.7-μm band on T_5 (Kudryavtsev and Novikov 1981). Taking into account the 10-15% error in i.r. radiation intensity measurements and the 10-30% error in gain coefficient determination, the accuracy of T_3, T_2, and T_5 calculation was evaluated as 5-10%.

Experimental Installation

Investigations were carried out on a shock tube driven GDL (Fig. 1) with a wedge supersonic nozzle at the end of the driven section. Vibrationally nonequilibrated gas parameters were measured at the nozzle exit. The nozzle could be isolated from the driven section by a thin aluminum valve to permit the evacuation of the test section to pressure $P \lesssim 10^{-5}$ atm. The starting time of the nozzle in our experimental conditions was less than 0.2 ms (Losev, 1977). The measurements of radiation intensity in the test section indicated that unsteady processes in the nozzle persisted for $t < 0.1$ ms.

The optical system permitted simultaneous observation in the same cross section of i.r. radiation in two spectral bands and small-signal gain coefficient. The optical systems of radiation measurements were identical and consisted of flat and spherical mirrors which focused i.r. radiation of the flow on the photoresistor sensitive element. Separation of required spectral bands was carried out by interference filters.

To calibrate the i.r. optical systems, the supersonic nozzle was removed and the cylindrical channel of the shock tube was extended to the test section. The sensitivity of the optical systems was determined from observations of radiation of the gas heated by the incident shock wave. The small-signal gain coefficient was measured by the probing-laser technique with a power stabilized discharge CO_2 laser. The radiation intensity of the probing-laser beam was determined just after the investigated mixture outflow process. An electrooptic modulator was employed to chop the laser beam.

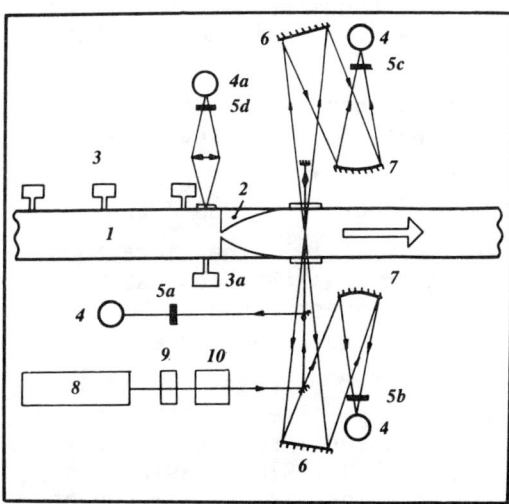

Fig. 1 Scheme of the experimental installation: 1) shock tube; 2) supersonic nozzle; 3) piezoelectric transducers; 4) photoresistors Ge - Au (77 K); 5) filters: a) $2 \times 10.6 \pm 0.6$ µm, b) 4.36 ± 0.5 µm, c) 4.75 ± 0.6 µm or 2.9 ± 0.5 µm, d) 6.3 ± 0.6 µm; 6) flat mirrors; 7) spherical mirrors; 8) discharge CO_2 laser; 9) polarizer; 10) electrooptic modulator.

Gas temperature T_0 and pressure P_0 in the stagnation region behind the reflected shock wave were calculated from shock jump conditions and the incident wave velocity. The latter is determined from the measurement of elapsed time between three stations. For this purpose three piezoelectric transducers and two time counters are used. Stagnation region pressure and chemical reactions were monitored with a piezoelectric transducer (3a) and an i.r. radiation intensity measurements with photoresistor (4a), respectively. Vibrational nonequilibrium parameters were measured at 0.2-0.3 ms reflection. For this time interval the conditions for hot gas in the stagnation region are undisturbed and a steady-state flow expansion persists (Anderson 1976; Losev 1977).

Investigations of CO_2 + N_2 + He Mixtures
_____

The feasibility of the method for determining CO_2 vibrational temperatures T_3 and T_2 from the radiation intensity measurements of 4.3- and 2.7-μm bands and calculation on this basis of the gain coefficient K_0 was demonstrated with the CO_2 GDL and a mixture of 0.1 CO_2 + 0.4 N_2 + 0.5 He. The employed supersonic nozzle was contoured by a circumference arc of 90 mm radius and was smoothly coupled with the circumference of 2.0 mm radius centered in the critical cross section. The throat height was $h_* = 1.5$ mm; the nozzle area ratio was $A/A_* = 38$.

Vibrational temperatures T_3 and T_2 deduced from radiation measurements made at different gas temperature T_0 and pressure P_0 are shown in Fig. 2 and the gain coefficient is shown in Fig. 3.

An increase in results in an increase of temperatures T_3 and T_2 (Fig. 2). Temperature T_3 is lower than the gas temperature T_* in the nozzle throat. The difference between T_* and T_3 is between 100 and 150 K at $T_0 = 1300$ K and rapidly increases with an increase in T_0. The vibrational temperature T_2 is 200-300 K higher than the flow gas temperature T^2 for the range of T_0 investigated. The vibrational temperatures T_3, T_2 and gain coefficient K_0 calculated from a procedure described by Losev (1977) are shown in Figs. 2 and 3, respectively. The results for the base case were obtained with vibrational relaxation rates recommended by Losev (1977), Anderson (1976), Taylor and Bitterman (1969), Simpson and Chandler (1970), Simpson et al. (1977) and Whitson and McNeal (1977). As the following relaxation processes have the strongest influence on CO_2-laser performance,

$$CO_2\ (01^10) + M\ \underset{\leftarrow}{\overset{Q_{20}}{\rightarrow}}\ CO_2\ (00^00) + M + \Delta E_1 \qquad (3)$$

$$CO_2\ (00^01) + M\ \underset{\leftarrow}{\overset{Q_{32}}{\rightarrow}}\ CO_2\ (11^10) + M + \Delta E_2 \qquad (4)$$

The effect of variation of the vibrational relaxation rate for these processes on vibrational temperature and the gain coefficient was by variation of Q_{20} and Q_{32}.

The presence of water molecules at $\xi_{H_2O} \cong 0.005$ is known to have a pronounced effect on the vibrational relaxation in $CO_2 + N_2$ mixture. The maximum water vapor concentration, which could be attributed to insufficient gas purification and water evaporation from the walls, was estimated to be $\xi_{H_2O} = 0.5\%$. As a consequence, the predicted effect of 0.5% H_2O molecules admixture on the gain coefficent and vibrational temperatures was determined. The vibrational

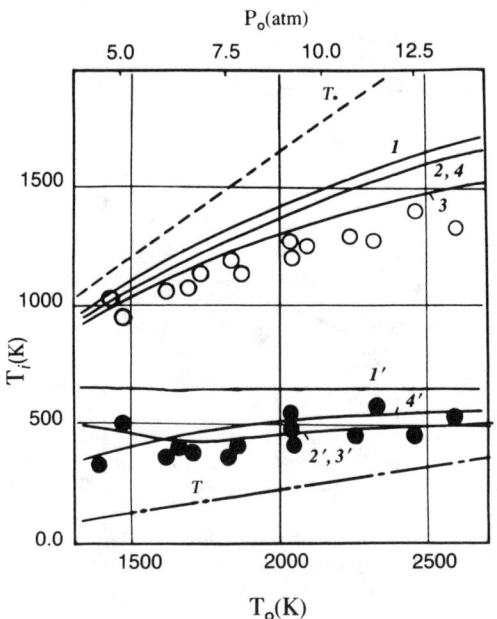

Fig. 2 Vibrational temperature of asymmeric-T_3 and combined-T_2 modes of carbon dioxide. Mixture compostion, x CO_2:n N_2:h He = 0.1:0.4:5. 0, T_3; and ●, T_2;(---): gas temperature at nozzle throat; (-··-): translation temperature at nozzle exit;(—): predicted vibrational temperatures. Case 1) base case relaxation rates; case 2) case 1 relaxation rates except Q_{20} and Q_{32} are doubled; case 3) 1 relaxation rates except $Q_{20} = (Q_{20})_1 \times 2$ and $Q_{32} = (Q_{32})_1 \times 3$; case 4) case 1 relaxation rates with 0.5% H_2O added to mixture. Unprimed numbers indicate T_3; primed numbers indicate T_2.

temperature T_3 for the base case coincides with the corresponding experimental values for temperatures $T_o \leq 1500$ K. For $T_o \gtrsim 1500$ K, the calculated values of T_3 exceed the experimental values. This discrepancy increases with T_o and is about 250-300 K at $T_o = 2500$ K. Neither the twofold increase of rate constant Q_{32} nor the effect of water impurity of 0.5% water molecules reduces the calculated values of T_3 for $T_o \gtrsim 2000$ K. Only for a fivefold increase of Q_{23} do the calculated vibrational temperatures T_3 with the corresponding experimental values for high T_o. The calculated value of T_2 does not seem particularly sensitive to the variations in relaxation rates in the addition of water.

The vibrational temperature T_2 measurements suggest a significant level of excitation of the combined CO_2 mode at the nozzle exit. For the range $1300 \leq T_o \leq 2700$ K, the vibrational temperature T_2 is 200-250 K higher than the translational temperature. The CO_2 calculated combined mode vibrational temperature for the base case are 200 K higher than the observed values of T_2. Both the twofold increase in the relaxation rate of the combined mode and the introduction of 0.5% water molecules into the mixture yield good agreement with the experiment.

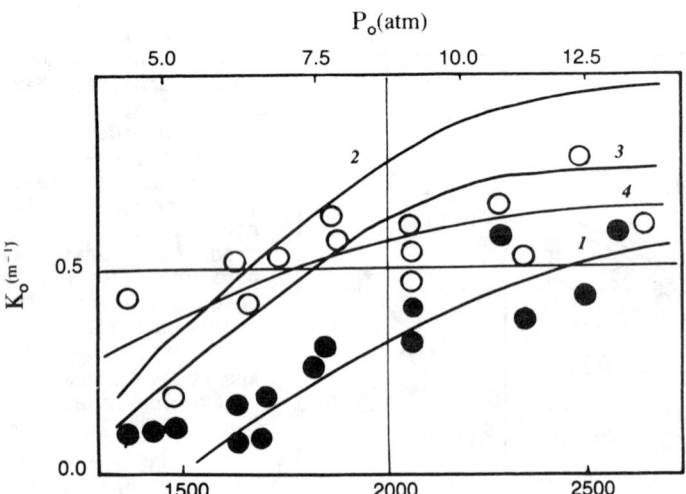

Fig. 3 Comparison of gain coefficient. ○: probing laser technique $K(L)$; ●: spontaneous radiation intensity measurements, $K_o(R)$; (——): theoretical calculations for cases 1-4 as indicated in Fig. 2. Experimental conditions correspond to Fig. 2.

The gain coefficients as determined by the probing-laser technique $K^{(L)}$ and by CO_2 radiation intensity measurements in CO_2 4.3- and 2.7-μm bands $K_o^{(R)}$ are scattered with $K_o^{(R)} > K_o^{(L)}$ (Fig. 3). If the probing laser is not frequency-stablized, the gain coefficient $K^{(L)}$ may be 20-30% smaller than $K_o^{(L)}$, which corresponds to the center of the rotational line $K_o^{(L)}$ (Soloukhin and Yakobi 1974). As a consequence, the discrepancy between gain coefficient measurements by the two methods should be smaller as $K_o^{(R)}$ is determined at the center of the rotational line P 20. For $K_o < 0.2$ m^{-1}, the accuracy of gain coefficient measured from spontaneous i.r. radiation intensity decreases because the differences of laser levels population become of the same order as the error of 00^01 and 10^00 levels population measurements.

The gain calculated gain coefficient for the base case is in good agreement with the results obtained for probing-laser technique $K^{(L)}$. However, it has been already noted that $K^{(L)}$ should be somewhat larger. The scatter in the data makes it difficult to state definitively as to whether any of the calculated cases should be preferred or whether one mode of measurement is more reliable.

On the basis of the results, it appears that for $T_o \geq$ 2000-2500 K, CO_2 asymmetric mode experiences a faster rate of relaxation. This observation cannot be explained by uncertainties in the choice of vibrational relaxation rate or by the precense of traces of water vapor. Probably, the increased rate of relaxation may be due to additional channels for energy exchange between the upper levels of asymmetric and combined CO_2 modes or the appearance in the mixture at high T_o of atomic components which enhance the rate of vibrational relaxation (Buchwald and Wolga 1975).

CO_2 GDL With Admixtures of Hydrogen

The addition of molecular hydrogen to the CO_2 laser mixture may enhance the gain, since the relaxation rate of the carbon dioxide combined mode is increased by the presence of H_2 molecules (Simpson and Chandler 1970). The population of the upper laser level in $CO_2 + N_2 + H_2$ mixture may be sufficient for an inversion if the hydrogen content

Table 2 Composition of CO_2 - GDL working fluids for Fig. 4

Mixture	Symbol	Mixture ratio x CO_2:y CO:z H_2:q N_2		
		x	y	z
1	○	1	0.8	0.1
2	◐	1	0.6	0.3
3	●	1	0.4	0.5

is $\xi_{H_2} \lesssim 0.1$ (Kudryavtsev et al. 1979c). When the stagnation region temperature is $T_o > 1200\text{-}1400$ K, the presence of H_2 results in the reaction:

$$CO_2 + H_2 \rightleftharpoons CO + H_2O - 12.8 \text{ kcal/mole} \qquad (5)$$

The carbon monoxide and especially the water molecules formed in reaction (5) affect the populations of the lasing levels (Losev 1977; Anderson 1976). Also, the reaction consumes the lasing carbon dioxide molecules. At $T_o = 1500$ K, the equilibrium CO_2 concentration is 20-30% and that of $H_2O \cong 70\%$ of the initial carbon dioxide content $\xi_{CO_2}^{(o)}$. As a consequence, this reaction reduces the gain coefficient (Rom and Stricker 1974; Kudryavtsev et al. 1976).

The addition of carbon monoxide to the lasing mixture will shift reaction (5) equilibrium to the left. In this work the effect on the GDL of the molecular hydrogen and carbon monoxide content mixtures were investigated (see Table 2). The wedge supersonic nozzle with half-angle $\phi = 15$ deg, throat height $h_* = 1.5$ mm, and area ratio $h/h_* = 23$ was employed. The results of gain coefficient measurements in mixtures 1-3 are shown in Fig. 4.

The water vapor content was calculated with the empirical relation of Brupbacher et al. (1976):

$$\xi_{H_2O} = \xi_{H_2O}^{(eq)} \left\{ 1 - \left[\exp - K(\xi_{H_2}^{(o)})^{0.3} [M]t \right] \right\}$$

$$K = 10^{20 \pm 0.2} \exp\left(- \frac{81400 \pm 2300}{RT}\right), \frac{cm^3}{mole \cdot s} \qquad (6)$$

$\xi_{H_2O}^{(eq)}$ is the equilibrium water vapor mole fraction; $\xi_{H_2}^{(o)}$ is the initial hydrogen content; and t is the time of the occurring chemical reaction. For the experimental conditions of this work, t = 0.2 ms. The water vapor content calculations according to Eq. (6) are in good agreement with the time-dependent concentration determined from radiation measurements of H_2O molecules in the 6.3-μm band in the stagnation region.

The maximum gain coefficient values ∿0.2 m^{-1} are obtained for mixture 1 with initial hydrogen content $\xi_{H_2}^{(o)}$ = 0.1 at temperature T_o = 1500 K and pressure P_o = 8 atm in the stagnation region. The increase of H_2 concentration results in a reduced gain coefficient. For mixture 3 with $\xi_{H_2}^{(o)}$ = 0.5, the maximum value of K_o does not exceed 0.1 m^{-1} at T_o = 1100 K and P_o = 5 atm. The sharp decrease in the gain coefficient at $T_o \geq$ 1500 K is due to water molecules appearance in the CO_2 + N_2 mixture. The CO_2 - T_3, T_2 and CO - T_5 vibrational temperatures deduced from experimental measurements and the calculated CO_2 and CO vibrational temperatures for the base case and a hydrogen-rich mixture are shown in Fig. 5. For all T_o investigated, the

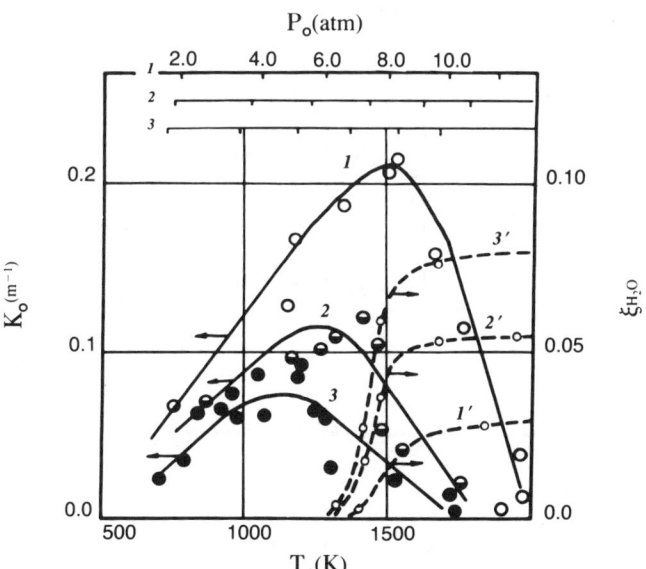

Fig. 4 Effect of H_2 and CO on the gain coefficient K_0 of a CO_2 GDL. (-o-): water vapor concentration calculated with Eq. (6); (——): predicted gain coefficients (see Table 2).

vibrational temperatures T_5 and T_3 practically coincide. For $T_0 \lesssim 900$ K, the values of T_5 and T_3 coincide with gas temperature at the nozzle throat T_*. The combined mode vibrational temperature T_2 = 300-450 K is much lower than T_3 and T_5 and is 100-150 K greater than the translational temperature T.

The gain coefficient decrease at $T_0 > 1100$ K is due to increased population of the CO_2 asymmetric mode and increased vibrational relaxation by CO. These results suggest that addition of carbon monoxide to the hydrogen-containing medium of CO_2 GDL is ineffective at the conditions of these experiments. However, it appears that an optimal condition for a high hydrogen concentration CO_2 GDL may exist at lower temperature T_0, which is necessary for "freezing" the troublesome chemical reactions.

At temperature T_0 = 1100-1250 K, the gain coefficient K_0 = 0.4-0.5 m^{-1} was observed (see Fig. 6). For such gas parameters, reaction (5) is "frozen" and molecular hydrogen influences only the vibrational relaxation kinetics of the mixture. The decrease of K_0 at high temperatures $T \geq 1300$ is due to the appearance of water molecules in the mixture.

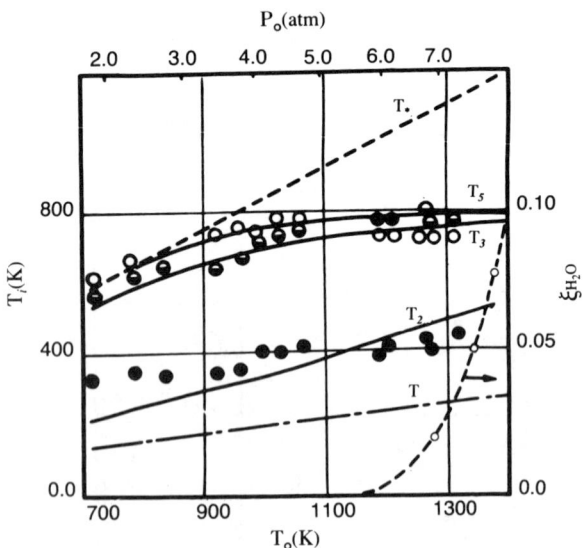

Fig. 5 Theoretical and experimental CO_2 - T_3, T_2 and CO - T_5 vibrational temperatures. Mixture composition, x CO_2:7 CO:z H_2 = 1:0.4:0.5. ○, T_5; ●, T_3; ●, T_2; (—): predicted vibrational temperature; (---): gas temperature at nozzle throat; (— · —): translation temperature at nozzle exit; (o--o): water concentration at stagnation region.

Significant increases in the gain coefficient for CO_2 GDL with $CO_2 + N_2 +$ He (H_2O) mixtures have been observed in a minimum length supersonic nozzle (Kuehn 1972). The gain coefficient observed in a hydrogen-rich mixture during expansion through such a nozzle is shown in Fig. 7. The throat height of the nozzle is $h_* = 1$ mm and the area ratio is $h/h_* = 23$. The supersonic part was contoured according to Ktalkherman et al. (1980). The measured maximum gain coefficient K_o (0.4-0.45 m^{-1}) at T_o = 1100-1250 K is not higher than that observed during expansion through a slit wedge nozzle (Fig. 6).

The vibrational temperatures T_3 and T_2 of carbon dioxide measured for the minimum length nozzle are shown in Fig. 8. At $T_o \leq 1200$ K, the asymmetric mode vibrational temperature is frozen at the nozzle throat, since $T_3 \cong T_*$. The CO_2 combined-mode vibrational temperature T_2 is about 150-180 K higher than the translational temperature.

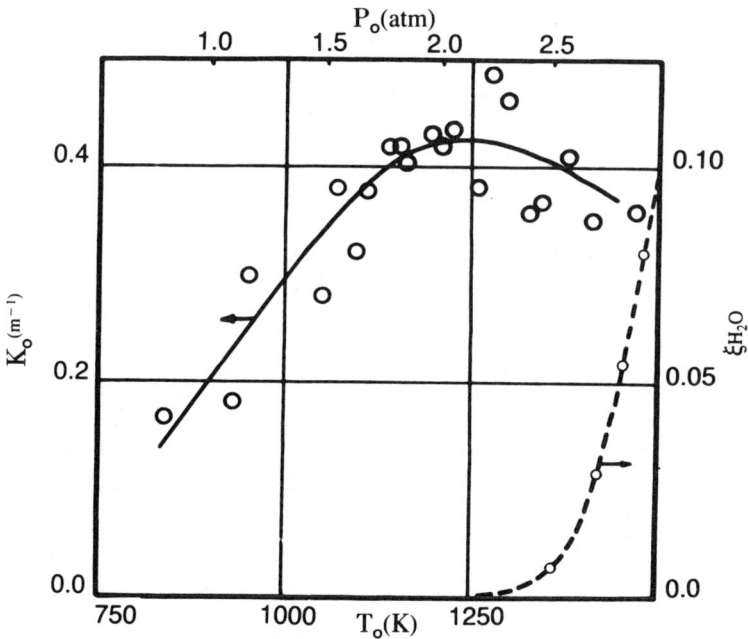

Fig. 6 Effect of reduced stagnation temperature on the gain coefficient of hydrogen-rich CO_2 GDL. ○: gain coefficient deduced from experimental observations; (——): predicted gain coefficient; (o--o) predicted water concentration; mixture composition, x CO_2:m N_2:zH_2 = 0.2:0.4:0.4.

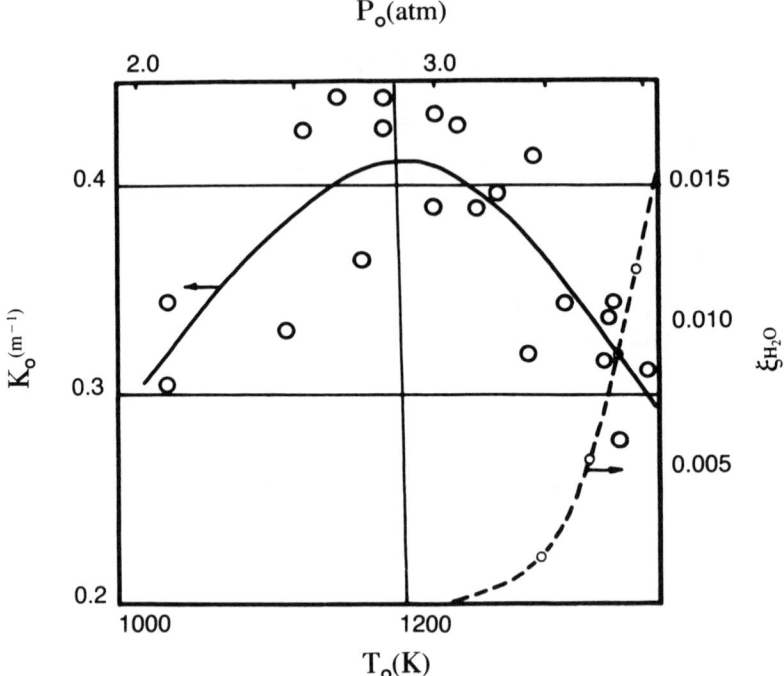

Fig. 7 Effect of a minimum length supersonic nozzle on the gain coefficient. o: gain coefficient deduced from experimental observation; (—): predicted gain coefficient; (o--o): predicted water concentration; mixture composition, x CO_2:m N_2:z H_2 = 0.2:0.4:0.4.

When $T_o \gtrsim$ 1200 K, the vibrational temperature T_3 becomes lower T_*, and T_2 tends to T. At this temperature range a considerable amount of water vapor arises in the CO_2 + H_2 mixture.

For $T_o \lesssim$ 1200 K, CO_2, the calculated vibrational temperatures correlate well with those deduced from the data. For $T_o \gtrsim$ 1200 K, the calculated T_3 and T_2 are 100-150 K greater than the measured ones. This may be accounted for by the strong influence of intermediate chemically active particles of reaction (5) on vibrational relaxation rate.

From the data of Figs. 6 and 7 it follows that for hydrogen content $\xi_{H_2}^{(o)}$ = 0.4 mixtures, the optimal temperature T_o = 1200 K and pressure P_o = 2.5 atm considerably smaller than the optimal conditions for CO_2 + N_2 + $He(H_2O)$ mixtures ($P_o \cong$ 20 atm, $T_o \cong$ 1500-1700 K) (Anderson 1976; Losev 1977).

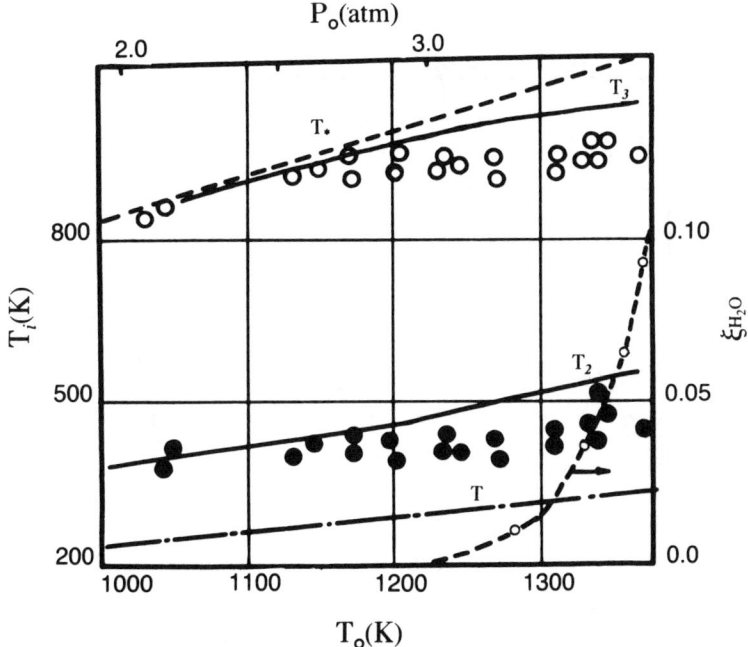

Fig. 8 Effect of a minimum length nozzle on the theoretical and experimental vibrational temperatures. **o**, T_3; **●**, T_2; (—): predicted vibrational temperature; (---): gas temperature at nozzle throat; (—·—): translation temperature at nozzle exit; (o--o): water concentration; mixture composition, x CO_2:m N_2:z H_2 = 0.2:0.4:0.4.

For practical applications in CO_2 GDL with high hydrogen content mixtures, the stagnation pressure should be higher. The results of investigations of mixture $0.2CO_2$:$0.4N_2$:$0.4H_2$ at high pressures are shown in Fig. 9. To compensate for increased vibrational relaxation rate, a contoured nozzle (Ktalkherman et al. 1980) with reduced throat height ($h_* = 0.4$ mm) was used. The area ratio of such a nozzle was $h/h_* = 50$.

The maximum gain coefficient $K_o = 0.3$-0.4 m^{-1} at temperature $T_o = 1000$-1100 K and pressure $P_o = 15$ atm was attained. Comparison of results represented in Figs. 7 and 9 makes it clear that at increased pressure in stagnation region P_o, maximum gain coefficient is obtained at low temperatures $T_o \cong 1100$ K.

These investigations of CO_2 GDL on high hydrogen content $CO_2 + N_2 + H_2$ mixtures have shown that the gain

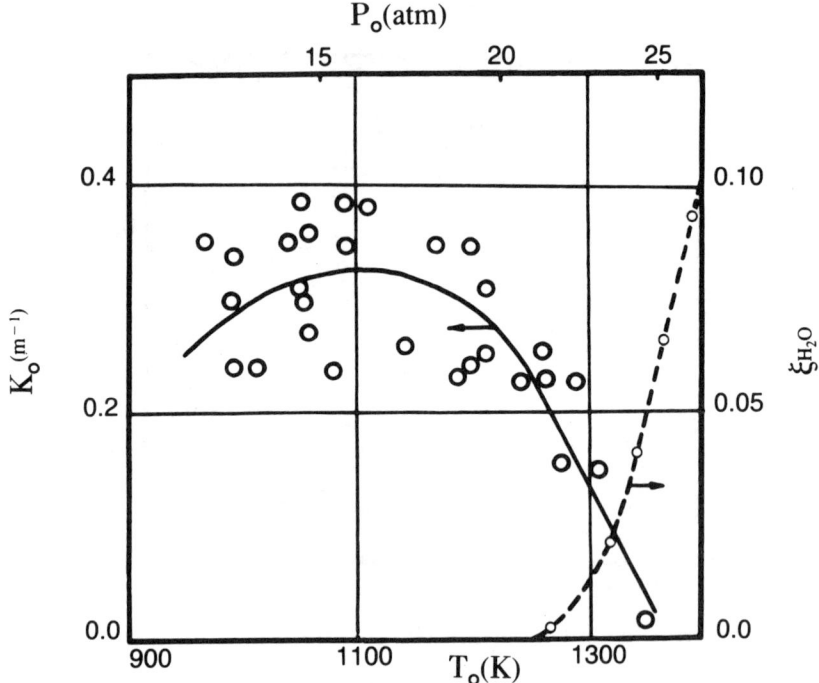

Fig. 9 Effect of pressure on the gain coefficient Ko and H2O of a CO2 GDL. (—) Predict gain coefficient, observed gain coefficient, (-o-) predicted water concentration; mixture composition x CO_2:m N_2:z H_2 = 0.2:0.4:0.4.

coefficient K_o = 0.4-0.5 m^{-1} sufficient for laser generation is possible. This considerably extends the gas mixtures employed in CO_2 gasdynamic lasers.

References

Anderson, J. D. (1976) <u>Gasdynamic Lasers: An Introduction</u> Academic Press, New York.

Brupbacher, J. M., Kern, R. D., and O'Grady, B. V. (1976) Reaction of hydrogen and carbon dioxide behind reflected shock waves. <u>J. Phys. Chem.</u> 80, 1031-1035.

Buchwald, M. I. and Wolga, G. J. (1975) Vibrational relaxation of CO_2 (001) by atoms. <u>J. Chem. Phys.</u> 62, 2828-2833.

Ktalkherman, M. G., Malkov, V. M., and Ruban, N. A. (1980) Experimental investigation of gasdynamic lasers nozzle flows. <u>Mech. Fluid Gases</u> 15, 178-182.

Kudryavtsev, N. N., Novikov, S. S., and Svetlychnyi, I. B. (1976) Effect of molecular hydrogen additives on CO_2 - laser radiation gain coefficient in expanding flow of carbon dioxide-nitrogen mixture. Phys. Combust. Explos. 12, 729-735.

Kudryavtsev, N. N., Novikov, S. S., and Svetlychnyi, I. B. (1979a) Experimental investigations of vibrational energy exchange in chemically active gas laser mixtures. Plasma Chem. Atomizdat. 6, 230-278.

Kudryavtsev, N. N., Novikov, S. S., and Svetlichnyi, I. B. (1979b) On a method for measurements of vibrational temperatures in CO_2 gas dynamic lasers. Quantum Electron. 6, 690-700.

Kudryavtsev, N. N., Novikov, S. S., and Svetlychnyi, I. B. (1979c) Vibrational temperatures of carbon dioxide in gas dynamic CO_2 - N_2 - H_2 laser. Phys. Combust. Explos. 15, 122-125.

Kudryavtsev, N. N. and Novikov, S. S. (1981) A study of infrared radiation of vibrationally excited CO in the 4.7 m band and CO_2 in the 4.3 m and 2.7 m bands. Rev. Phys. Appl. 16, 49-66.

Kuehn, D. M. (1972) Importance of nozzle geometry to high pressure gasdynamic lasers. Appl. Phys. Lett. 21, 112-118.

Losev, S. A. (1977) Gasdynamic Lasers, Nauka, Moscow, USSR.

Makarov, V. N. and Losev, S. A. (1975) On the Influence of admixtures on the optical gain coefficient for the gas relaxing in supersonic nozzle. Phys. Combust. Explos. 11, 804-811.

Rom, J. and Stricker, J. (1974) Effects of chemical reactions on the performance of gas dynamic lasers. Acta Astron. 1, 1101-1117.

Simpson, C. J. and Chandler, T. (1970) A shock tube study of vibrational relaxation in pure CO_2 and mixtures of CO_2 with inert gases-nitrogen, deuterium and hydrogen. Proc. R. Soc. London Ser. A 317, 265-304.

Simpson, C. J., Gait, R. D., and Simmie, J. M. (1977) The vibrational deactivation of the bending mode of CO_2 by O_2 and N_2. Chem. Phys. Lett. 47, 133-136.

Soloukhin, R. I. and Yakobi, Y. A. (1974) On gain coefficient measurements. J. Appl. Mech. Tech. Phys. 15, 3-12.

Taylor, R. L. and Bitterman, S. (1969) Survey of vibrational relaxation data for the processes important in the CO_2 - N_2 laser system. Rev. Mod. Phys. 41, 26-47.

Whitson, M. B. and McNeal, R. J. (1977) Temperature dependence of the quenching of vibrational excited N_2 by NO and H_2O. J. Chem. Phys. 66, 2696.

Time-Dependent Nozzle and Base Flow/Cavity Model for CW Chemical Laser Flowfield

N.L. Rapagnani* and D.W. Lankford†
Kirtland Air Force Base, N. Mex.

Abstract

A detailed calculation of the flow emerging from the primary (fluorine) and secondary (deuterium) nozzles into the cavity base region of a deuterium fluoride (DF) continuous wave chemical laser is examined. A two-dimensional time-dependent numerical method, the APACHE code, is used to predict the flow starting at the nozzle throats and proceeding approximately twenty base widths into the cavity. The APACHE code is an outgrowth of the Los Alamos RICE code and was developed to give greater numerical efficiency and geometric flexibility than RICE. This analysis was performed to ascertain the effect of the base region on the upstream nozzle flow and to investigate the mixing process in a laminar cavity flow with lateral pressure gradients. A thorough description of the flowfield is provided by contour, profile, and carpet plots of primary and derived variables including velocities, pressure, density, Mach number, temperature, and chemical species concentrations. The base region for this case is shown to influence the upstream nozzle flow, but not drastically enough to cause separation. Although, chemical reactions are releasing heat in the base region the maximum temperature and pressure in the cavity remain moderate and are controlled by the nozzle wall temperature. The reaction zone remains narrow to 0.5 cm beyond the nozzle base and is shown to be limited by the diffusion rate. Therefore, macroscopic induced mixing, such as caused by injection (trip) jets, would be expected to greatly perturb and enhance the reaction zone.

Presented at the 8th ICOGER, Minsk, USSR, Aug. 23-26, 1981. This paper is declared a work of the U.S. Government and therefore is in the public domain.
*Chief, Computational Group, Chemical Laser Branch.
†Member, Computational Group, Chemical Laser Branch.

Introduction

Background

Chemical laser cavity flows are the result of complex physical phenomena including interaction between chemical reactions, mass diffusion, heat transfer, viscous stresses, radiation, and convection processes. These flows are predicted in varying levels of detail from simple one-dimensional premixed models to complex multidimensional, chemically reacting models (Zelazny et al. 1978). Increasingly complex computations continue to be performed as uneasiness grows over approximations which may not be valid, or as problems occur which cannot be addressed by simplified models. The analysis of laser nozzles has proceeded from inviscid equations, to the fully viscous boundary-layer approximation, to complete two-dimensional time-dependent Navier-Stokes solution (Cline 1976). The analysis of the cavity flow has followed a similar develop-

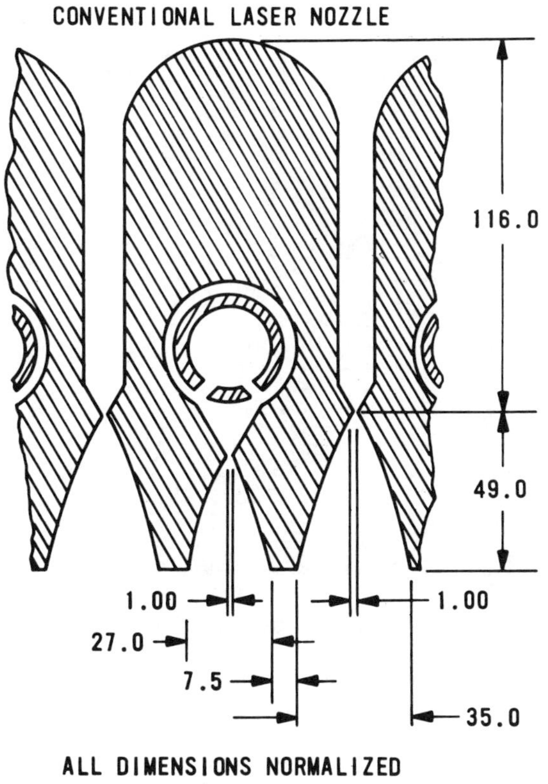

Fig. 1 CL-X1 nozzle geometry.

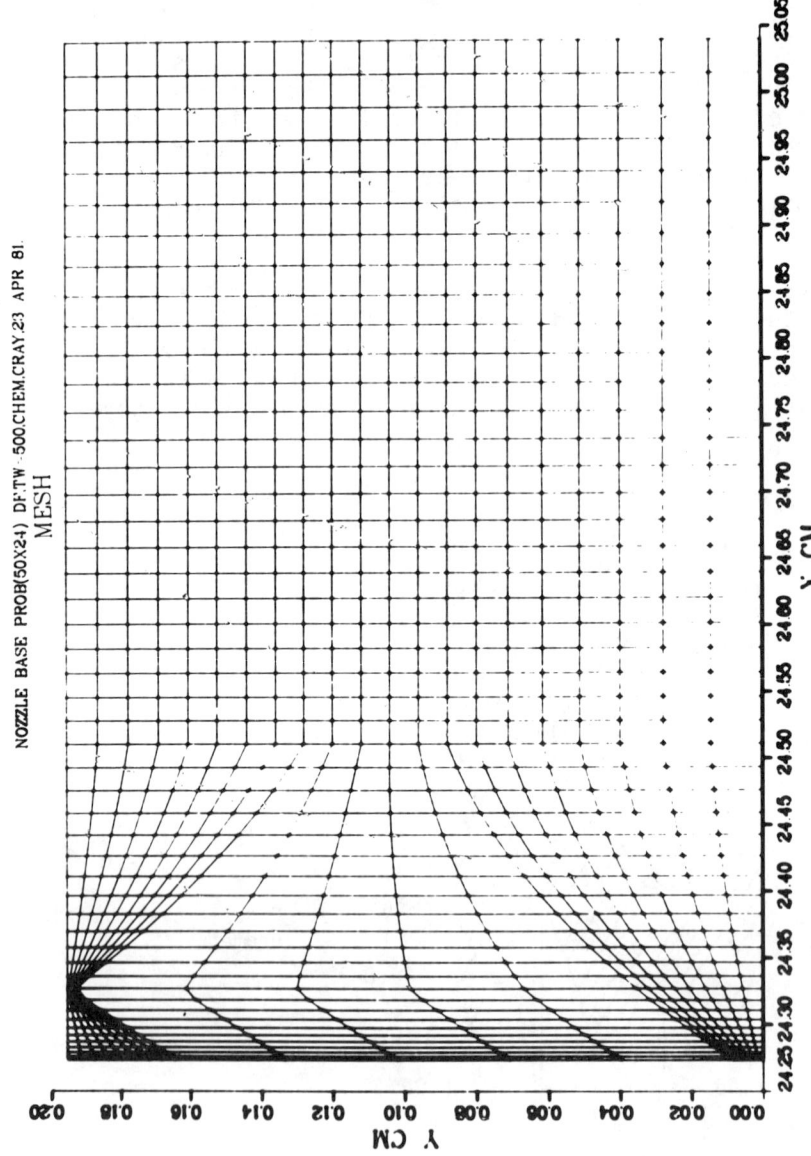

Fig. 2 calculational mesh used by the APACHE code.

ment. Generally this progression has led to more information about the flow, greater numerical complexity, and substantially more expensive calculations. It is essential to maintain a proper prespective of these codes in order to apply them effectively and economically. Extremely important results are obtained by performing a larger number of parametric cases using simplified models, but the only way to completely understand viscous nozzle flow and cavity mixing is to perform detailed experiments in conjunction with detailed analysis. This detailed understanding can then be applied to simplified analysis so that it is known what approximations are valid and what error they introduce.

Objective

The objective of this investigation was to obtain an in-depth description of the flow in the cavity base region of a typical laser nozzle. Of particular concern was the interaction of the base pressure with the upstream nozzle flow and any communication between the primary and secondary nozzles. The possible existence of recirculation in the cavity and nozzles was to be examined. Mixing and kinetic processes in laminar flow were to be investigated in order to determine the reacting region and concentrations levels of vibrationally excited species. Injection jet flow interactions were beyond the scope of the present investigation because of their three-dimensional aspects.

The Computer Model

Computational Region

Figure 1 shows the geometric configuration of the nozzles which were analyzed. In this figure, the primary nozzles are formed by aligning nozzles side-by-side of one another, whereas the secondary nozzles are the smaller ones contained within the nozzle itself. Initially separate calculations for the primary and secondary nozzles were performed starting in the subsonic converging sections. Once these two calculations were completed, the combined nozzles/cavity base case was performed using the results of the separate nozzles calculation as input and initial conditions. The geometric configuration and computational mesh for the base case are shown in Fig. 2. Half the primary nozzle is on the bottom and half the secondary nozzle is on the top. It should be noted that the mesh and subsequent contour plots are distorted by a factor of 3 in the Y direction for ease of understanding the results. The calculation was started just downstream of the nozzle throats and proceeded 0.5 cm beyond the nozzle exit plane (NEP). The calculation is bound on bottom and top by the nozzle centerlines. The depth dimension, perpendicular to the plane in Fig. 2, was varied to represent a radial expansion. A variable cell size was used in order to resolve points in areas where large gradients were expected and in order to obtain good resolution in the throat and base regions.

Governing Equations

This investigation was performed using the APACHE code (Ramshaw and Dukowicz 1979). APACHE, revised from the RICE (Rivard, Farmer and Butler 1975) code, is a time-marching finite-difference code that solves a set of equations for transient viscous multicomponent reactive fluid dynamics that includes the complete two-dimensional Navier-Stokes equations using a variable depth formulation. The variable depth formulations makes the solution "quasi-" three-dimensional. This numerical procedure and code, which has been verified against classical fluid dynamics problem, are described in detail by Ramshaw and Dukowicz (1979). The governing equations are presented here for reference.

Continuity

$$\frac{\partial \hat{\rho}}{\partial t} + \nabla \cdot (\hat{\rho} \underline{U}) = 0$$

where $\hat{\rho} = \rho R$, ρ is the density; R is the variable depth; and $\underline{U}$ is the velocity.

Momentum

$$\frac{\partial}{\partial t}(\hat{\rho}\underline{u}) + \nabla \cdot (\hat{\rho}\underline{u}\underline{u}) = -R \nabla p + \nabla \cdot (R\underline{\underline{\sigma}}) + \sigma_0 \nabla R$$

where p is the pressure, and $\underline{\underline{\sigma}}$ and σ_0 are viscous stresses given by

$$\underline{\underline{\sigma}} = \mu[(\nabla \underline{u}) + (\nabla \underline{u})^T] + (\lambda/R) \nabla \cdot (R\underline{u}) \underline{\underline{I}}$$

$$\sigma_0 = (1/R) [2\mu \underline{u} \cdot \nabla R + \lambda \nabla \cdot (R\underline{u})]$$

Internal Energy

$$\frac{\partial}{\partial t}(\hat{\rho} e) + \nabla \cdot (\hat{\rho} e \underline{u}) = -p \nabla \cdot (R\underline{u}) + R\underline{\underline{\sigma}} : \nabla \underline{u} + \sigma_0 \underline{u} \cdot \nabla R$$

$$- \nabla \cdot (R\underline{F}) + R\dot{Q}_C + R\dot{Q}_R$$

where e is the internal energy per unit mass; $\dot{Q}_C$ and $\dot{Q}_R$ are the chemical and radiative heat source terms, respectively; and $\underline{F}$ is the heat flux vector.

Species Continuity

$$\frac{\partial \rho_k}{\partial t} + \underline{\nabla} \cdot (\hat{\rho}_k \underline{u}) = - \underline{\nabla} \cdot (R \ \underline{J}_k) + R(\dot{\rho}_k)_C + R(\dot{\rho}_k)_R$$

where $\hat{\rho}_k = \rho_k R$, ρ_k being the mass density of species k; J_k is the mass diffusion flux of species k; and $(\dot{\rho}_k)_C$ and $(\dot{\rho}_k)_R$ are the chemical and radiative mass exchange terms, respectively.

Inlet and Initial Conditions

Analysis of the combined nozzle/cavity case was preceded by calculating the flow through the separate nozzles. The subsonic nozzle inlet conditions were set by specifying a fixed pressure, temperature, and species concentrations. Initial velocity at the inlet was specified to produce sonic conditions at the geometric throat. These initial conditions are given in Table 1. Results of these calculations were used as the inlet and initial conditions for the combined nozzle/cavity case. The combined case was started just downstream of the nozzle throats where the centerline Mach number was 2.20 for the secondary nozzle and 1.94 for the primary nozzle. The cavity initial conditions were set by extending the nozzle exit conditions throughout

Table 1 Pressure, temperature, and gas composition used in this calculation

	Primary	Secondary
Pressure (stagnation)	1176.24 kPa (170.6 psia)	875.63 kPa (127 psia)
Temperature (stagnation)	1890 K	540 K
Species, mixture ratios		
Combustor diluent ratio $\frac{\text{moles diluent}}{\text{moles fluorine as all } F_2} = 15$		
Cavity diluent ratio $\frac{\text{moles total diluent}}{\text{moles fluorine as all } F_2} = 30$		
$\frac{\text{Moles cavity fuel } (D_2)}{\text{Moles fluorine as all } F_2} = 3.0$		

the cavity. The base region was initially set to contain only helium.

Boundary Conditions

Applications of the no-slip conditions made the velocity equal to zero on the nozzle walls. The temperature was set at 500 K on these surfaces from which the corresponding energy was calculated. The pressure on the walls was equal to the pressure one cell away from the wall. Then a new density was calculated from the state equation $p = (\gamma-1)\rho e$. Symmetry was applied along the nozzle centerlines, giving

$$\frac{\partial u}{\partial y} = \frac{\partial \rho}{\partial y} = \frac{\partial \rho_k}{\partial y} = \frac{\partial e}{\partial y} = v = 0$$

Since the outflow boundary was completely supersonic, the exit conditions were arbitrarily set to satisfy

$$\frac{\partial u}{\partial x} = \frac{\partial v}{\partial x} = \frac{\partial e}{\partial x} = \frac{\partial \rho}{\partial x} = \frac{\partial \rho_k}{\partial x} = 0$$

DF Chemistry Model

A set of 40 chemical reactions was employed for the analysis and is listed in Table 2. This set includes pumping of the first four vibrational levels of DF by the cold reaction ($F + D_2$). A single hot reaction ($D + F_2$) for pumping DF(4) was included to account for the higher heat release. A number of single and multiquantum deactivation reactions by catalytic species were included. Rate constants for these reactions are primarily those reported by Cohen (1977) and were arrived at after a careful study of some 236 reactions selecting the dominant rates (Rapagnani and Anderson, 1980).

Results of Calculation

An overwhelming quantity of information is obtained from any complex computer code such as APACHE. The figures in this section were selected to give an overview of the processes occurring in the cavity. Several types of plots are used for this purpose. Contour plots of primary variables are presented to identify high and low value regions of a quantity and how a parameter varies across the entire computation. They were also used to determine when the calculation approached a steady-state solution. Carpet

Table 2 Reaction rates used in the APACHE code for this code for this calculation

Reaction	Forward Rate cc/mole-s
1. $F + D_2(0) \rightarrow DF(1) + D$	1.0 E13* exp (-986.0/T)
2. $F + D_2(0) \rightarrow DF(2) + D$	3.5 E13* exp (-986.0/T)
3. $F + D_2(0) \rightarrow DF(3) + D$	1.2 E14* exp (-986.0/T)
4. $F + D_2(0) \rightarrow DF(4) + D$	6.9 E13/$T^{0.1}$* exp (-198.0/T)
5. $DF(1) + F \rightarrow DF(0) + F$	4.0 E12* exp (1610.0/T)
6. $DF(2) + F \rightarrow DF(0) + F$	4.0 E12* exp (-1610.0/T)
7. $DF(2) + F \rightarrow DF(1) + F$	8.0 E12* exp (-1610.0/T)
8. $DF(3) + F \rightarrow DF(2) + F$	1.2 E13* exp (-1610.0/T)
9. $DF(3) + F \rightarrow DF(1) + F$	6.0 E12* exp (-1610.0/T)
10. $DF(3) + F \rightarrow DF(0) + F$	4.0 E12* exp (-1610.0/T)
11. $DF(4) + F \rightarrow DF(3) + F$	1.6 E13* exp (-1610.0/T)
12. $DF(4) + F \rightarrow DF(2) + F$	8.0 E12* exp (-1610.0/T)
13. $DF(4) + F \rightarrow DF(1) + F$	5.3 E12* exp (-1610.0/T)
14. $DF(1) + D \rightarrow DF(0) + D$	5.0 E11* exp (-1010.0/T)
15. $DF(2) + D \rightarrow DF(1) + D$	1.0 E13* exp (-1010.0/T)
16. $DF(2) + D \rightarrow DF(0) + D$	1.0 E13* exp (-1010.0/T)
17. $DF(3) + D \rightarrow DF(2) + D$	1.0 E13* exp (-1010.0/T)
18. $DF(3) + D \rightarrow DF(1) + D$	1.0 E13* exp (1010.0/T)
19. $DF(3) + D \rightarrow DF(0) + D$	1.0 E13* exp (-1010.0/T)
20. $DF(4) + D \rightarrow DF(3) + D$	1.0 E13* exp (-1010.0/T)
21. $DF(4) + D \rightarrow DF(2) + D$	1.0 E13* exp (-1010.0/T)
22. $DF(4) + D \rightarrow DF(1) + D$	1.0 E13* exp (-1010.0/T)
23. $DF(1) + M \rightarrow DF(0) + M^a$	8.0 E14/$T^{1.3}$ + 1.1 E4* $T^{2.37}$
24. $DF(2) + M \rightarrow DF(1) + M$	K23* 6.0
25. $DF(2) + M \rightarrow DF(0) + M$	3.5 E2* $T^{2.8}$* exp (-191.0/T)
26. $DF(3) + M \rightarrow DF(2) + M$	K23* 12.0
27. $DF(3) + M \rightarrow DF(1) + M$	3.5 E2* $T^{2.8}$* exp (-191.0/T)
28. $DF(3) + M \rightarrow DF(0) + M$	K27
29. $DF(4) + M \rightarrow DF(3) + M$	K27* 20.0
30. $DF(4) + M \rightarrow DF(2) + M$	K27
31. $DF(4) + M \rightarrow DF(1) + M$	K27
32. $DF(1) + He \rightarrow DF(0) + He$	4.0 E6* $T^{4.75}$
33. $DF(2) + He \rightarrow DF(1) + He$	8.0 E6* $T^{4.75}$
34. $DF(3) + He \rightarrow DF(2) + He$	1.2 E5* $T^{4.75}$
35. $DF(4) + He \rightarrow DF(3) + He$	1.6 E5* $T^{4.75}$
36. $DF(1) + D_2(0) \rightarrow DF(0) + D_2(0)$	1.4 E3* $T^{2.4}$
37. $DF(2) + D_2(0) \rightarrow DF(1) + D_2(0)$	2.8 E3* $T^{2.4}$
38. $DF(3) + D_2(0) \rightarrow DF(2) + D_2(0)$	4.2 E3* $T^{2.4}$
39. $DF(4) + D_2(0) \rightarrow DF(3) + D_2(0)$	5.6 E3* $T^{2.4}$
40. $D + F_2 \rightarrow DF(4) + F$	1.24E13* exp (-1260.0/T)

[a] $M = DF + 0.5*$ HF.

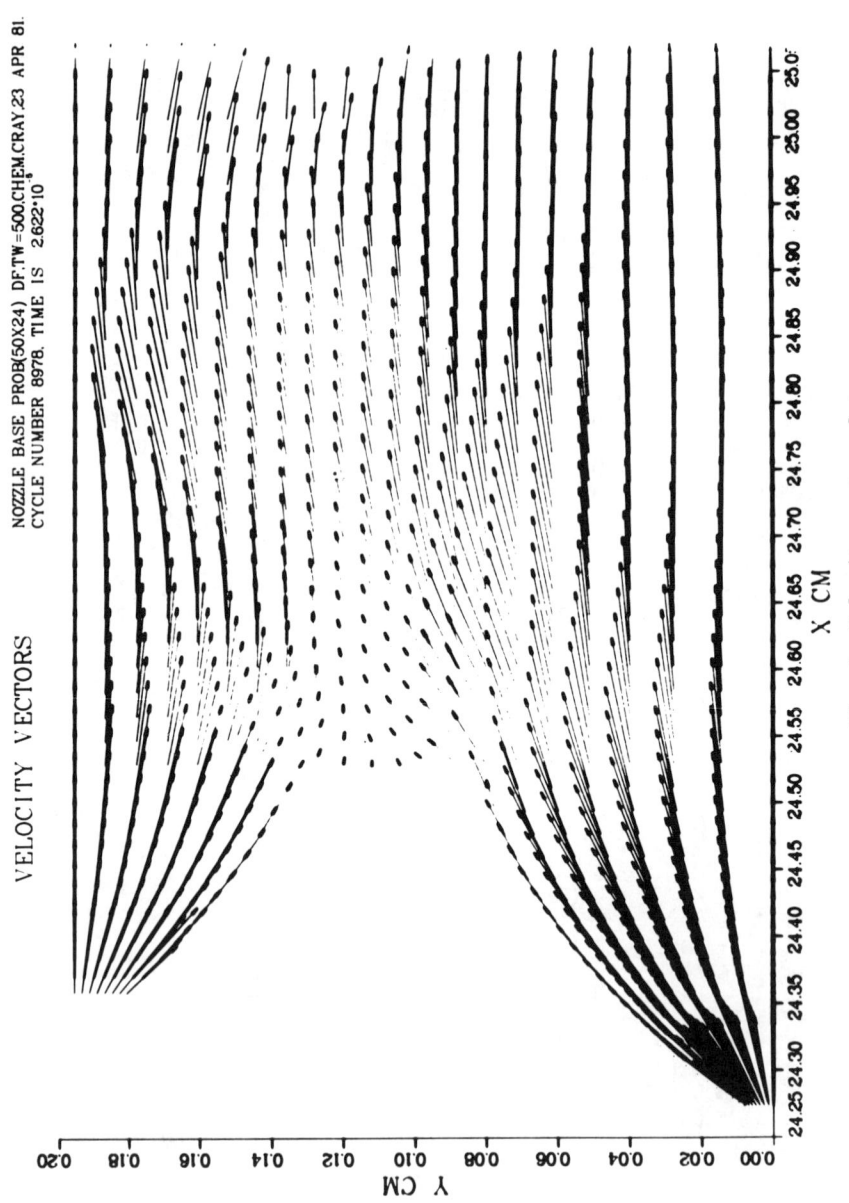

Fig. 3 Velocity vector plot.

CW CHEMICAL LASER FLOWFIELDS

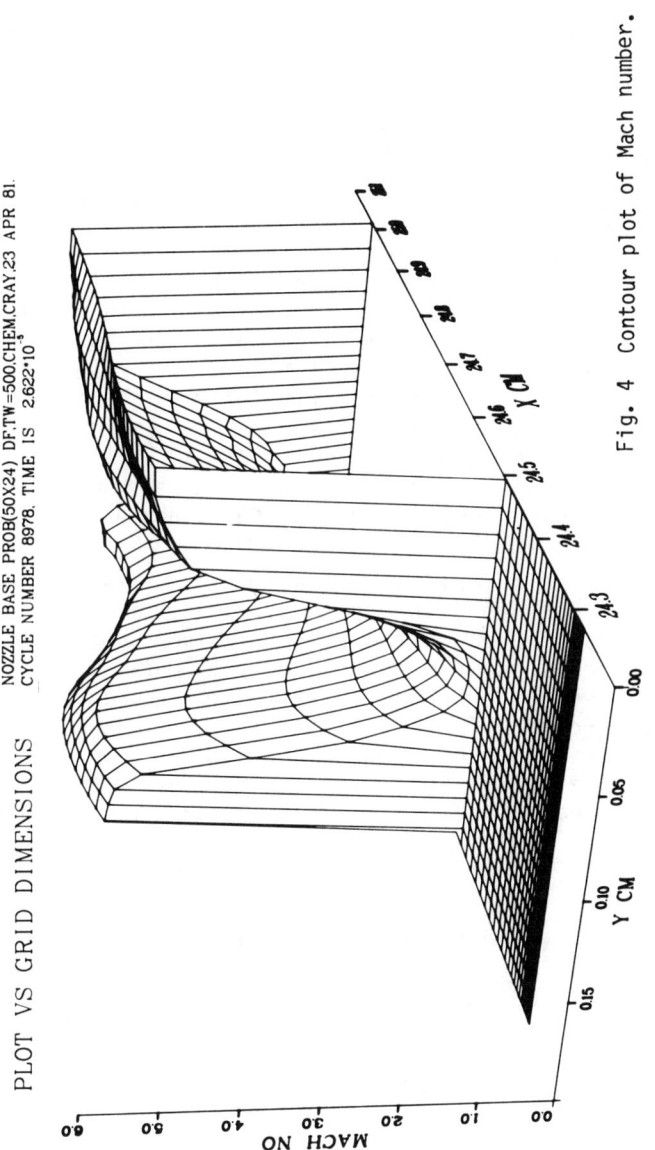

Fig. 4 Contour plot of Mach number.

plots, which are three-dimensional plots of the parameter vs the two space variables, are presented to give a heuristic impression of what is physically occurring in the cavity. Profile plots are presented to identify trends and the actual value of a quantity as it varies across the cavity.

Figure 3 is a velocity vector plot where the length of the vector is linearly proportional to the magnitude of the velocity. One observes peak velocities on the nozzle centerlines and small velocities in the wake of the base. A base recirculation zone that might be expected is not present. The absence of recirculation may be due to the low local Reynolds number in this area and/or insufficient resolutions of the base flow. The Reynolds number of this flowfield was from 10^3 to 10^4/cm, with the cell Reynolds number in the base region being close to 10.

The Mach number is an important parameter in this analysis. Of particular interest is the location of the sonic line. Figure 4 shows that the sonic line extends 1.6 base regions past the exit plane. This identifies the communicating region between the base and nozzles. Obviously, the base pressure will always influence the nozzle flow and vice versa. The boundary layer should be thin near the exit when the base pressure is low and should be expected to thicken as the base pressure increases. This leads to a question which remains unanswered. What conditions will make the boundary layer separate? Another important bit of information obtained from the Mach number contours is that the exit was completely supersonic. This alleviates concern about the arbitrary manner in which the boundary conditions were set. Figure 5 shows a Mach number trough proceeding down the cavity. The nozzle region values in the foreground were set to zero so as not to interfere with the data presentation of the cavity.

Both nozzle exit centerlines were close to Mach 5, and the flow exited with a peak Mach number of 4. The minimum Mach number at the downstream boundary was 1.5. Figure 6 shows that the velocity varied similarly to the Mach number. The maximum axial velocity was 2.5×10^5 cm/s, and the minimum trough velocity accelerated from 0.0 to 1.5×10^5 cm/s.

Figures 7 and 8 show the pressure variation for this mixing case. The contour plots (Fig. 7) show steep pressure gradients in the nozzles, but large pressure fluctuations do not occur in the base wake. The carpet plot (Fig. 8) indicates some pressure waves throughout the cavity. The peak pressure at the primary nozzle exit appears to traverse the cavity to the secondary side. This is corroborated by

CW CHEMICAL LASER FLOWFIELDS 347

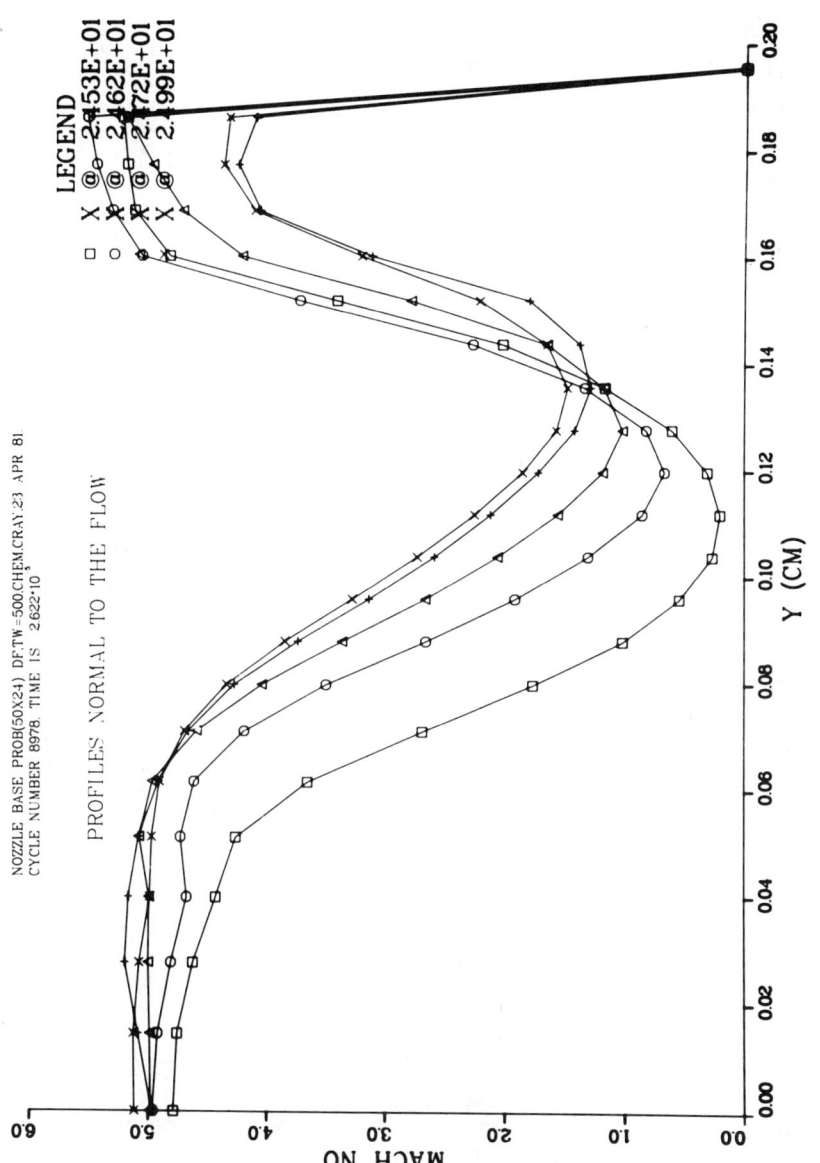

Fig. 5 Carpet plot of Mach number.

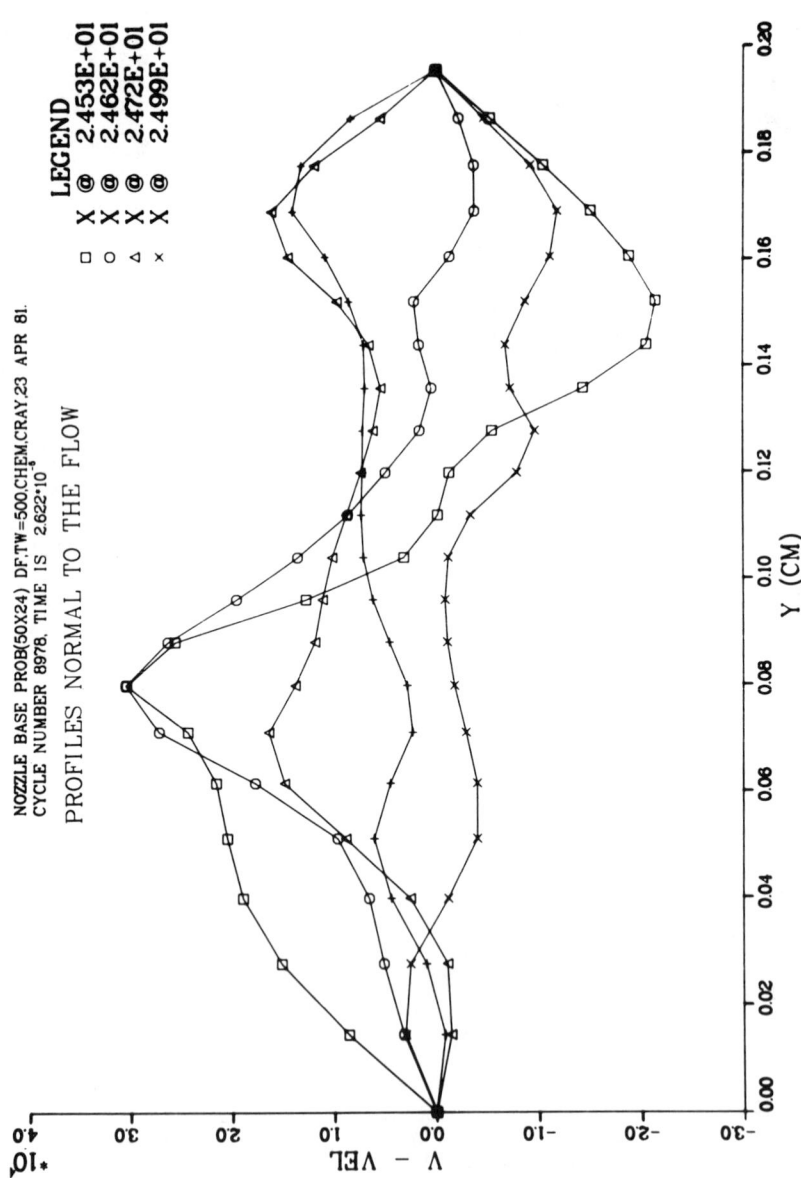

Fig. 6 Profile plot of axial velocity.

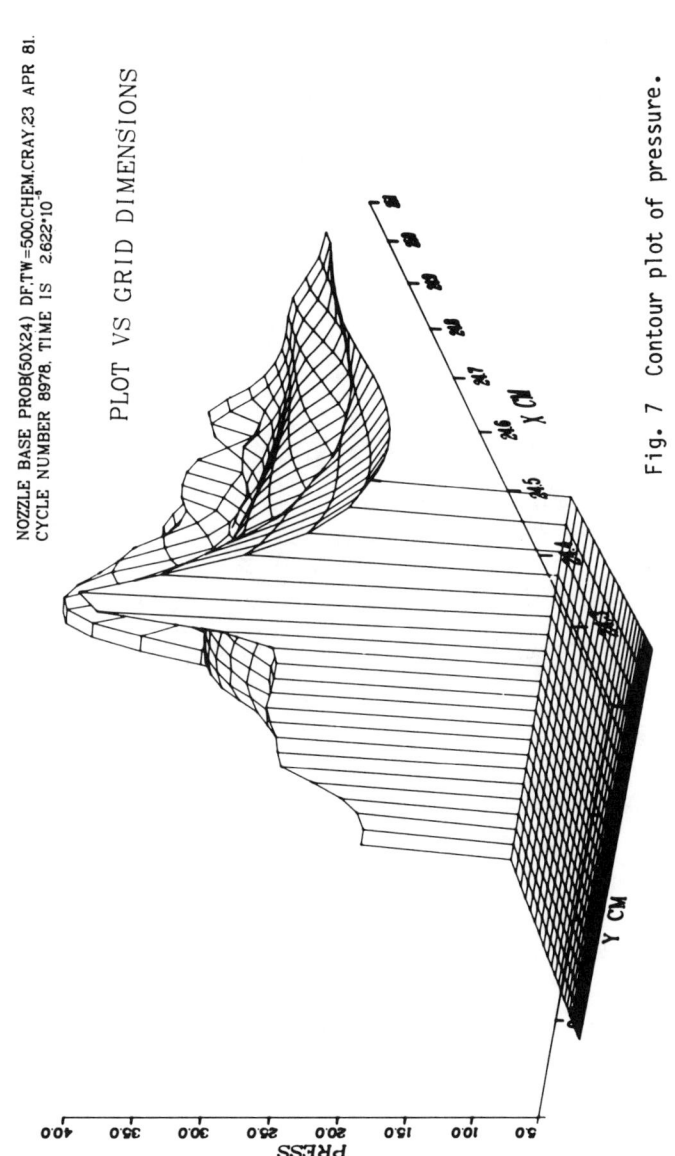

Fig. 7 Contour plot of pressure.

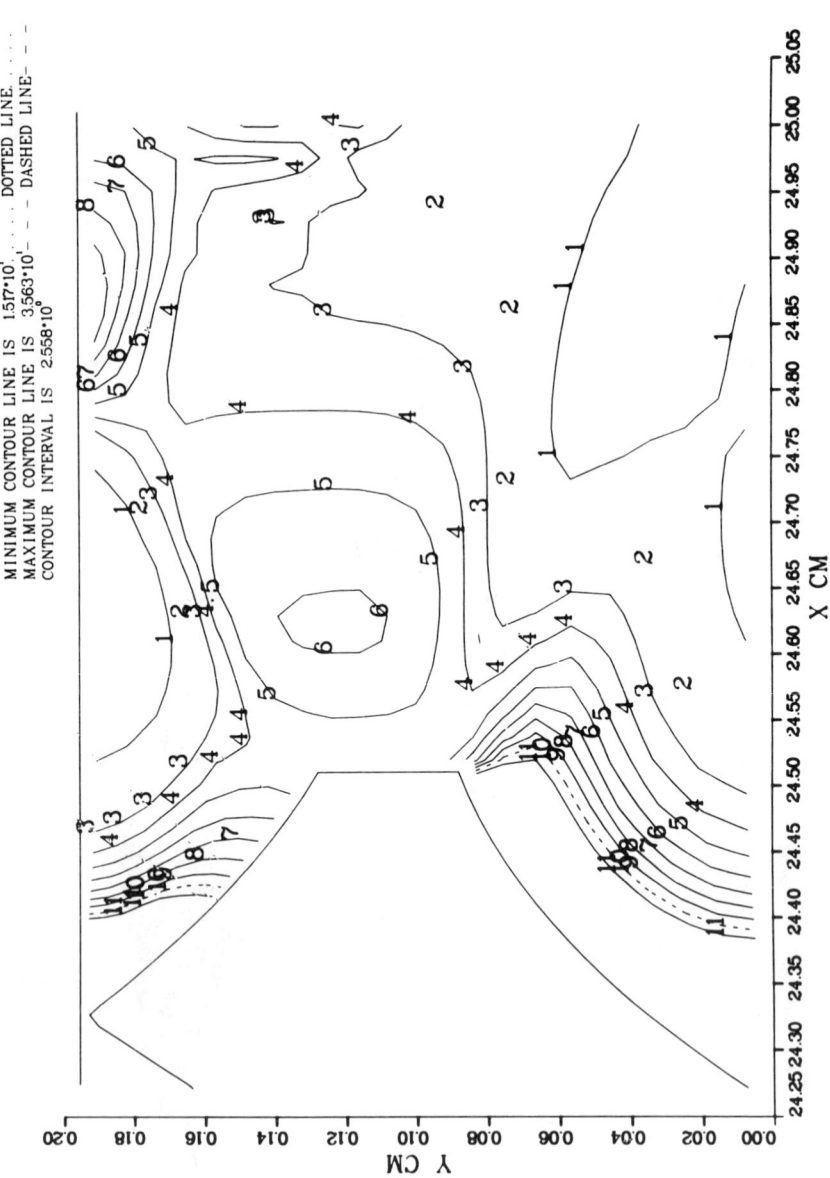

Fig. 8 Carpet plot of pressure.

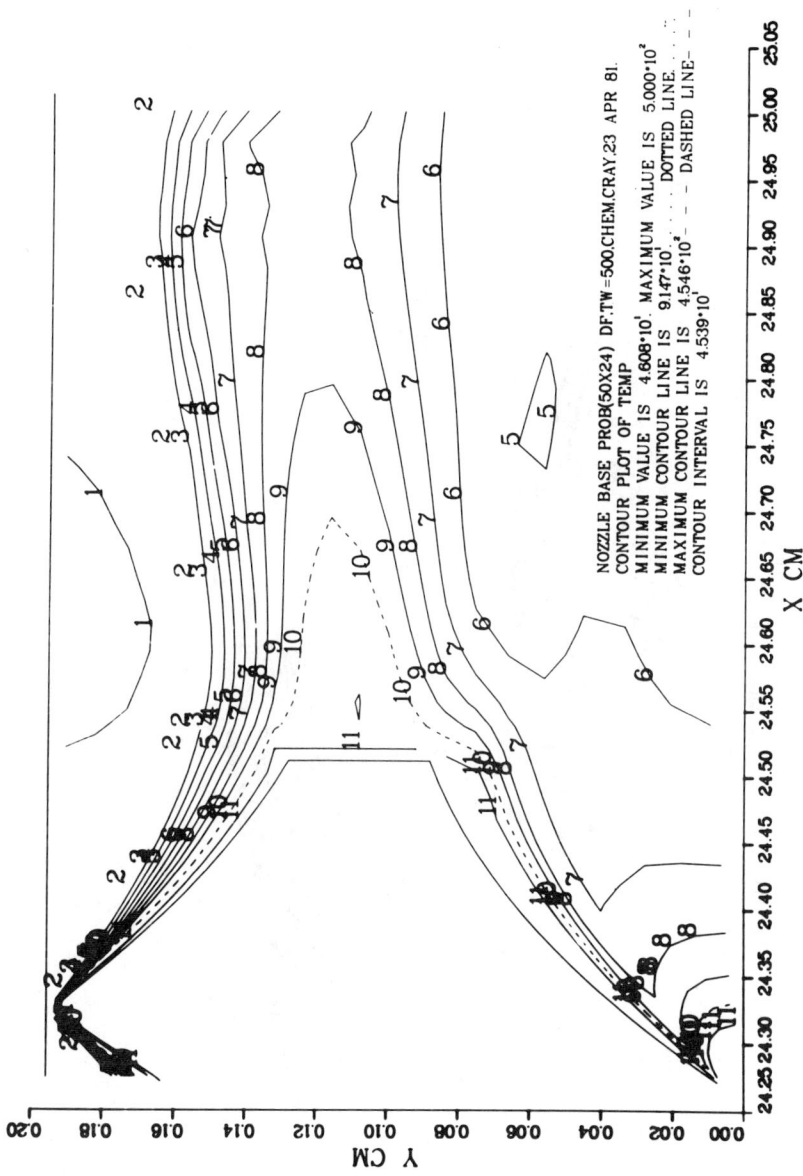

Fig. 9 Contour plot of temperature.

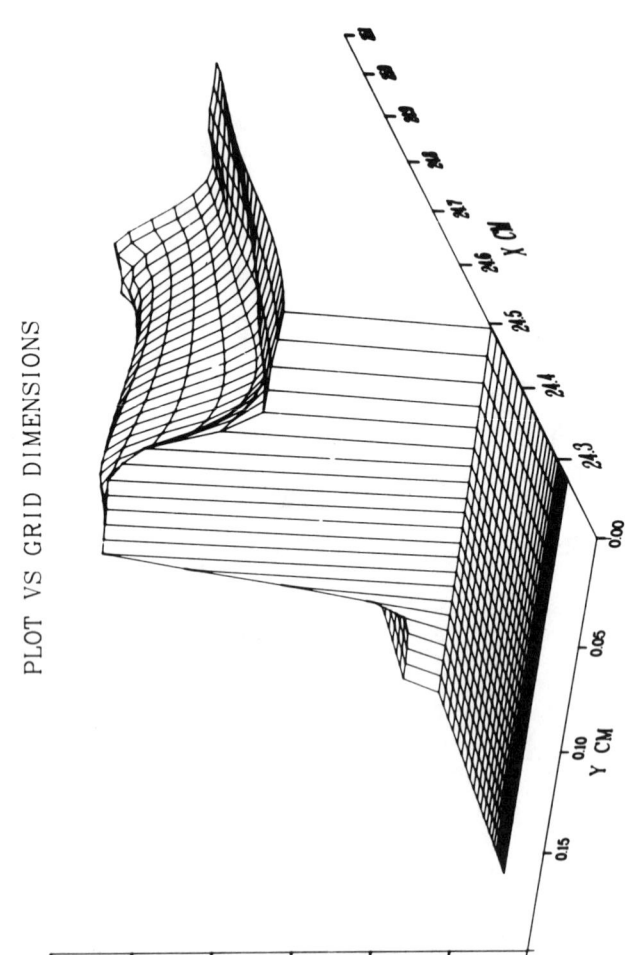

Fig. 10 Carpet plot of temperature.

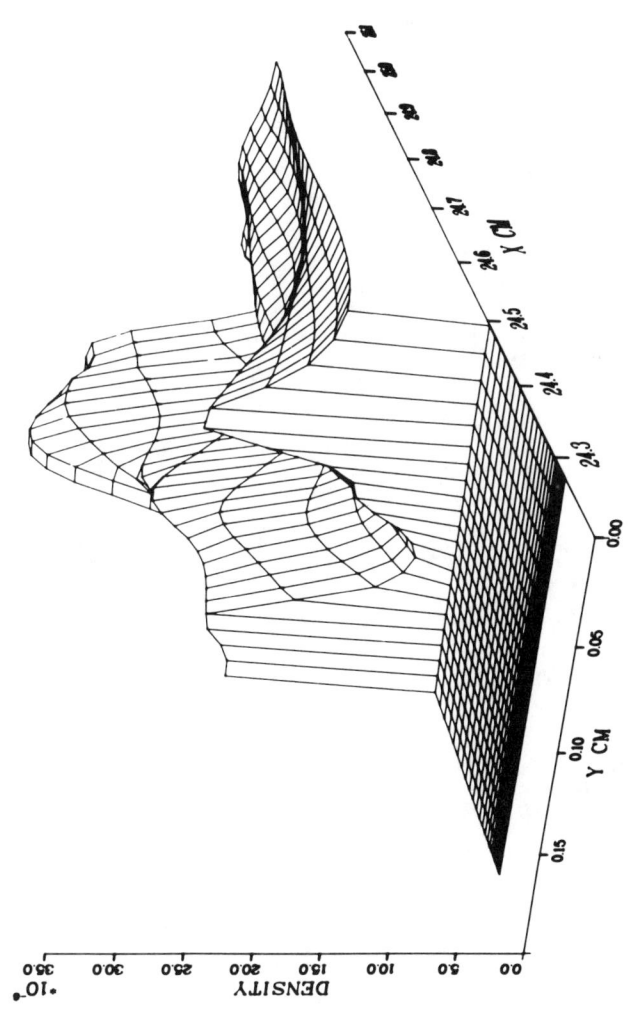

Fig. 11 Contour plot of density.

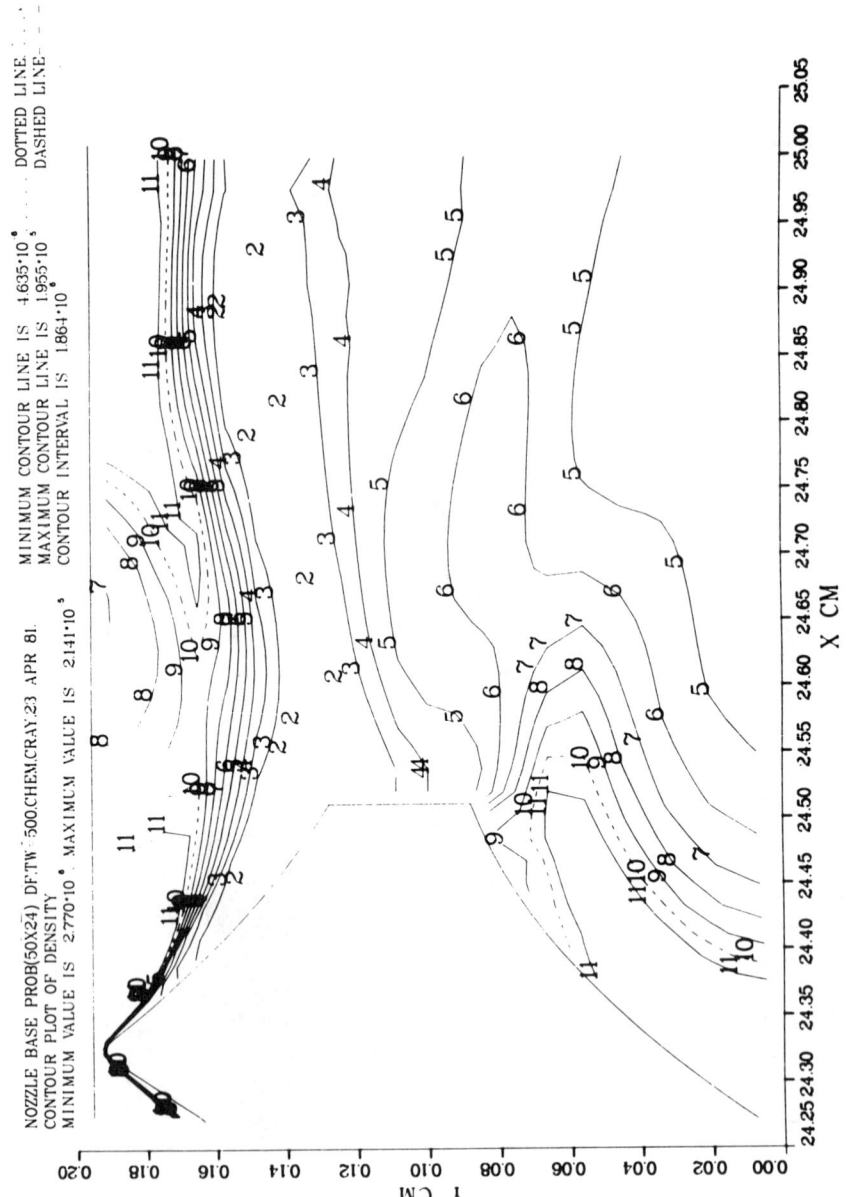

Fig. 12 Carpet plot of density.

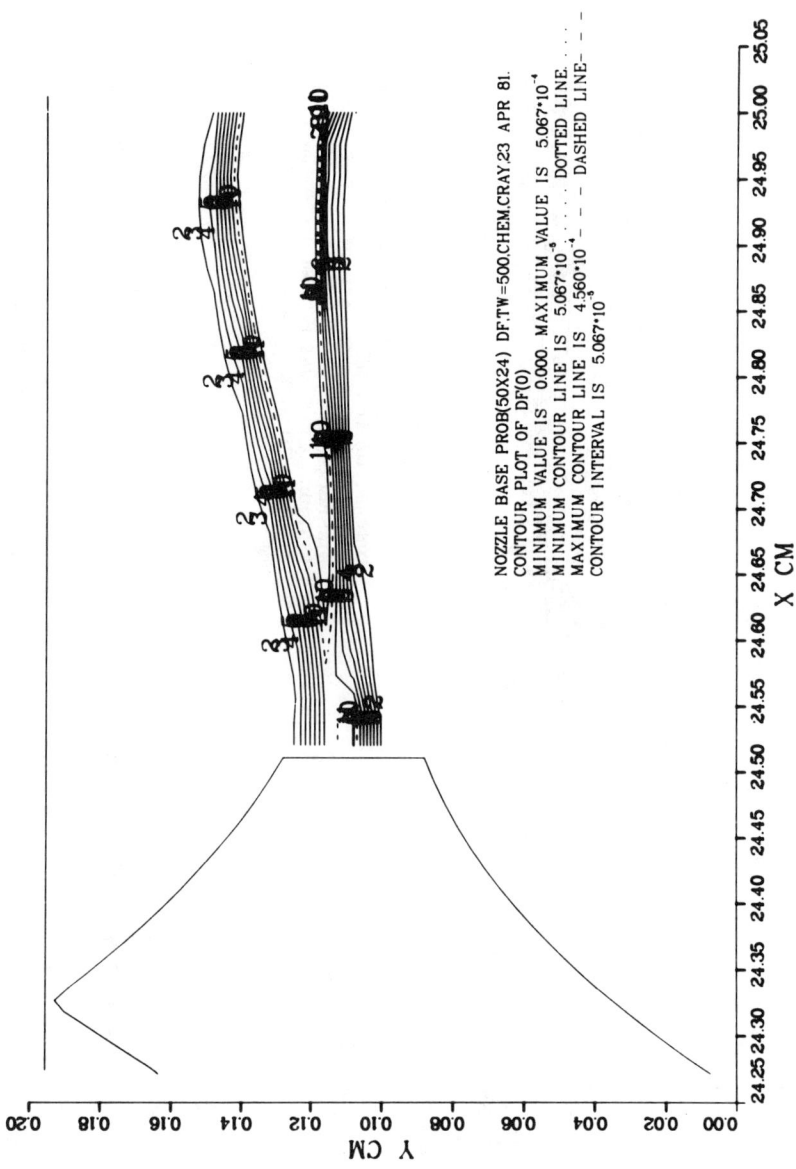

Fig. 13 Contour plot of density DF(0).

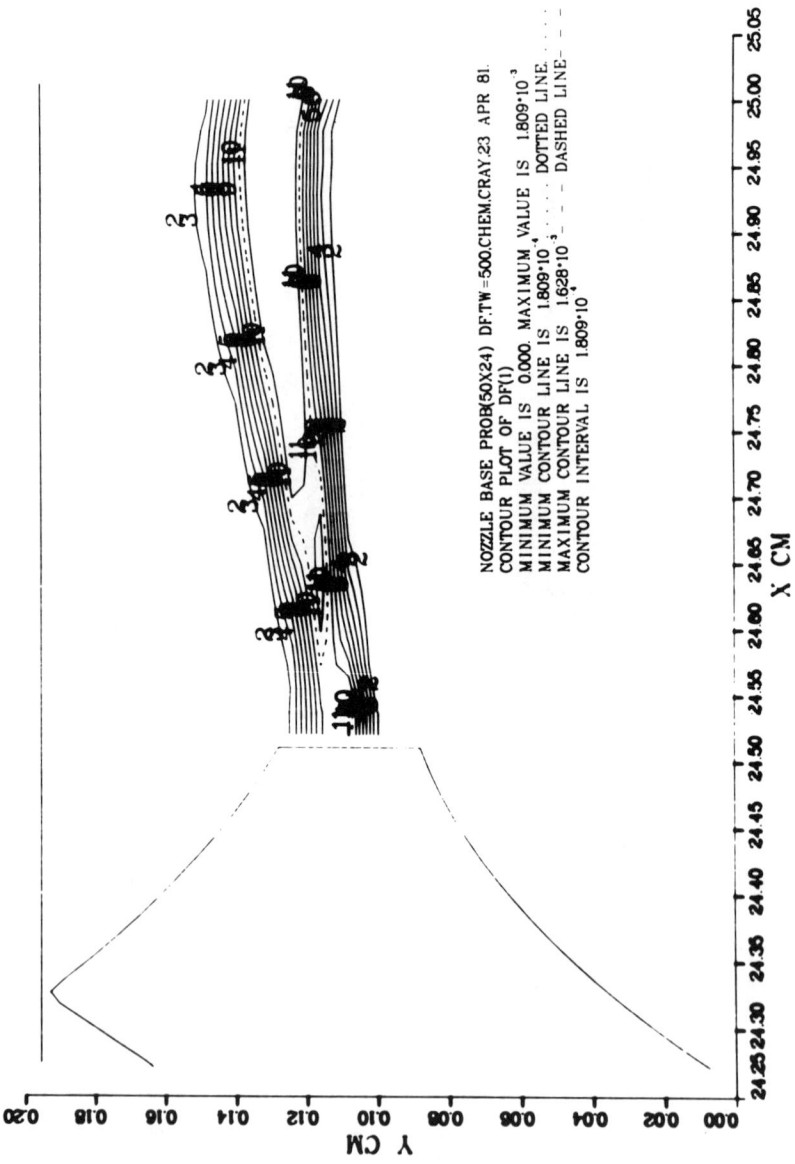

Fig. 14　Contour plot of density DF(1).

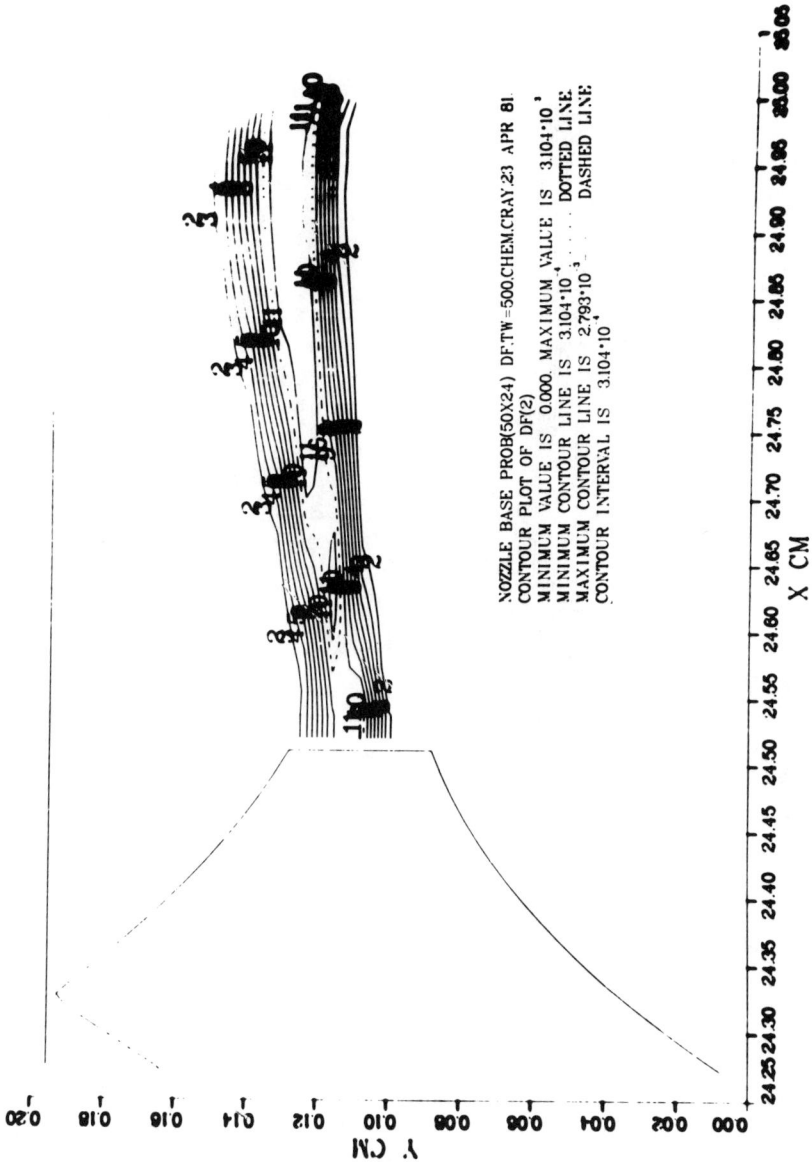

Fig. 15 Contour plot of density DF(2).

358 N.L. RAPAGNANI AND D.W. LANKFORD

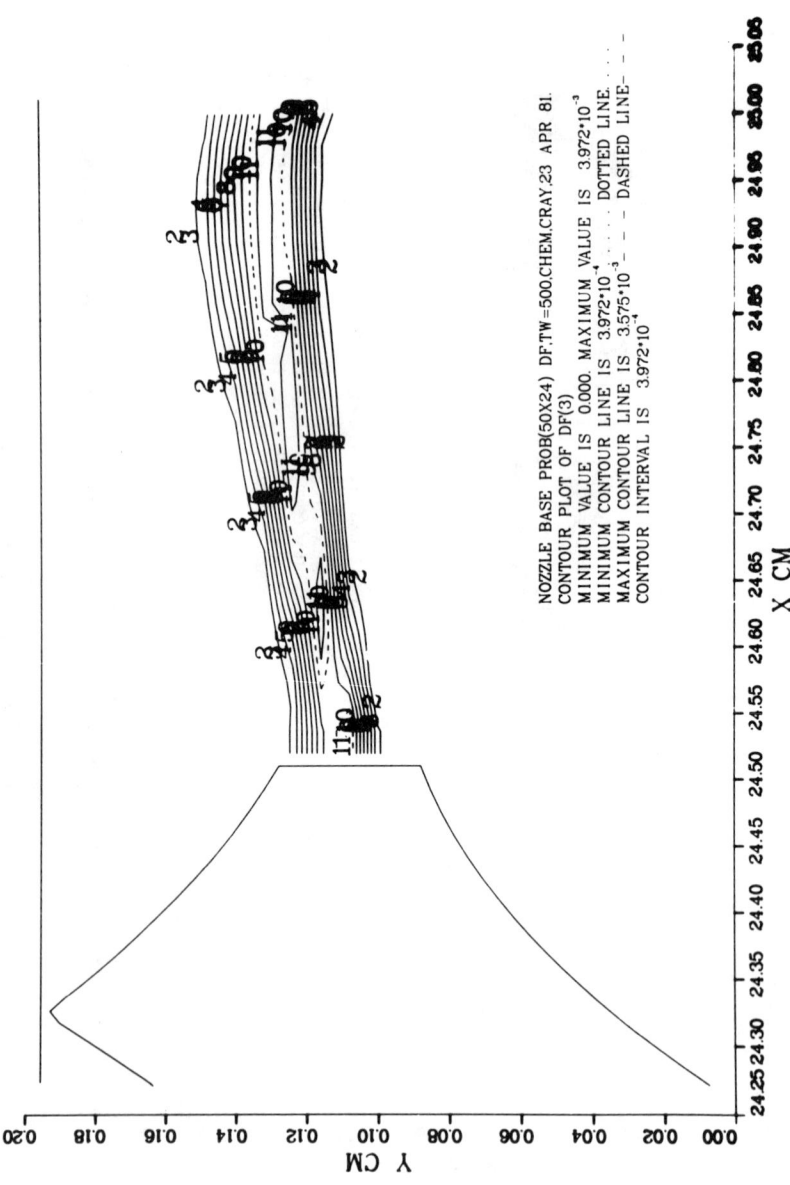

Fig. 16 Contour plot of density DF(3).

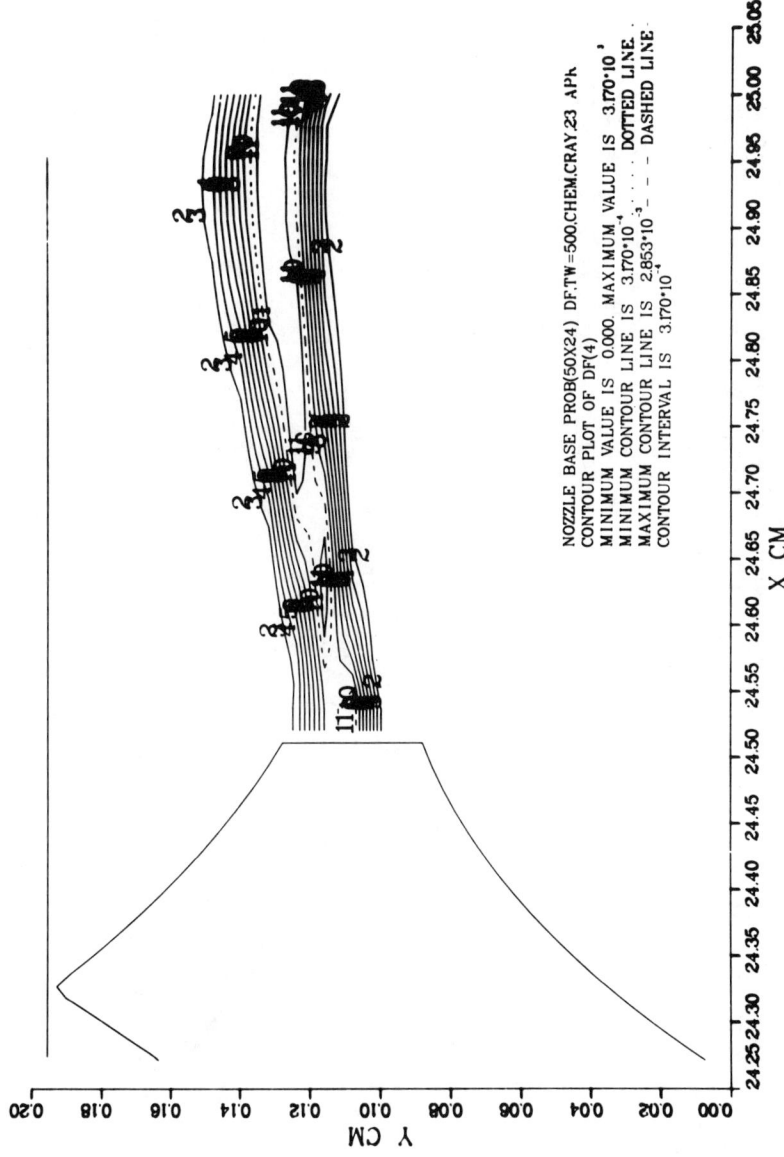

Fig. 17 Contour plot of density DF(4).

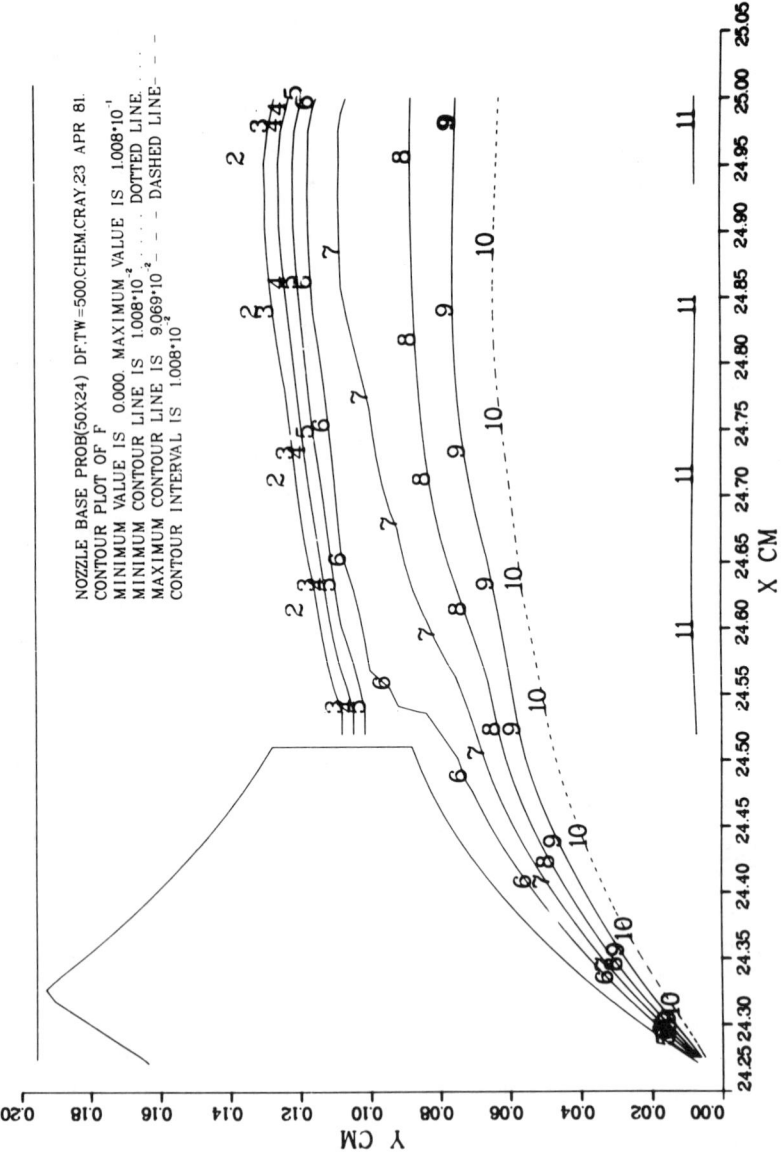

Fig. 18 Contour plot of density F.

Fig. 3, which shows the velocity vectors directed from the primary to the secondary pressure peak. The profile plots show a pressure peak of 35 Torr in the primary stream at an $\underline{x}$-position that is 0.03 cm beyond the base. The centerline pressure of both nozzles was 20 Torr at the NEP. The pressure gradient is seen to relax in the flow direction, and the peak is seen to travel from centerline to centerline. The wall pressure for these nozzles was experimentaly measured to be 23 Torr, which is the same as the calculated value.

The temperature contour plot, Fig. 9, shows the wall region to be the hottest. Reactions in the base wake were insufficient to overcome the effect of the base relief, and the temperature dropped gradually in the streamwise direction. The carpet plot, Fig. 10, shows this temperature ridge in the wake. The secondary nozzle centerline is seen to be the low-temperature area. It should be noted that no temperature overshoot occurred in the fluid near the wall and that it was controlled by the nozzle wall temperature. The temperature profiles are seen to relax in the downstream direction, the maximum being 400 K.

The high-density location corresponding to the low-temperature location is located along the secondary centerline, Fig. 11. The low-density region occurs just above the base wake region. A density trough is noted in the base wake, Fig. 12. A considerable momentum defect is expected to exist here since a velocity trough also occurs in the wake. These figures show the density profile becoming smoother as the flow progresses down the cavity, but the wake remains a substantial low-density region.

The chemical reactions producing vibrationally excited DF occur in this wake region. Figures 13-17 show the production of the various levels of DF. As can be seen from these figures, the DF concentrations are greatest in the mixing zone. These concentration levels are strongly influenced by the cold reaction, and this by the amount of F atoms immediately available for reaction. These calculations were made with Jumper's (1975) wall recombination model, which is a strong function of wall pressure and temperature. The F and F_2 contour plots are presented in Figs. 18 and 19. The effects of wall recombination can be seen in Fig. 19 where the F_2 profiles build up near the wall. The F atoms react with D_2 in the secondary stream, producing excited DF and D atoms. Thus the D atoms can be thought of as the summation of all the cold reactions. The D_2 and D atom contours are shown in Figs. 20 and 21. It is

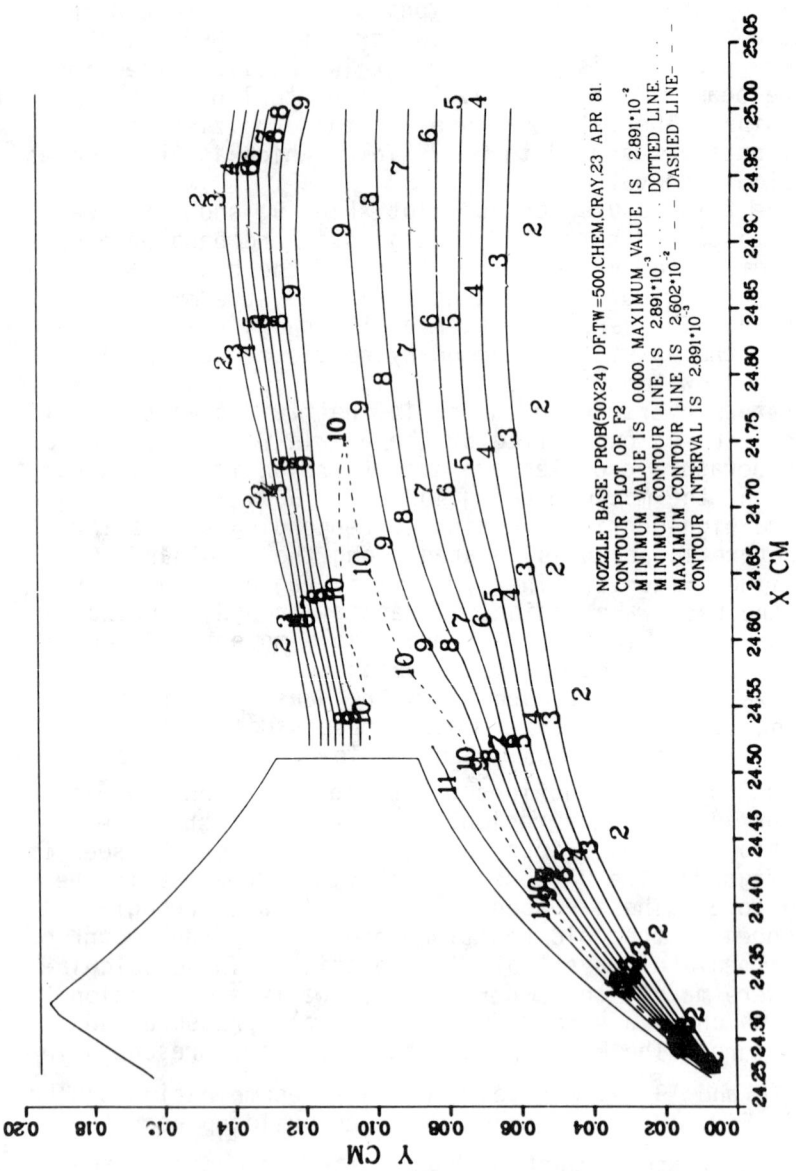

Fig. 19 Contour plot of density F_2.

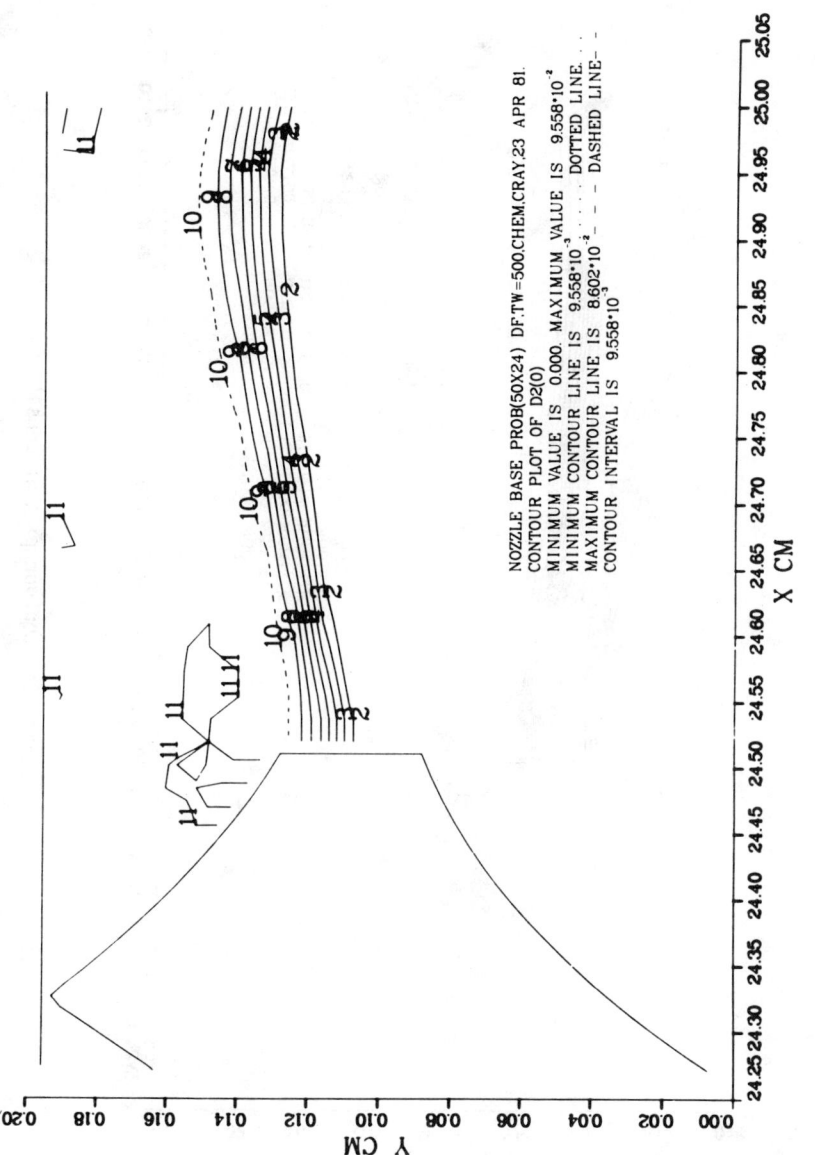

Fig. 20 Contour plot of density D_2.

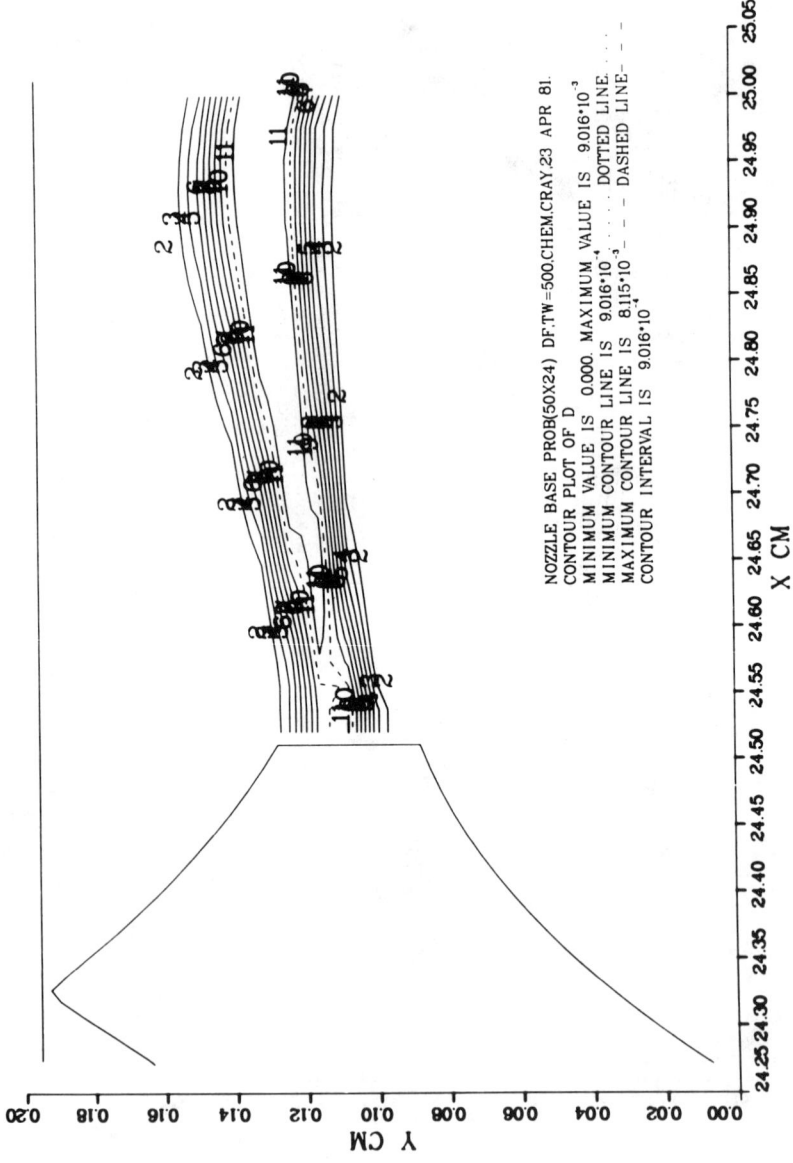

Fig. 21 Contour plot of density D.

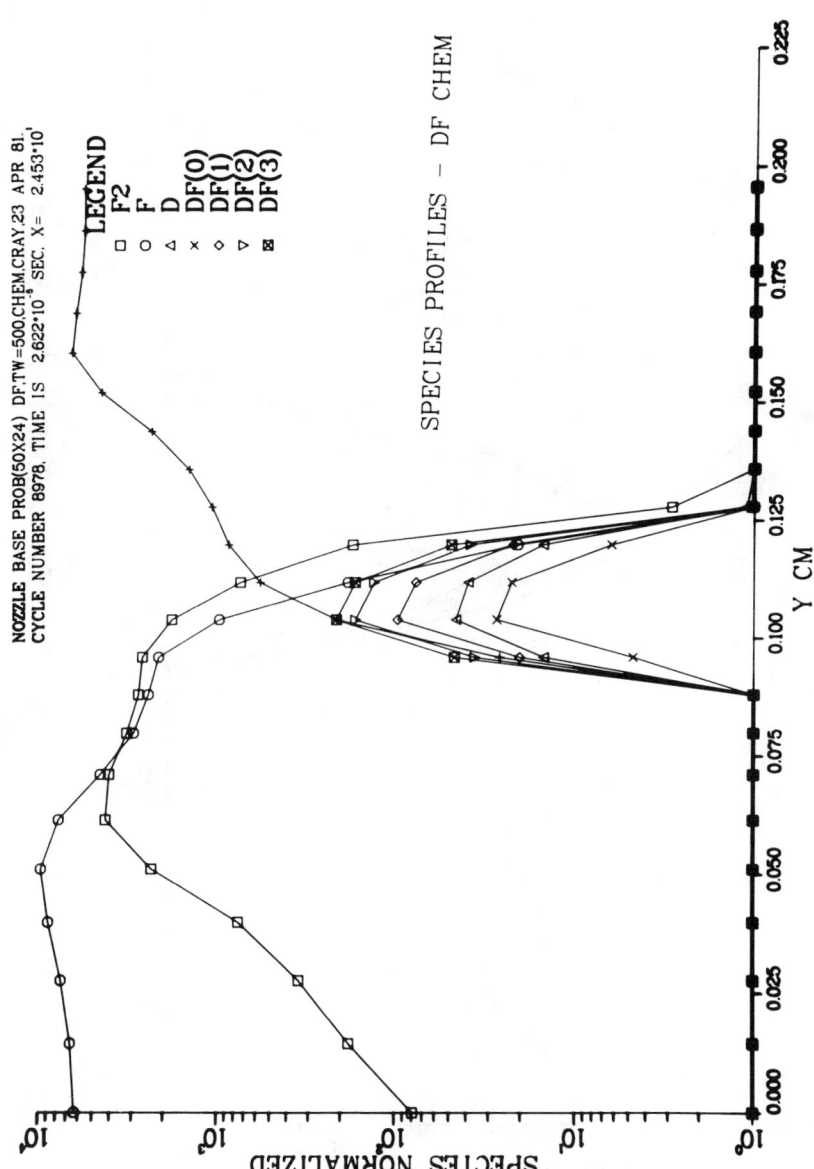

Fig. 22 Profile plot of major species at nozzle exit plane.

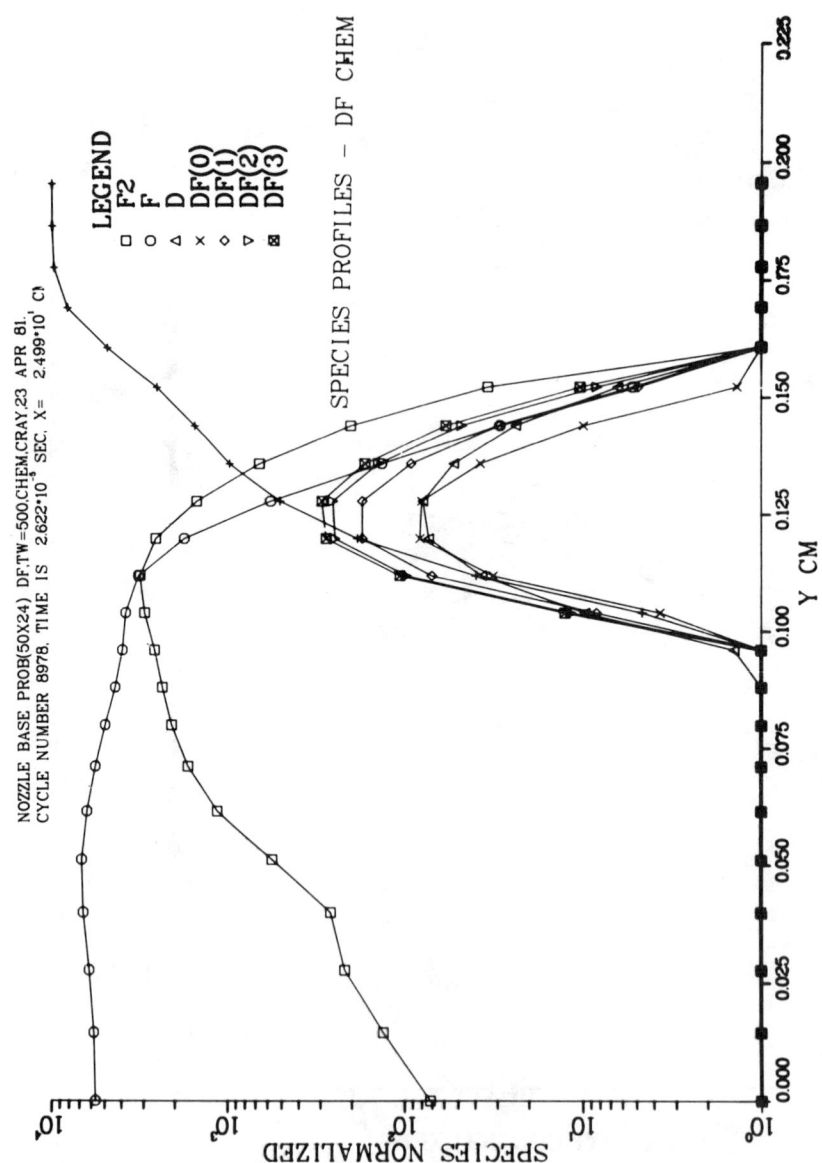

Fig. 23 Profile plot of major species at the exit plane of the calculation.

shown in Fig. 21 that by the time the flow has gone 0.5 cm downstream, the laminar mixing zone has spread only over 30% of the centerline to centerline distance. This is shown again, rather dramatically, in Figs. 22 and 23. These figures are profile plots of all the principle species at the NEP and at 0.5 cm from the NEP. The initial mixing is done because the pressure is higher in the primary nozzle (see Fig. 7) and by the time the flow has gone 0.5 cm downstream, the mixing has become diffusional because the mixing zone is growing faster into the primary nozzle than into the secondary nozzle, as is seen in Fig. 23. The pressure peak in the primary nozzle has displaced the mixing zone towards the secondary side. Further downstream, the pressure peak in the secondary nozzles forces the mixing zone backtowards the primary side. The slow growth of the mixing zone indicates diffusion limited mixing. The initial thicknesss of the mixing zone is probably a result of the low velocity in the base region, making the diffusion relatively fast.

Conclusions

A detailed calculation of the flow emerging from the primary and secondary nozzles into the cavity base region of a DF cw chemical laser is examined. The location of the initial mixing zone has been shown to be initially controlled by the pressure mismatch between the primary and secondary nozzles and eventually is controlled by diffusional effects. As the mixing zone was calculated to reamin narrow, macroscopic mixing phenomena such as vortices or turbulence would be expected to greatly enhance the mixing rate, causing large perturbations in the flowfield. No hot spots were seen in the flowfield owing to the heat release from the chemical reactions. A study of heat release due to chemical reaction will be the subject of an additional paper.

References

Cline, M. C. (1976) Computation of two-dimensional, viscous nozzle flow. AIAA J. 14, 295-296.

Cohen, N. (1977) A review of rate coefficients in the $D_2 + F_2$ chemical laser system. SAMSO TR-77-152, The Aerospace Corporation, Los Angeles, Calif.

Jumper, E. J. (1975) A model for fluorine recombination at a metal surface. Ph.D. Thesis, Air Force Institute of Technology, Dayton, Ohio; see also (1975) DS/ME-75-2 Wright-Patterson Air Force Base, Ohio.

Ramshaw, J. D. and Dukowicz, J. K. (1979) APACHE: a generalized-mesh eulerian computer code for multicomponent chemically reactive fluid flow. LA-7427, Los Alamos Scientific Laboratory, Los Alamos, N. Mex.

Rapagnani, N. L. and Anderson, D. A. (1980) Dominant kinetic rates in a typical flowing DF chemical laser. AFWL-TR-80-4, Kirtland Air Force Base, N. Mex.

Rivard, W. C., Farmer, O. A., and Butler, T. D. (1975) RICE: a compter program for multicomponent chemically reactive flows at all speeds. LA-5812, Los Alamos Scientific Laboratory, Los Alamos, N. Mex.

Zelazny, S. W. et al.(1978) Modeling DF/HF cw lasers: an examination of key assumptions. AIAA J. 16, 297-304.

Numerical Modeling of a Chemically Driven H_2-HCl Transfer Laser

V.K. Baev* and V.I. Golovichev†
USSR Academy of Sciences, Novosibirsk, USSR
and
H. Guenoche‡ and C. Sedes§
Université de Provence, Marseille, France

Abstract

This study is concerned with the modeling of a chemically driven H_2-HCl energy transfer laser in which hydrogen is previously vibrationally excited in a supersonic mixing-controlled device. The analysis complements a previous theoretical study for a H_2-HF system by Golovichev (1979a) and utilizes the same physical models. The numerical technique has been modified to include the capability for calculation of unbalanced supersonic flows. The aim of the investigation was to estimate the operation efficiency of a chemical HCl laser in the presence of vibrationally excited hydrogen in a test gas. In this case the process of cold "pumping," $Cl + H_2 \rightleftarrows HCl(v) + H$, is assumed to be rapid, and the process of nonresonance V-V exchange between vibrationally excited molecules $H_2(v)$ and $HCl(v)$ is assumed to be effective. At these conditions, vibrational energy transfer from $H_2(v)$ molecules excited in a glow discharge to $HCl(v)$ molecules produced in chemical reactions should be possible. The validity of investigation depends primarily on the

---

Presented at the 8th ICOGER, Minsk, USSR, Aug. 23-26, 1981. Copyright © American Institute of Aeronautics and Astronautics, Inc., 1982. All rights reserved.

*Professor and Deputy Director, Institute of Pure and Applied Mechanics.

†Senior Scientist, Institute of Pure and Applied Mechanics.

‡Professor and Director, Laboratoire de Dynamique et Thermophysique des Fluides.

§Research Scientist, Laboratoire de Dynamique et Thermophysique des Fluides.

kinetic rate coefficients for the above processes. They were estimated and then changed as parameters. The energy deposited in vibration of hydrogen has been also considered as a model parameter. The efficiency of the laser operation was predicted to be a 40-50-J/g mixture at pressures and flow parameters in the cavity typical of chemical HF laser. The total flow rate of toxic component Cl_2 can be decreased proportionally to the content of vibrationally excited H_2 in the mixture.

Introduction

Orayevsky et al. (1978) emphasized that combined methods of working gas vibrational excitation in laser media may efficiently produce partial inversion states by V-V exchange between molecules of an "energy carrier" gas and a lasing gas. A glow discharge can be used for selective "heating" of vibrational modes of an energy carrier gas which is then expanded through a supersonic nozzle and mixed with lasing gas molecules. High efficiency of vibrational-to-coherent radiation energy conversion in flow systems similar to mixing gasdynamic lasers with selective thermal excitation was demonstrated by Taylor et al. (1976) for a number of binary mixtures diluted by monoatomic gas.

In this study the laser system is chemically driven because a chemical reaction mechanism is of importance for laser operation, and energy transfer is realized between vibrationally excited H_2 molecules and V-R states of hydrogen-chloride product. For this system V-V interaction between $H_2(v)$ and $HCl(v)$ molecules is a principal mechanism of pumping, while the chemical mechanism is in this sense a subsidiary one and can be considered mainly as a source of vibrationally excited molecules of a lasing gas.

The analysis is limited to the two-dimensional flow produced by alternative supersonic jets of reactants. One nozzle injects vibrationally excited hydrogen, while the other contains thermal dissociation products of diluted molecular chlorine. In general the jet static pressures are not matched and the pressure gradients induced by compression and expansion waves as well as by chemical heat release highly influence the energy conversion process. The latter phenomenon is a matter of paramount importance for HCl-chemical laser operation, since vibrationally excited $HCl(v)$ molecules can be produced only in a "hot" branch of a chemical mechanism and the degree of Cl_2 dissociation cannot be arbitrarily increased. The use of a nozzle bank with the base relief has proved to be an effective means for preventing the flow "thermal blockage" due to heat release in a mixing zone (Grohs and Emanuel 1976). The domain of interest is the region between the centerlines of the adjacent nozzles, including the near wake region behind the nozzles bank with base relief.

In principle, the system of complete Navier-Stokes equations supplemented with the mass and energy conservation laws in the form of the relevant partial differential equations can be used to model the flow of multicomponent mixture with finite-rate physical and chemical processes. These equations were solved for the near wake region to investigate the effects of base injection conditions and heat release modeled by one-step chemical reaction on the pressure field and mixing quality. A detailed chemical mechanism was included into consideration on the basis of the so-called "parabolized" Navier-Stokes equations, which are advantageous when describing the ducted supersonic reacting gas flows at high Reynolds numbers in a distant wake region. The advanced k-ε model of turbulence (Launder and Spalding 1974) was adopted to account for the turbulent mixing effects. The efficiency of the energy conversion process was estimated on the same theoretical basis.

Mathematical Formulation

Governing Equations

The general conservation equations that govern the fluid dynamics, species mixing, and chemical reactions written in cylindrical coordinates with axial symmetry and zero azimuthal velocity in standard notation are continuity,

$$\frac{\partial \rho}{\partial t} + \frac{1}{r}\frac{\partial}{\partial r}(\rho v r) + \frac{\partial}{\partial z}(\rho u) = 0 \tag{1}$$

streamwise momentum,

$$\frac{\partial(\rho u)}{\partial t} + \frac{1}{r}\frac{\partial}{\partial r}(\rho u v r) + \frac{\partial}{\partial z}(\rho u^2) = -\frac{\partial p}{\partial z} + \frac{1}{r}\frac{\partial}{\partial r}(\sigma_{rz} r) + \frac{\partial \sigma_{zz}}{\partial z} \tag{2}$$

lateral momentum,

$$\frac{\partial(\rho v)}{\partial t} + \frac{1}{r}\frac{\partial}{\partial r}(\rho v^2 r) + \frac{\partial}{\partial z}(\rho u v) = -\frac{\partial p}{\partial r} + \frac{1}{r}\frac{\partial}{\partial r}(\sigma_{rr} r) + \frac{\partial \sigma_{rz}}{\partial z} - \frac{\partial \sigma_{\psi\psi}}{r} \tag{3}$$

energy,

$$\frac{\partial(\rho E)}{\partial t} + \frac{1}{r}\frac{\partial}{\partial r}(\rho v E r) + \frac{\partial}{\partial z}(\rho u E) = \frac{1}{r}\frac{\partial}{\partial r}(pvr) - \frac{\partial}{\partial z}(pu)$$

$$+ \frac{1}{r}\frac{\partial}{\partial r}(\sigma_{rr} v + \sigma_{rz} u) + \frac{\partial}{\partial z}(\sigma_{rz} v + \sigma_{zz} u)$$

$$- \frac{1}{r}\frac{\partial}{\partial r}(q_r r) - \frac{\partial}{\partial z}(q_z) + \dot{q}_c + \dot{q}_{rad} \quad (4)$$

species density,

$$\frac{\partial \rho_\ell}{\partial t} + \frac{1}{r}\frac{\partial}{\partial r}(\rho_\ell v r) + \frac{\partial}{\partial z}(\rho_\ell u) = \frac{1}{r}\frac{\partial}{\partial r}[\rho D_\ell r \frac{\partial}{\partial r}(\rho_\ell/\rho)]$$

$$+ \frac{\partial}{\partial z}[\rho D_\ell \frac{\partial}{\partial z}(\rho_\ell/\rho)] + (\dot{\rho}_\ell)_c + (\dot{\rho}_\ell)_{rad} \quad \ell = 1, \ldots, N_c \quad (5)$$

Mixture properties are defined by the following algebraic and differential relations:

$$\sigma_{zz} = 2\frac{\partial u}{\partial z} + \lambda \operatorname{div} \vec{v}$$

$$\sigma_{rr} = 2\frac{\partial v}{\partial r} + \lambda \operatorname{div} \vec{v}$$

$$\sigma_{rz} = \mu(\frac{\partial u}{\partial r} + \frac{\partial v}{\partial z})$$

$$\sigma_{\psi\psi} = 2\mu \frac{v}{r} + \lambda \operatorname{div} \vec{u}$$

$$\operatorname{div} \vec{v} = \frac{1}{r}\frac{\partial}{\partial r}(vr) + \frac{\partial u}{\partial z} \qquad \lambda = -\frac{2}{3}\mu$$

$$q_r = -k\frac{\partial T}{\partial r} - \Sigma_\ell \rho h_\ell(T)\frac{\partial}{\partial r}(\frac{\rho_\ell}{\rho})$$

$$q_z = -k\frac{\partial T}{\partial z} - \Sigma_\ell \rho h_\ell(T)\frac{\partial}{\partial z}(\frac{\rho_\ell}{\rho})$$

$$E = J + \tfrac{1}{2}(u^2 + v^2)$$

$$p = (\gamma-1)\rho J \quad \gamma = C_p/C_v$$

$$C_p = \sum_\ell (\rho_\ell/\rho) C_{p_\ell}(T) \quad C_v = \sum_\ell (\rho_\ell/\rho) C_{v_\ell}(T)$$

$$\rho = \sum_\ell \rho_\ell \quad \dot{q}_c = q^0 \dot{\omega} \quad \dot{q}_{rad} = (\dot{\rho}_\ell)_{rad} = 0$$

where μ and λ are the first and second viscosity coefficients; k is the thermal conductivity; $h_\ell(T)$ is the specific enthalpy of ℓ species; D_ℓ is the effective binary diffusion coefficient; ρ_ℓ is the density of ℓ species; p is the pressure; T is the static temperature; J is the specific internal energy; γ is the isentropic exponent; σ_{rr}, σ_{rz}, σ_{zz}, and $\sigma_{\psi\psi}$ are components of the viscous stress tensor; and q_z and q_r are components of energy flux due to the heat conduction and species diffusion.

The mathematical statement of the problem suggests the specification of kinetic equations in the form of phenomenological mass action law in accordance with the mechanism of kinetic processes or chemical reactions, reaction rate coefficients, and thermal properties of mixture components.

The chemical reaction scheme is set by stoichiometric relations

$$\sum_\ell \nu'_{\ell j} A'_\ell \quad \sum_\ell \nu''_{\ell j} A''_\ell$$

where $\nu'_{\ell j}$ and $\nu''_{\ell j}$ are the stoichiometric coefficients for the forward and backward j reaction, respectively; A'_ℓ and A''_ℓ are symbols of chemical species taking part in the reaction, $A'_\ell \neq A''_\ell$, $\ell \in N$; and k_f^j and k_B^j are the rate coefficients for forward and backward reactions, respectively.

The rate of the ℓ species density change from a chemical reaction may be written as follows:

$$(\dot{\rho}_\ell)_c = M_\ell \sum_{j=1}^{N_r} (\nu''_{\ell j} - \nu'_{\ell j}) \dot{\omega}_j \quad (6)$$

where

$$\dot{\omega}_j = k_f^j(T) \pi_\ell (\rho_\ell/M_\ell)^{\nu'_{\ell j}} - k_B^j(T) \pi_\ell (\rho_\ell/M_\ell)^{\nu''_{\ell j}}$$

Table 1 Kinetic model of chemical H_2/HCl laser

								A1	A2	A3
Cl	+ H_2(0)	+	= H	+ HCl(0)	+		1	0.480E 11	0.00	5.26
Cl	+ H_2(0)	+	= H	+ HCl(1)	+		2	0.480E 11	0.00	13.51
Cl	+ H_2(0)	+	= H	+ HCl(2)	+		3	0.480E 11	0.00	21.46
Cl	+ H_2(0)	+	= H	+ HCl(3)	+		4	0.480E 11	0.00	29.12
HCl(0)	+ HCl(2)	+	= HCL(1)	+ HCl(1)	+		5	0.300E 10	0.50	2.40
HCl(1)	+ HCl(2)	+	= HCl(0)	+ HCl(3)	+		6	0.100E 10	0.50	6.00
HCl(2)	+ HCl(2)	+	= HCl(1)	+ HCl(3)	+		7	0.900E 10	0.50	2.40
HCl(3)	+ MA	+	= HCl(2)	+ MA	+		8	0.540E-02	3.00	0.00
HCl(3)	+ MB	+	= HCl(2)	+ MB	+		9	0.150E 03	2.14	0.00
HCl(2)	+ MA	+	= HCl(1)	+ MA	+		10	0.360E-02	3.00	0.00
HCl(2)	+ MB	+	= HCl(1)	+ MB	+		11	0.100E 03	2.14	0.00
HCl(1)	+ MA	+	= HCl(0)	+ MA	+		12	0.180E-02	3.00	0.00
HCl(1)	+ MB	+	= HCl(0)	+ MB	+		13	0.500E 02	2.14	0.00
HCl(0)	+ M	+	= H	+ Cl	+ M		14	0.113E 19	-2.00	102.30
HCl(1)	+ M	+	= H	+ Cl	+ M		15	0.113E 19	-2.00	110.55
HCl(2)	+ M	+	= H	+ Cl	+ M		16	0.113E 19	-2.00	118.50
HCl(3)	+ M	+	= H	+ Cl	+ M		17	0.113E 19	-2.00	126.20
H_2(0)	+ M	+	= H	+ H	+ M		18	0.240E 17	0.00	61.50
Cl_2	+ M	+	= Cl	+ Cl	+ M		19	0.615E 19	-2.00	57.30
H	+ Cl_2	+	= Cl	+ HCl(0)	+		20	0.172E 09	0.00	2.40
H	+ Cl_2	+	= Cl	+ HCl(1)	+		21	0.172E 11	0.00	2.40
H	+ Cl_2	+	= Cl	+ HCl(2)	+		22	0.344E 11	0.00	2.40
H	+ Cl_2	+	= Cl	+ HCl(3)	+		23	0.579E 11	0.00	2.40
H	+ Cl_2	+	= Cl	+ HCl(4)	+		24	0.116E 11	0.00	2.40
Cl	+ H_2(0)	+	= H	+ HCl(4)	+		25	0.480E 11	0.00	36.48
HCl(0)	+ HCl(4)	+	= HCl(1)	+ HCl(3)	+		26	0.800E 09	0.50	7.60
HCl(1)	+ HCl(4)	+	= HCl(2)	+ HCl(3)	+		27	0.266E 11	0.50	6.00
HCl(2)	+ HCl(4)	+	= HCl(3)	+ HCl(3)	+		28	0.180E 11	0.50	2.40
HCl(4)	+ MA	+	= HCl(3)	+ MA	+		29	0.720E-02	3.00	0.00
HCl(4)	+ MB	+	= HCl(3)	+ MB	+		30	0.200E 03	2.14	0.00
Cl	+ H_2(1)	+	= H	+ HCl(0)	+		31	0.960E 11	0.00	5.26
Cl	+ H_2(1)	+	= H	+ HCl(1)	+		32	0.960E 11	0.00	13.51
Cl	+ H_2(1)	+	= H	+ HCl(2)	+		33	0.960E 11	0.00	21.46
Cl	+ H_2(1)	+	= H	+ HCl(3)	+		34	0.960E 11	0.00	29.12
Cl	+ H_2(1)	+	= H	+ HCl(4)	+		35	0.960E 11	0.00	36.48
HCl(1)	+ H_2(0)	+	= HCl(0)	+ H_2(1)	+		36	0.534E 11	-1.00	0.00
HCl(2)	+ H_2(0)	+	= HCl(1)	+ H_2(1)	+		37	0.107E 12	-1.00	0.00
HCl(3)	+ H_2(0)	+	= HCl(2)	+ H_2(1)	+		38	0.160E 12	-1.00	0.00
H_2(1)	+ M	+	= H_2(0)	+ M	+		39	0.250E-06	4.30	0.00

[a]Rate coefficients in a form kf = A1 T A2 (-a3/t) in liter, mole, and second (T* = T/1000.0*R).

[b]Reaction rate coefficients for stages 36-38 have non-Arrhenius form.

[c]Ma = He, MB = H, Cl, H_2, Cl_2, HCl(V), V = 1, ..., 4, M= MA + MB.

The rate of energy release from a chemical reaction is given by

$$\dot{q}_{cj} = \dot{q}_j^o \omega_j \qquad (7)$$

where q_j^o is the heat of reaction.

In the presence of several chemical reactions the net rates of species densities change and energy release are obtained by summation of contributions to each reaction.

The kinetic model that governs the operation of HCl-chemical laser is listed in Table 1 (Sêdes 1977). A complete reaction scheme used in the calculations (39 elementary stages) may be divided as follows: the reaction of "cold" pumping between atomic chlorine and molecular hydrogen (excited and in the ground state) (stages 1-4, 25, 31-35); V-V processes of molecular energy redistribution between vibrational modes of an active molecule HCl(v) (stages 5-7, 26-28); V-T processes of HCl(v) and H_2(v) collisional deactivation (stages 8-13, 39); processes of nonequilibrium dissociation of excited molecules HCl(v) and H_2(v) (stages 14-18); reactions of "hot" pumping between atomic hydrogen and molecular chlorine (stages 20-24); V-V exchange processes between HCl(v) and H_2(v) molecules (stages 36-38).

The forward reaction rate coefficients are taken in the Arrhenius form

$$k_f^j(T) = A_j T^{n_j} \exp(-E_j/R_o T)$$

Those for the backward reaction rates can be defined using the concept of detailed equilibrium,

$$k_B^j(T) = k_f^j(T)(1/R_o T)^{\nu_j}/k_p^j(T) = k_f^j(T)/k_c^j(T)$$

where $\nu_j = 0$ for bimolecular reactions, and $\nu_j = 1$ for thermolecular ones.

Rate coefficients for V-V exchange processes between H_2(v) and HCl(v) molecules were approximated by the non-Arrhenius relations:

$$k_f^j(H_2^{u \to u-1}, HCl^{v-1 \to v}) = uv[8.89 \times 10^{-11} T^{-1} \exp(-1.633 \times 10^{-3} \Delta E'')]$$

$$k_B^j(H_2^{u-1 \to u}, HCl^{v \to v-1}) = k_f^j(H_2^{u \to u-1}, HCl^{v-1 \to v})$$

$$[\exp(-1.4388 \Delta E''/T)]$$

where u = 1 and v = 1, 2, 3 are vibrational quantum numbers, and

$$\Delta E'' = (\omega_e - 2u\omega_e x_e)_{H_2} - (\omega_e - 2v\omega_e x_e)_{HCl}$$

It was assumed for the reaction rate coefficients of "cold" pumping with participation of $H_2(v)$ that vibrational excitation of one reactant either does not lead to a change of the elementary process rate or increases it by changing the relative efficiency of $H_2(v)$ molecules in collisions.

The perfect gas equations of state are used in the forms:

$$c_p = c_p(T) \quad c_v = c_v(T)$$

$$\ln(k_c^j) = \Delta G_0^j / R_o T - \ln(R_o T) \nu_j$$

$$\Delta G_0^j = \sum_\ell M_\ell (\nu''_{\ell j} - \nu'_{\ell j}) \mu_\ell^0 \tag{8}$$

where ΔG_0 is the standard Gibbs free energy of j reaction, and μ_ℓ^0 is the chemical potential of ℓ species at the temperature T and the unit pressure.

In the laminar limit numerical solutions of the time-dependent Navier-Stokes equations and species conservation equations for a simplified chemical mechanism are useful for the estimation of the supersonic reactive flow parameters. For turbulent mixing and detailed kinetic mechanism, however, storage and computer time required exceed the capacity of the available computers. It is therefore expedient to search for another mathematical approach to solve the problem more realistically than on the basis of complete Navier-Stokes equations.

The parabolic approximation of Navier-Stokes equations can be formally obtained for the steady-state ($\partial/\partial t = 0$) form of Eqs. (1-5) through the elimination of the second derivatives of unknown functions with respect to the spatial variable which coincides with the direction of a main flow. The resulting equations can be easily set in numerical form. Numerical solutions are sufficiently economical, since a marching scheme of numerical integration can be constructed. The numerical model can incorporate, when applied to the analysis of two-dimensional turbulent flows, the basic concepts of semiempirical turbulence theories. For the purposes of the present work the supplementary conservation equations for the kinetic energy of turbulence and for its effective dissipation rate were solved:

$$\rho u \frac{\partial k}{\partial z} + \rho v \frac{\partial k}{\partial r} = \frac{1}{r}\frac{\partial}{\partial r}\left(\frac{\mu_e^r}{\sigma_k} \cdot \frac{\partial k}{\partial r}\right) + S_k \qquad (9)$$

$$\rho u \frac{\partial \varepsilon}{\partial z} + \rho v \frac{\partial \varepsilon}{\partial r} = \frac{1}{r}\frac{\partial}{\partial r}\left(\frac{\mu_e^r}{\sigma_\varepsilon} \frac{\partial}{\partial r}\right) + S_\varepsilon \qquad (10)$$

The latter quantities are the averaged scalar "analogs" of the appropriate tensor pulsation modes:

$$k = \langle u'_\alpha u'_\alpha \rangle \qquad \varepsilon_{\alpha\beta} = 2\, \nu_{\ell am} \left\langle \frac{\partial u_\alpha}{\partial x_\beta} \frac{\partial u_\alpha}{\partial x_\beta} \right\rangle$$

The source terms in Eqs. (9,10) are defined as

$$S_k = G_k - \rho\varepsilon \qquad S_\varepsilon = (\varepsilon/k)(c_1 G_k - c_2 \rho\varepsilon)$$

$$G_k = \mu_t \left\{ 2\left[\left(\frac{\partial u}{\partial z}\right)^2 + \left(\frac{\partial v}{\partial r}\right)^2 + \left(\frac{v}{r}\right)^2 + \left(\frac{\partial u}{\partial r} + \frac{\partial v}{\partial z}\right)^2 \right] \right\} \qquad (11)$$

The turbulence model is "closed" by the functional dependence of the effective turbulent viscosity μ_e on k and ε that specify the pulsation motion:

$$\mu_e = \mu_{\ell am} + \mu_t(k,\varepsilon) = \mu_{\ell am} + c\mu\rho k^2/\varepsilon$$

The model contains five "universal" constants: $c_\mu = 0.09$; $c_1 = 1.43$; $c_2 = 1.92$; $\sigma_k = 0.9$; and $\sigma_\varepsilon = 2.72$. Here $k = 0.4$ is the von Kármán constant.

If the initial and boundary conditions are specified, the mathematical problems considered can be numerically analyzed.

Numerical Solution

The finite-difference analogs of the general conservation laws in terms of the complete Navier-Stokes equations are solved by the partially implicit (ICE) method (Rivard et al. 1975) that employs the concept of operator splitting with respect to physical processes. The splitting of the difference scheme does not decrease the order of approximation, but increases the computational efficiency of the method, since there is a possibility of taking into account the physical processes proceeding at essentially different rates in a consecutive order. Separation of the gasdynamic

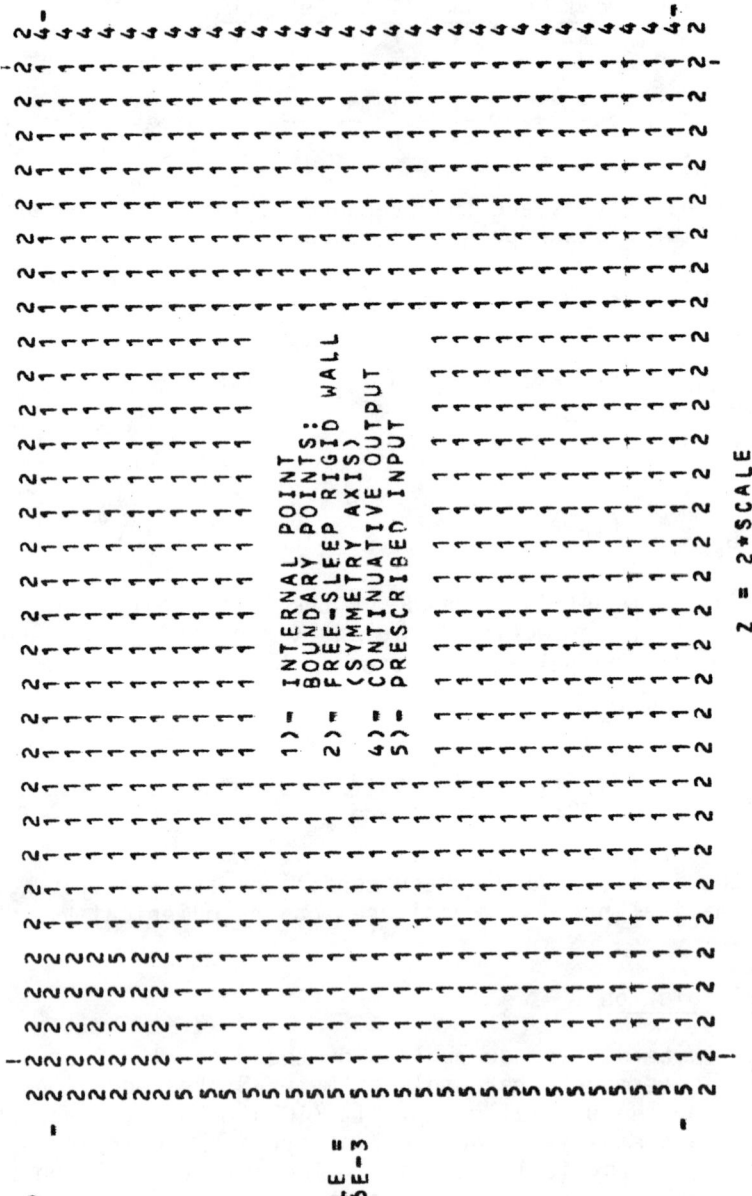

Fig. 1 Flow geometry and boundary conditions for the sample problem.

processes from the chemical ones and calculation of the latter with the help of implicit difference schemes are of importance. It is necessary to split up various terms and solve a smaller set of equations by the implicit method, especially when chemical changes proceed much faster than gasdynamic ones. The RICE method is the simplest operator splitting when the individual operators are both implicit or explicit, but explicitly coupled.

For the numerical solution of the problem, the computational domain according to the flow geometry (see Fig. 1) is divided into a great number of elementary cells with dimensions δr and δz. Following Rivard et al. (1975), the finite-difference approximation of the initial equations in the predictor-corrector form for the equations of continuity and momentum are

$$\rho_{ij}^{n+1} = \tilde{\rho}_{ij} + \theta \Delta t \left\{ \frac{1}{r_j \delta r} \left[(\rho v)_{i,j-\frac{1}{2}}^{n+1} r_{j-\frac{1}{2}} - (\rho v)_{i,j+\frac{1}{2}}^{n+1} r_{j+\frac{1}{2}} \right] \right.$$
$$\left. + \frac{1}{\delta z} [(\rho u)_{i-\frac{1}{2},j}^{n+1} - (\rho u)_{i+\frac{1}{2},j}^{n+1}] \right\} \quad (12)$$

where Δt is the time interval corresponding to the integration step, and θ is the parameter used for varying the relative time centering of the convection terms. The value $\tilde{\rho}_{ij}$ is determined by the values at the n-th time

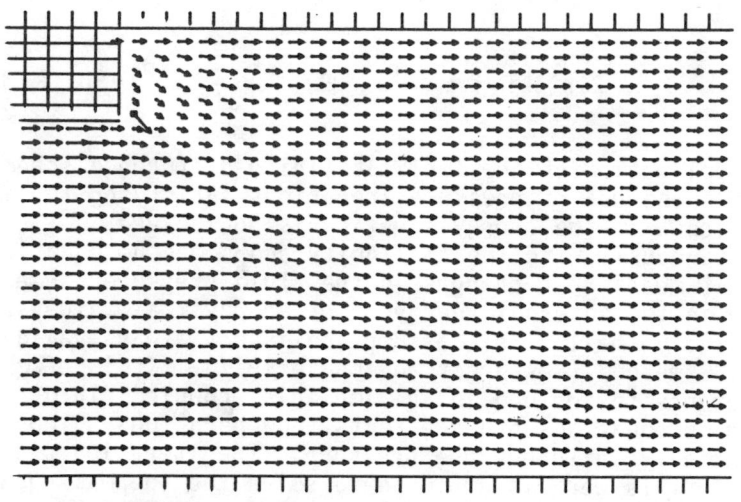

Fig. 2 Velocity vectors in the near wake flow with the H_2 injection along the flow centerline.

level:

$$\tilde{\rho}_{ij} = \rho^n_{ij} + \Delta t \left\{ (1-\theta) \left[-\frac{1}{r_j \delta r} (\rho v)^n_{i,j-\frac{1}{2}} r_{j-\frac{1}{2}} - (\rho v)^n_{i,j+\frac{1}{2}} r_{j+\frac{1}{2}} \right] \right.$$

$$\left. + \frac{1}{\delta z} \left[(\rho u)^{n+1}_{i-\frac{1}{2},j} - (\rho v)^{n+1}_{i+\frac{1}{2},j} \right] \right\} \quad (13)$$

The numerical approximation of the momentum equations is written down in the following form:

$$(\rho v)^{n+1}_{i,j+\frac{1}{2}} = (\tilde{\rho} v)_{i,j+\frac{1}{2}} \pm \psi \frac{\Delta t}{\delta r} \delta \overline{P}_{ij}$$

$$(\rho u)^{n+1}_{i\pm\frac{1}{2},j} = (\tilde{\rho} u)_{i\pm\frac{1}{2},j} \pm \psi \frac{\Delta t}{\delta z} \delta \overline{P}_{ij} \quad (14)$$

where the intermediate values $\tilde{\rho} u$ and $\tilde{\rho} v$ are determined by relations similar to Eq. (13). The expression for $\tilde{\rho} v$ is

$$(\tilde{\rho} v)_{i,j+\frac{1}{2}} = (\tilde{\rho} v)^n_{i,j+\frac{1}{2}} + \Delta t \left(\frac{1}{r_{j+\frac{1}{2}} \delta r} \left\{ r_j [(\rho v^2)^n_{ij} - (\sigma_{rr})^n_{ij}] \right. \right.$$

$$\left. - r_{j+1} (\rho v^2)^n_{i,j+1} - (\sigma_{rr})^n_{i,j+1} \right\} + \frac{1}{\delta z} [(\rho uv)^n_{i-\frac{1}{2},j+\frac{1}{2}}$$

$$- (\sigma_{rz})^n_{i-\frac{1}{2},j+\frac{1}{2}} - (\rho uv)^n_{i+\frac{1}{2},j+\frac{1}{2}} + (\sigma_{rz})^n_{i+\frac{1}{2},j+\frac{1}{2}}]$$

$$\left. + \frac{1}{2\delta r} (p^n_{i,j-1} - p^n_{i,j+1}) - \frac{1}{r_{j+\frac{1}{2}}} (\sigma_{\psi\psi})_{i,j+\frac{1}{2}} \right) \quad (15)$$

The application of the implicit finite-difference scheme assumes the use of pressure iterations satisfying the mass equation. If the energy equation is not solved simultaneously with the continuity and momentum equations, then it means that changes in the fluid pressure caused by those in the internal energy or other physical values are relatively small. This assumption is not always valid in the ICE method (Westbrook and Chase 1977). It is of importance to include in the implicit, coupled fluid dynamics equations other effects and operations which could essentially influence the pressure. The iteration procedure used in this work is employed to solve the thermal equation of state with a prescribed accuracy. The solutions for ρ^{n+1}_{ij}, $(\rho v)^{n+1}_{i,j+\frac{1}{2}}$, and $(\rho u)^{n+1}_{i+\frac{1}{2},j}$ are accomplished by iterating

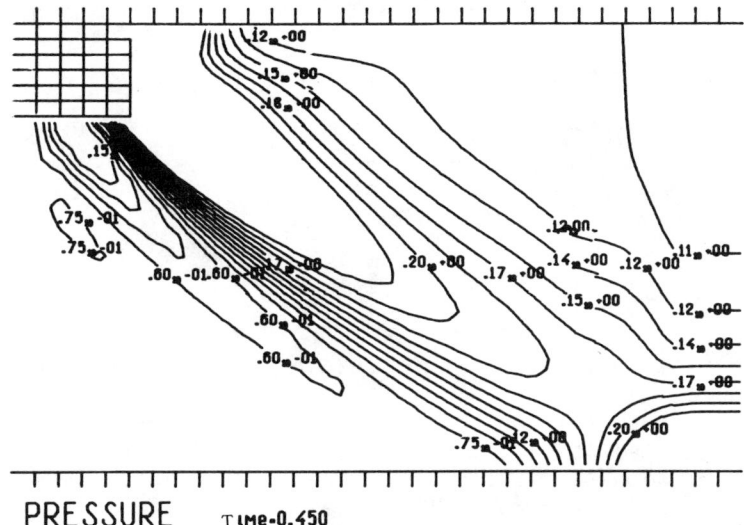

Fig. 3 Pressure isolines for the case corresponding to Fig. 2.

over the pressure $\overline{p}_{ij} = p^n_{ij} + \delta\overline{p}_{ij} + (\gamma-1)\delta(\rho\theta)_{ij}$. The internal energy is updated by the pdV work term

$$\frac{\partial(\rho\theta)}{\partial t} = -\psi p \text{ div } \vec{v} \qquad (16)$$

where the role of ϕ factor is similar to that of θ in Eq. (12). Details of computational procedure have been given by Brandt (1980). The energy equation is integrated then with the help of the explicit scheme of Eq. (13) with the updated values of density, pressure, velocities, and internal energy. The solution procedure for the species transport equations contains three stages. Convection, diffusion, and chemistry terms are treated separately and their corresponding contributions are added successively in order to yield the final species density at the time level (n + 1). The chemistry step uses a one-step backward differentiation formula. It ensures that the species concentrations will not be negative during the time step Δt. A special algorithm is used to impose the constraint law in the form:

$$\sum_\ell \left\{ \frac{1}{r}\frac{\partial}{\partial r}[\rho D_\ell \, r \frac{\partial}{\partial r}(\rho_\ell/\rho)] + \frac{\partial}{\partial z}[\rho D_\ell \frac{\partial}{\partial z}(\rho_\ell/\rho)] \right\} = 0 \qquad (17)$$

This is achieved by monitoring the diffusional fluxes across the boundaries of the computing cell to satisfy Eq. (17).

Final values of the species densities and total energy density are calculated by evident means:

$$(\rho_\ell)_{ij}^{n+1} = (\tilde{\rho}_\ell)_{ij} + \sum_{r=1}^{N_r} \delta \dot{\omega}_r M_r (\nu''_{\ell r} - \nu'_{\ell r})$$

$$(\rho E)_{ij}^{n+1} = (\tilde{\rho} E)_{ij} + \sum_{r=1}^{N_r} q_r^0 \delta \dot{\omega}_r \qquad (18)$$

A computational cycle is completed by calculating the final pressure.

The boundary conditions according to the flow geometry are presented in Fig. 1. The numerical integration of the parabolized Navier-Stokes equations uses a more simple technique (Golovichev 1979b; Zelazny et al. 1978). The von Mises transformation is employed to convert the governing equations to the form which is convenient for the numerical technique. The conservation laws in transformed coordinates can be expressed as a generalized parabolic equation:

$$\rho u \frac{\partial f}{\partial z} = \rho u \frac{\partial}{\partial \psi} (\rho u \mu e \sigma_f^{-1} r^2 \frac{\partial f}{\partial \psi}) + S_f \qquad (19)$$

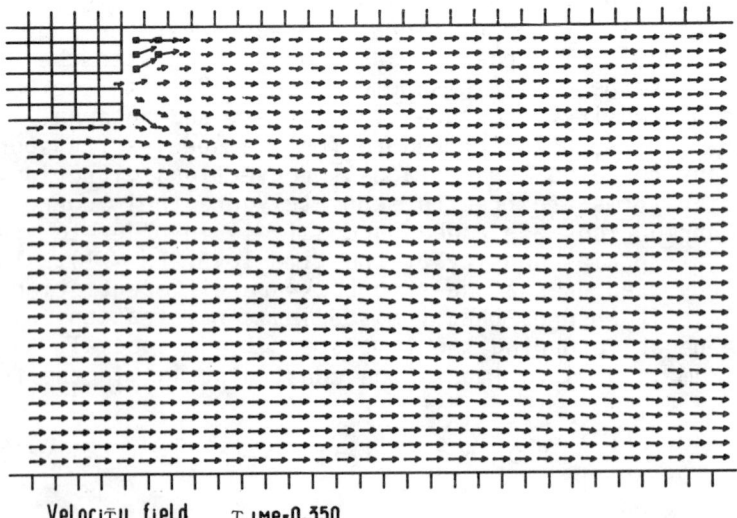

Fig. 4 Velocity vectors in the near wake flow with the H_2 injection in the circumference of a backward step.

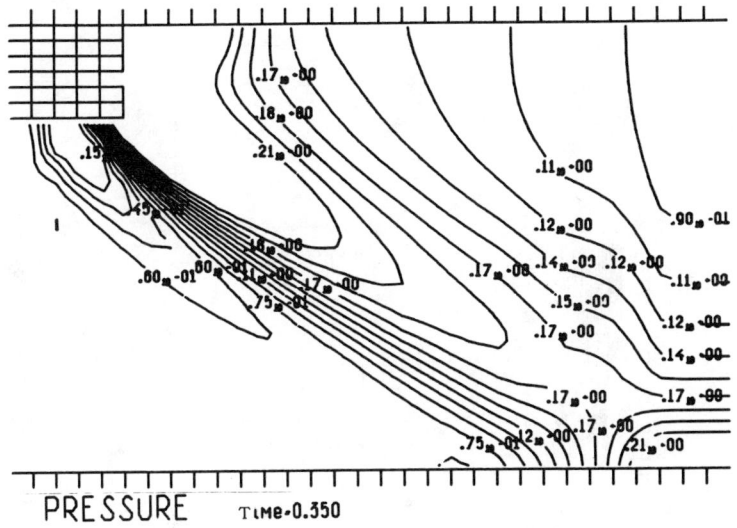

Fig. 5 Pressure isolines for the case corresponding to Fig. 4.

where $\vec{f} \equiv (u,v,H,\rho_\ell,\ldots,k,\varepsilon)^T$ is the vector of unknown functions, and S_f is the source term in the appropriate balance equations. The marching numerical procedure used to solve Eq. (19) is based both on the monotonic implicit finite-difference scheme of the second order and on the quasilinearization technique. Shock waves encountered in the calculation of the mixing of unbalanced pressure jets were treated as smeared isentropic compressions. The numerical model described does not take into account the reverse flow in the predominant flow direction and must contain a special technique to suppress the counterflow disturbance propagation in the subsonic region.

The numerical results were used to estimate the optical properties of the medium for the zero-power gain for selected P transitions. The population of each excited vibrational-rotational state was calculated with the assumption that the rotational and translational temperatures of the mixture were in equilibrium. Standard formulas were used to compute the line-broadening parameters for Doppler broadened lines. During the amplification mode, the available coherent power is estimated under the assumption of a total inversion destruction of the inverted transitions. In the oscillator mode the calculation of the laser output parameters is reduced to the solution of Eq. (19) for the lasing species, consistent with the threshold

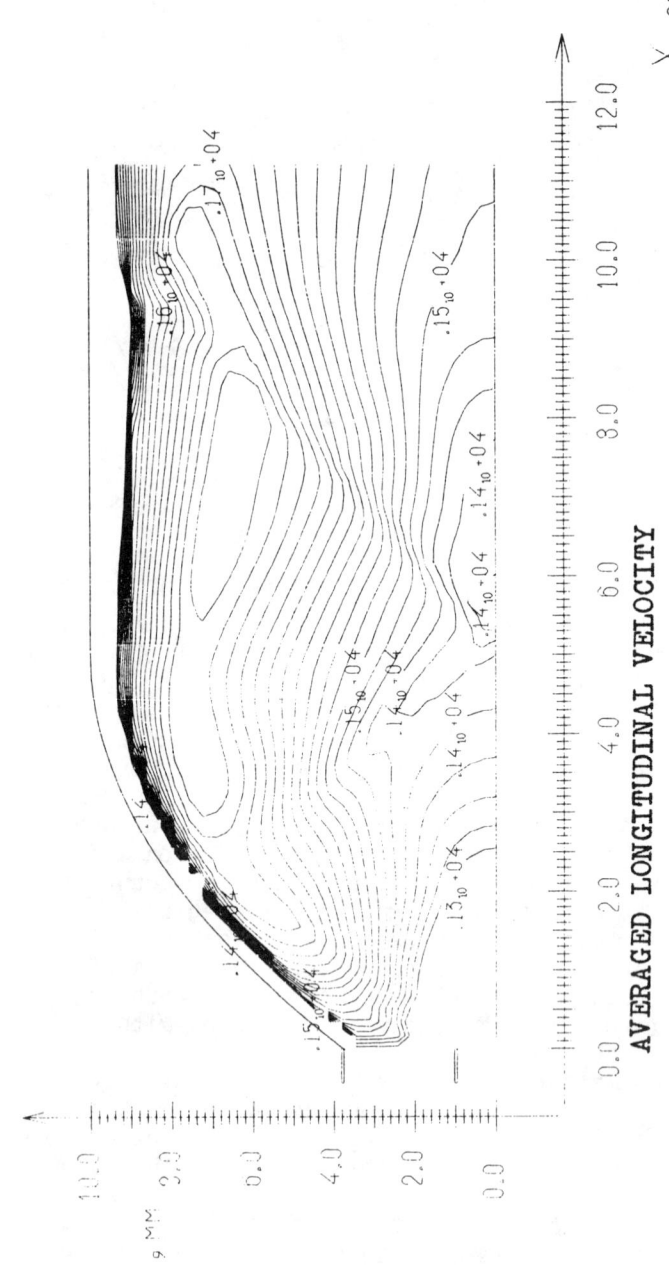

Fig. 6 Averaged velocity isolines in the distant wake flow of the H_2-HCl laser.

condition for the gain coefficient:

$$\bar{g}_{v,J} = \frac{1}{\psi_{L_c}} \int_0^{\psi_{L_c}} g_{v,J}(\psi)\, d\psi = \frac{1}{2L_c} \ln(r_1 r_2) \tag{20}$$

where L_c is the jet width between mirrors; and r_1, r_2 are the mirror reflectivities. The efficiency of the energy conversion process is characterized by terms $\sum_v \sum_J g_{v,J} I_{v,J}$ entering into the equation of energy.

Numerical Results

As an example the numerical techniques were used to model a HCl chemical diffusion laser in the base region produced by the nozzle trailing edges with planar geometry. While a detailed kinetic description of inversion was not attempted in this example, the influence of mixing and heat release effects on the gasdynamic parameters together with the effect of geometrical factors was considered. The presence of a base-relief region favors the reductions of pressure because of the interactions of the expansion form with the reacting flow immediately downstream of the base. The products of chlorine thermal dissociation diluted with an inert heat absorbing gas (N_2) enter the lasing cavity at Mach number 4.3, a static temperature of 650 K, and pressure of 7.5 Torr. This flow produces the near wake behind the backward step. Partially excited hydrogen is injected under sonic conditions and at a pressure of 250 Torr through a slit near the bottom of the backward step. These conditions were used by Rivard et al. (1975) to model a chemical HF-laser. The mixing design is discussed in further detail by Grohs and Emanuel (1976). This design suffers from the shortcoming that the pressure imbalance causes the fuel (hydrogen) to be pushed into the Cl/Cl_2 stream within the oxidizer nozzle. As a consequence, some of the chemical reactions occur within the oxidizer nozzle. Numerical results presented in Figs. 2 and 3 support this observation. The steady-state velocity vector field and the contours of equal pressure values are plotted in these figures. The flow conserves the supersonic jetlike nature. The formation of recirculation zones when the Mach number is less than unity reduces the efficiency of computations, but Brandt (1980) has proposed a method to handle the recirculation zone. Another scheme of mixing suggested by Grohs and Emanuel (1976) was also analyzed. In this case the base-relief nozzles have finite-size dividing walls between reactant jets, and a flow area between two hydrogen-

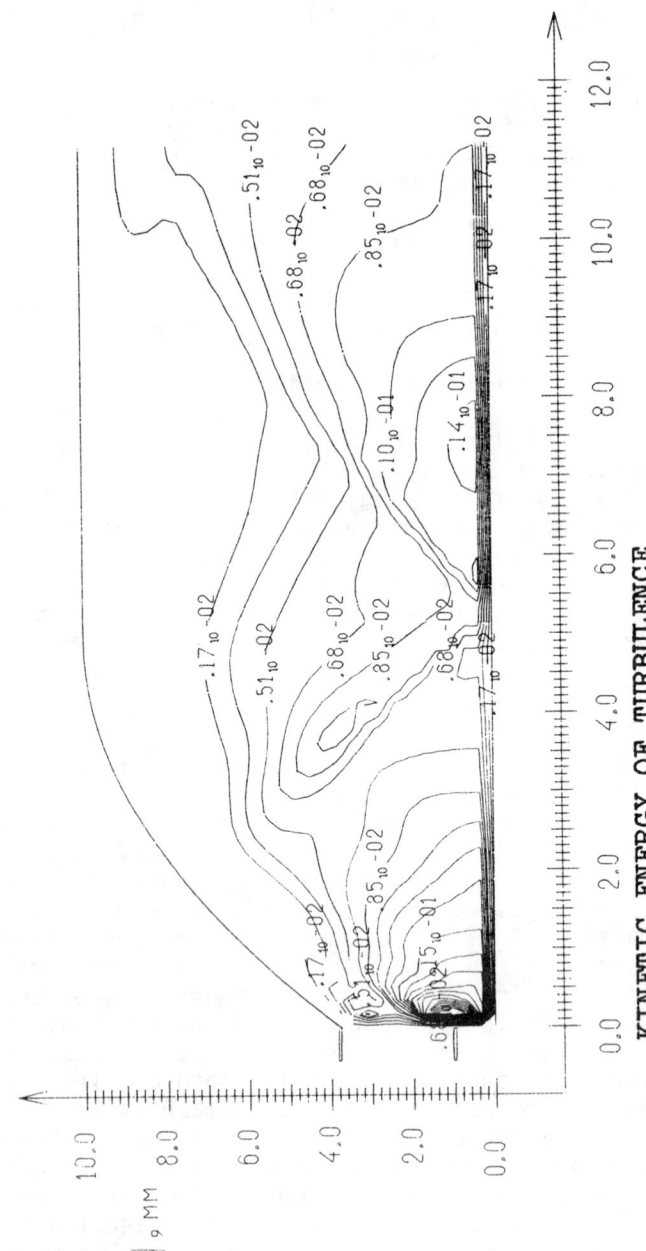

Fig. 7 Turbulent kinetic energy isolines for the case corresponding to Fig. 6.

injection slits is introduced as shown in Figs. 4 and 5. Numerical results indicate that base pressure is reduced because of the flow area between adjacent hydrogen jets, but the principal nature of the flow does not change significantly.

The parabolized Navier-Stokes equations have been used to model a far wake. This model satisfactorily describes the transverse flow motion induced by the oblique shock that forms the characteristic diamond-shaped structure. The gasdynamic nature of the flow is illustrated by Figs. 6 and 7 in which characteristics of the averaged and fluctuating fluid motion are shown. A considerable increase in the kinetic turbulence energy occurs in the vicinity of the oblique shock as a result of lateral and longitudinal pressure gradients. This phenomenon leads to a noticeable enhancement of the mixing process.

The calculated mixture composition, optical gain coefficients on selective P transitions and output parameters of the laser system are presented in Figs. 8-10. The longitudinal distributions of the population densities

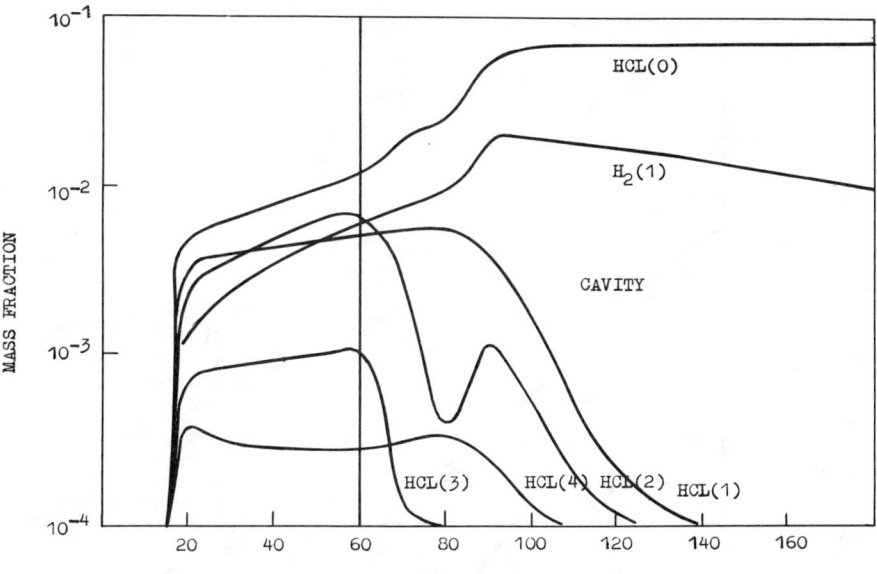

Fig. 8 The centerline profiles of the H_2-HCl-laser mixture composition; location $Z/2r_j = 60$ corresponds to the cavity exit plane; the value of $H_2^j(1)$ mass fraction should multiplied by 10; the maximum pressure level in the cavity is equal to 30.4 Torr; $r_1 = 1$, $r_2 = 0.95$.

of vibrationally excited HCl(v) molecules in Fig. 8 are attributed to the fact that cavity radiation field deactivates the excited states on the 1/0 and 2/1 bands, since the vibration modes 4/3 and 3/2 serve as an "energy container" which is evacuated only by the V-V and V-T processes. A sudden change in the longitudinal distributions at $Z/2r_j = 60$ takes place owing to the location of a cavity exit plane at this point. Transverse profiles of the optical gain coefficients for selected P transitions presented in Fig. 9 are indicative of a rather high quality of the laser system medium produced by chemical reaction and V-V exchange processes. As the V-V exchange processes are more significant than the chemical pumping processes downstream of the flow, transverse profile "wings" are produced. Figure 10 illustrates that the H_2/Cl_2 system is less sensitive to the pressure increase than the H_2/F_2 system. Even the lasing "cutoff" position is slightly influenced by the pressure level in mixture. A comparison between the curves 1 and 1' makes possible the estimation of the energy conversion process efficiency.

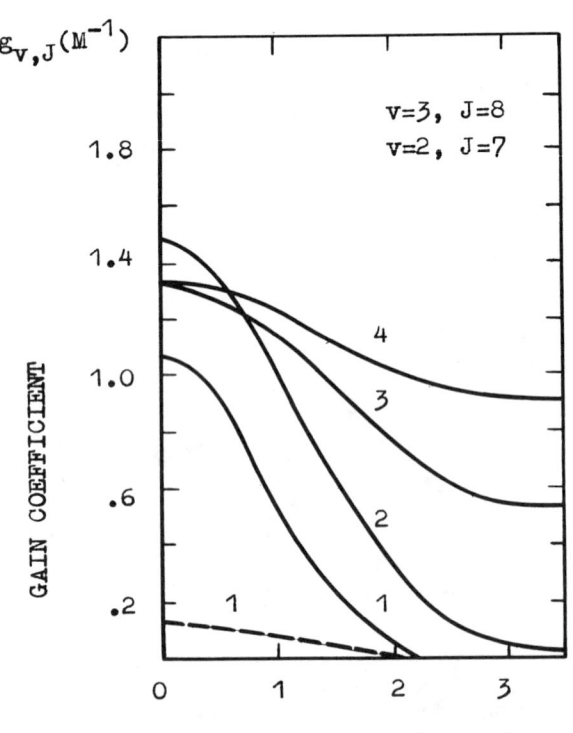

Fig. 9 Transverse profiles of the optical gain coefficients for different flow cross sections: 1 - $Z/2r_j = 50$; 2 - $Z/2r_j = 100$; 3 - $Z/2r_j = 150$. The maximum pressure level in the cavity equals 7.6 Torr.

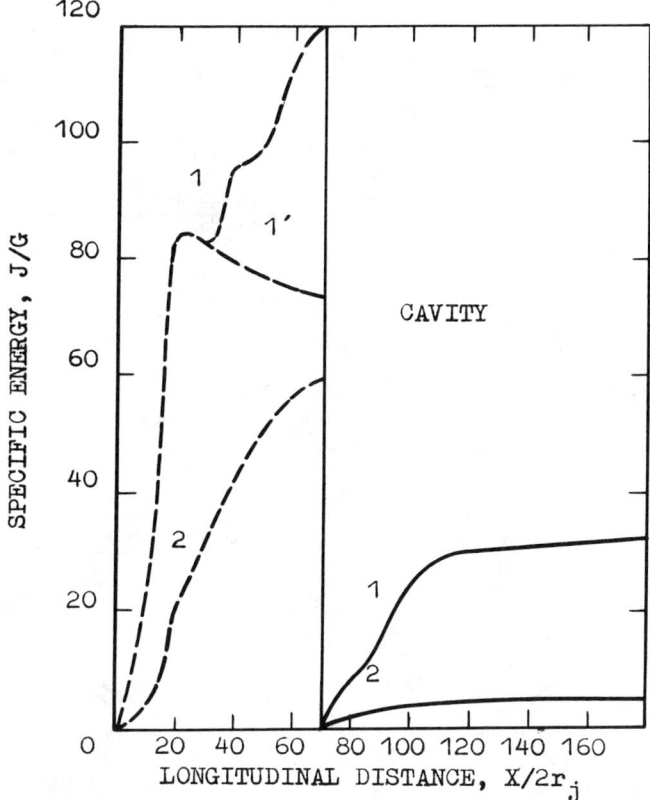

Fig. 10 Power characteristics of the H_2-HCl laser in the amplifier and oscillator regimes. Curve 1 correponds to the maximum pressure level in cavity equal to 7.6 Torr; curve 1' to a purely chemical HCl laser; curve 2 to the maximum pressure level equal to 30.4 Torr.

Conclusions

The illustrations present the informative results on the kinetics of vibrational relaxation under the conditions of chemical activity and working gas motion. The modeling suggests that demonstration of chemically driven $H_2(v)/HCl(v)$ transfer laser is possible. Preliminary operating conditions, mixing and injection schemes are determined. This information can be used for the construction of an experimental design.

Acknowledgments

The authors are indebted to N. N. Yanenko for support of the work and to R. I. Soloukhin for stimulating discussions.

References

Brandt, A., Dendy, J. E. Jr., and Ruppel, H. (1980) The multigrid method for semi-implicit hydrodynamics codes. J. Computat. Phys. 34, 348-370.

Golovichev, V. I. (1979a) Power estimates of diffusion type chemically driven HF-gasdynamic laser. Issledovaniye Rabochego Protsessa Gasodinamicheskikh i Khimicheskikh Laserov, pp. 146-158. Novosibirsk, USSR.

Golovichev, V. I. (1979b) Numerical modeling of turbulent mixing process of supersonic free and ducted flows of reactive gases. Gasodinamika Goreniya v Sverkhzwukovykh Potokakh, pp. 27-52. Novosibirsk, USSR.

Grohs, G. L. and Emanuel, G. (1976) Gas dynamics of supersonic mixing lasers. Handbook of Chemical Lasers (edited by Gross, R. W. F. and Bott, J. F.), pp. 263-388. John Wiley & Sons, New York.

Launder, B. E. and Spalding, D. B. (1974) The numerical computation of turbulent flows. Computat. Math. Appl. Mech. Eng. 3, 269-289.

Orayevsky, A. N., Rodionov, N. B., and Shcheglov, V. A. (1978) Thermal partial inversion gasdynamic lasers. Zh. Tekh. Fiz. 48, 1432-1441.

Rivard, W. C., Farmer, O. A., and Butler, T. D. (1975) RICE: A computer program for multicomponent chemically reactive flows at all speeds. LA-5812, Los Alamos Scientific Laboratory, Los Alamos, N. M.

Sèdes, C. (1977) Contribution à l'etude des méchanismes de création d'inversions de population entre niveaux de vibration de molécules formées par réactions chimiques au sein d'un écoulement. Ph.D. Thesis, pp. 76-89, Université de Provence, Marseille, France.

Taylor, R., Caledonia, G., Lewis, P., Wu, P., Teare, J. D., and Cronin, J. (1976) Analytic modeling of electrically excited D_2/HCl and HCl laser experiments. PSI TR-58, Physical Science Inc., Andover, Me.

Westbrook, C. K. and Chase, L. K. (1977) A one-dimensional combustion model. UCRL-52297, Lawrence Livermore Laboratory, Livermore, Calif.

Zelazny, S. W., Driscoll, R. J., Raymonda, J. W., Blauer, J. A., and Solomon, W. C. (1978) Modeling DF/HF CW lasers: An examination of key assumptions. AIAA J. 16, ,297-304.

A Gasdynamic Laser Using Products of Acetylene Explosions

A.B. Britan,* A.N. Khmelevskii,† V.A. Levin,‡ S.A. Losev,‡
V.V. Lugovskoi,§ G.D. Smekhov,† and A.M. Starik†
Moscow University, Moscow, USSR

Abstract

A combustion pulse explosive gasdynamic laser (GDL) has been used to investigate the characteristics of acetylene-air fuel combustion products flow. The 2-liter GDL spherical explosion chamber was connected to the nozzle grid by a flat widening adapter. The chamber and adapter volumes were separated with a copper diaphragm. The explosion products escaped through the nozzle grid to the evacuated 200-liter damper tank with probe windows on the side walls. The nozzle grid (500 mm wide) consisted of ten variable nozzle blades. To obtain a uniform flow in the outlet section, the nozzles were profiled (0.35 mm nozzle throat and 140 area ratio). Gain measurement was carried out in two sections behind the nozzle grid edge. Experimental values of the gain, obtained over the pressure region from 5 to 100 atm at 1800 and 2500 K, is in satisfactory agreement with calculations. Acetylene, nitrogen, and nitrous oxide mixtures produced the best energy characteristics over the investigated experimental region. The specific stored energy values in such mixtures can be as high as 120 J/g, with the pressure from 6 to 10 atm and a temperature of 3100 K.

Presented at the 8th ICOGER, Minsk, USSR, Aug. 23-26, 1981. Copyright © American Institute of Aeronautics and Astronautics, Inc., 1982. All rights reserved.
*Senior Research Scientist, Institute of Mechanics.
†Senior Research Engineer, Institute of Mechanics.
‡Professor, Institute of Mechanics.
§Research Engineer, Institute of Mechanics.

Introduction

Development of a gasdynamic laser (GDL) with optimum main parameters is of utmost importance in GDL applications. One of the main characteristics of the laser system is a gain, K_ν, which depends on the degree of inversion and laser level population and determines the radiation power generated by a laser. Comparison of experimental and calculated values of K_ν allows evaluation of the facilities and validity of the calculational procedures (Losev 1977).

These experiments were concerned primarily with a laser mixture formed directly by combustion of fuels. These experiments were carried out on the laboratory plant and provided information in good agreement with the results of tests with real GDL's.

However, in this case the use of the results and parameter range is limited by the choice of fuel type (Ktalkherman and Malkov 1979; Eutyukhin et al. 1978; Ivonov 1978).

The employment of an additional energy source [shock waves (Britan and Starik 1980; Volkov et al. 1979) or pulse discharge in a closed volume (Odintsov et al. 1976)] makes it possible to investigate a new class of noncombustible mixtures. Composition and parameters of working mixtures in these experiments are independently varied in a wide range, and this allows modeling of the combustion product flowing for a number of laser fuels (Losev 1977).

At the same time it is worthwhile to combine a number of different-type experimental systems in one investigation. Gradual complication of gas composition from the simplest model mixtures to the multicomponent combustion products of real fuels greatly expands the field of investigation and leads to the determination of the effect of individual components on GDL working medium characteristics (Britan et al. 1974).

If a theoretical analysis is carried out using a unique calculation procedure and a fixed set of constants, a comparison of experiment and theory makes it possible to check, with a high degree of reliability, the validity of the mathematical flow model.

This paper is a continuation of previous shock-tube experiments (Britan and Starik 1980; Britan et al. 1974; Britan et al. 1980), and is devoted to the analysis of the inversion and energetic characteristics of GDL on acetylene-air combustion products in which along with air-oxygen, nitrous oxide was used as an oxidizer.

LASER USING PRODUCTS OF ACETYLENE EXPLOSIONS

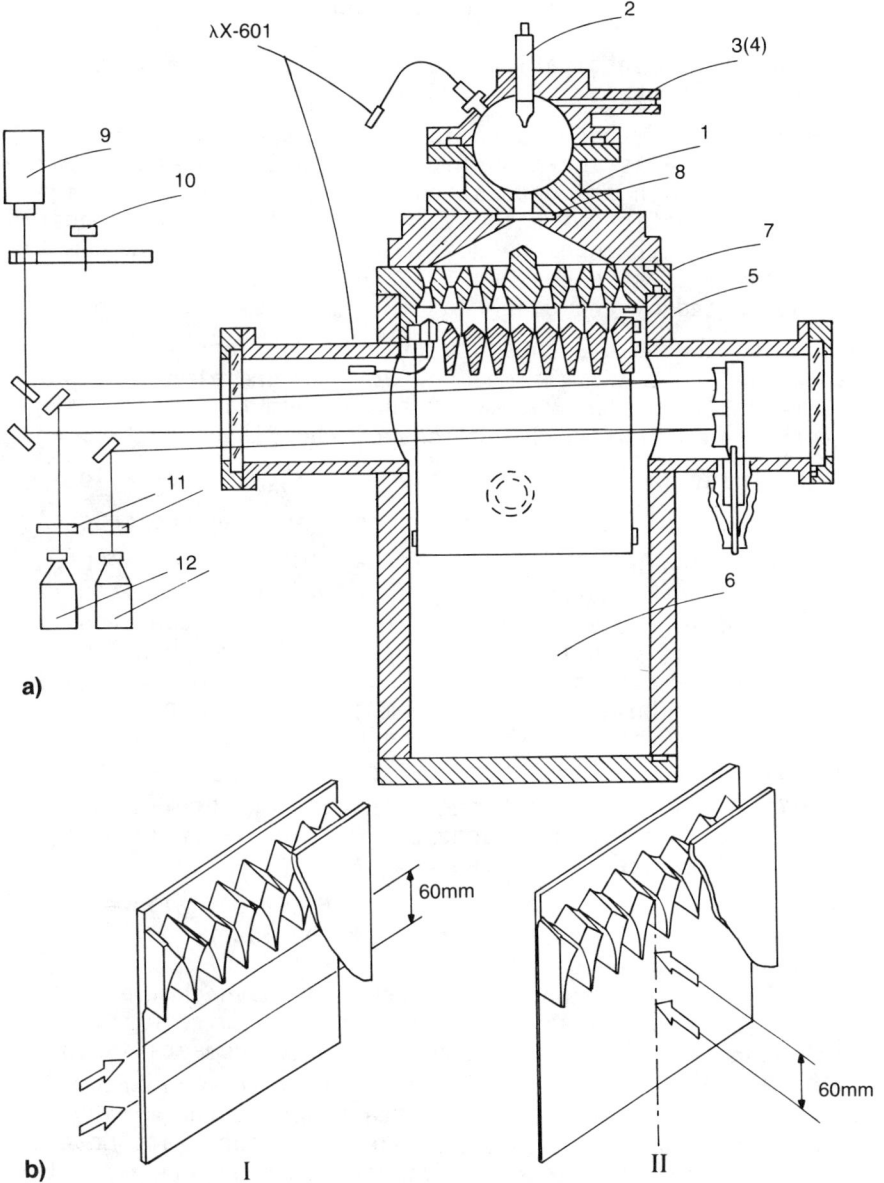

Fig. 1a) Gasdynamic laser unit, view in the plane of symmetry of the nozzle grid. Gain measurement system: probe laser (9); modulator (10); 10.6-m filter (11); photodetectors (12); X-601 - pressure gages positioned in the explosive chamber and at the grid exit. 1b) Various nozzle grid positions relative to the plane of the probe laser beam. → - direction of the probe beam; I - measurements in the nozzle grid plane; II - measurements for a single nozzle.

Experimental Setup

In this investigation the GDL system was designed with ultimate parameters: P_0 = 200 atm, T_0 = 3000 K. The general view of the gasdynamic unit in the nozzle grid symmetry plane is shown on Figs. 1a and 1b. The unit face is provided with the detonation ignitor (2) and the union (3) for connection with the vacuum line (4). The combustion products flow through the nozzle grid (5) to the vacuum tank (6). The explosive chamber (1) and the vacuum tank volumes are separated by the diaphragm (8). The system may be evacuated and hold a pressure of 2×10^{-2} Torr. The explosive and prenozzle chamber volumes are almost identical (2 liters), and the volume of the vacuum tank (6) is 200 liters. The flow cross sections of the diaphragm, S_1, and of the holes in the upper flange (7), S_2, have a ratio, S_1/S_2 = 1.6 and are considerably greater than the cross section of the nozzle grid (S_2/S^* = 720).

The length and supersonic section of the nozzle were calculated in a one-dimensional formulation (Losev and Makarov 1976), with the output radiation power maximum at a given mixture composition and stagnation parameters. The nozzle throat height (h^* = 0.35 mm) was chosen on the basis of the two-dimensional calcuation of such a flow whose parameters were optimum (Makarov and Tunik 1978). The nozzle opening angle was ε = 140 deg. In gain measurements, use was made of the standard scheme (see Fig. 1a) with a continuous electrically excited CO_2 laser $\Lambda\Gamma$ - 22. The capacitance filter in the supply unit circuit allowed reduction of the fluctuation level of the laser radiation power to the value of $\sim$3%.

The mixture detonation in the explosive chamber was initiated by the explosion of a constantan wire with resistance of $\sim$ 100 Ω. The pressure measurements in the explosion chamber allowed the control of the diaphragm break moment and the volume to be constant during the mixture combustion. This condition is necessary for exact determination of the combustion product composition and parameters. Experimental pressure data in the explosion chamber P_K, and before the nozzle grid entrance were measured at various initial mixture pressures, P_h, and are presented in Fig. 2. The pressure values obtained from thermodynamic calculations (solid line) (Smekhov and Fotiev 1978) are higher than the experimental ones.

The calculated values exceed those of the explosion products temperatures as is clear from the comparison of the calculated values T_0 (see Table 1) with the measured temperatures (Ivanov 1978).

This difference is obviously due to incomplete combustion of acetylene (El'patanov and Srizhevskii 1968). According to Ivanov (1978), for a mixture containing 4% C_2H_2, T_0 = 1850 K, and for a mixture containing 6.54% C_2H_2, T_0 = 2500 K. Calculations show that combustion of the above

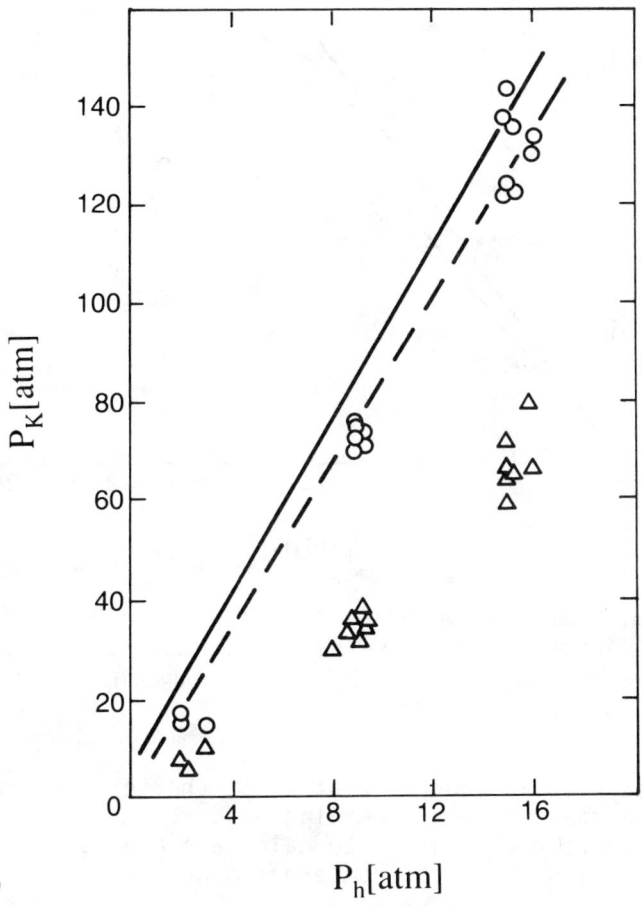

a)

Fig. 2a) Variation of explosive products pressure P_K with initial pressure P_h for 6.54% C_2H_2-air mixture: △ - pressure at the nozzle grid plane; o - pressure in the explosive chamber; —— thermodynamic calculation data; ---- average experimental data.

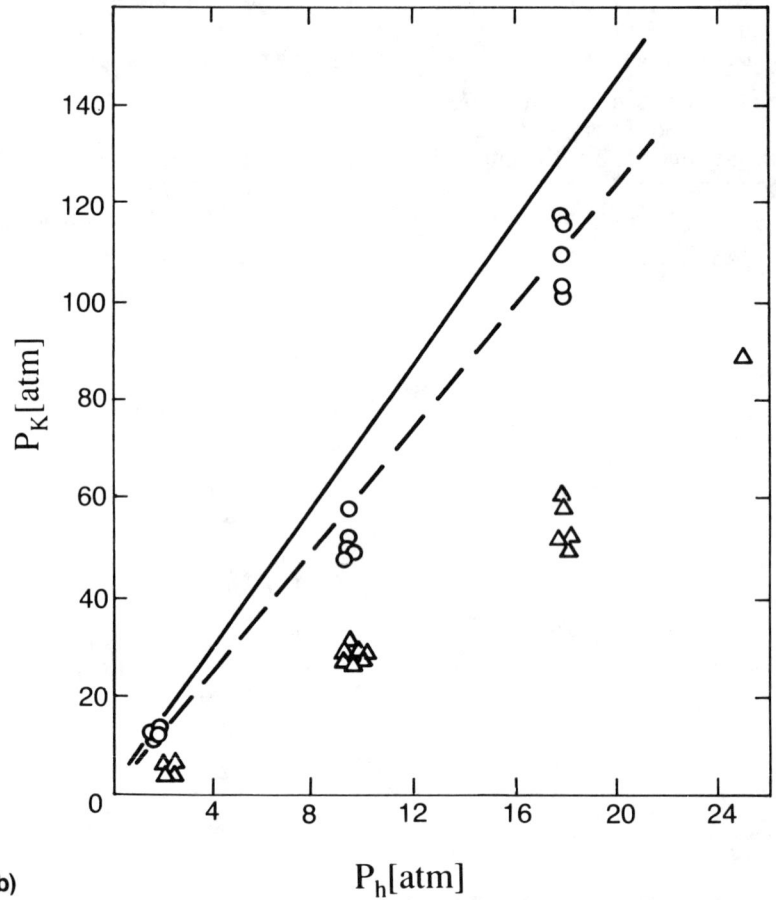

Fig. 2b (cont.) Variation of explosive products pressure P_K with initial pressure P_h for 4% C_2H_2-air mixture. Δ - pressure at the grid exit; o - pressure in the explosive chamber; —— thermodynamic calculation data; ---- average experimental data.

mixtures leads to a slight variation of the total number of molecules in the gas (not exceeding 3%).

It is therefore possible to calculate a pressure rise in the combustion products from experimental values of T_0 (Ivanov 1978)

$$\frac{P_K}{P_h} = \frac{T_K}{T_h} \frac{\mu_h}{\mu_K} \qquad (1)$$

where μ is the molecular mixture weight, and the subscripts h and K correspond to the initial and final mixture parameters, respectively.

Table 1 Data of thermodynamics calculation for different initial composition of mixtures

P_h atm	P_K atm	T_0 K	% CO_2	% H_2O	% N_2	% O_2	% CO	% OH	% O	% H	% H_2	% NO
\multicolumn{12}{c}{6.54% C_2H_2 + air}
4	37.1	2851	11.1	6.18	74.5	3.2	1.68	0.7	0.16	0.05	0.12	1.55
6	57.2	2875	11.77	6.19	76.6	3.2	1.65	0.7	0.157	0.05	0.12	1.58
8	76.8	2875	11.86	6.22	74.6	3.16	1.57	0.67	0.146	0.04	0.11	1.58
10	91.47	2750	12.57	6.4	75.06	3.07	0.92	0.47	0.08	0.088	0.067	1.31
\multicolumn{12}{c}{4% C_2H_2 + air}
4	28.8	2157	8.21	4.06	76.9	9.84	0.092	0.096	0.01	0.37×10^{-3}	0.2×10^{-2}	0.8
6	43.3	2157	8.21	4.06	76.9	9.85	0.017	0.087	0.97×10^{-2}	0.27×10^{-3}	0.16×10^{-2}	0.8
8	57.7	2157	8.22	4.06	76.9	9.84	0.015	0.081	0.84×10^{-2}	0.22×10^{-3}	0.15×10^{-2}	0.8
10	72.1	2157	8.21	4.07	76.9	9.85	0.014	0.077	0.75×10^{-2}	0.18×10^{-3}	0.13×10^{-2}	0.8
\multicolumn{12}{c}{6.67% C_2H_2 + 33.33% N_2O + 60% N_2}
1	12.	3000	6.46	4.55	78.9	1.6	4.88	1.02	0.41	0.3	...	1.35
2	24.2	3039	6.95	4.73	79.1	1.43	4.48	0.92	0.31	0.22	...	1.34
4	49.1	3086	7.33	4.86	79.4	1.29	4.14	0.841	0.244	0.17	...	1.34
6	73.9	3101	7.63	4.96	79.6	1.19	3.86	0.775	0.2	0.14	...	1.31
10	124.	3133	7.92	5.04	79.7	1.08	3.59	0.71	0.16	0.11	...	1.31

P_K values, calculated from Eq. (1), are in a good agreement within experimental error and are presented in Fig. 2 (dotted line). The data in Fig. 2 show that the expansion of the system volume, after the diaphragm breaks, leads to pressure drops; therefore, in most of the experiments, $P_0 = (0.5-0.7)P_K$, where P_0 and P_K are the maximum values of the signal amplitude from the pressure gages mounted before the nozzle grid and in the explosion chamber.

Theoretical Analysis

Thermodynamic calculations (see Table 1) indicate that CO, NO, H_2, OH, O, H molecules are present in significant concentrations as well as CO_2, N_2, O_2, and H_2O. Any small quality of the former species complicates the kinetics of the vibrational energy exchange in comparison to the CO_2 + N_2 + O_2 + H_2O mixture flow model. In this case it is necessary to take into consideration the new energy exchange channels (Genich et al. 1979; Biryukov 1975; Egorov and Komarov 1975; Kulagin 1979; Kozlov et al. 1979; Levin and Starik 1980) in addition to those examined by Britan and Starik (1980):

$CO_2(00^00) + CO \rightleftarrows CO_2(00^01) + CO\ (V=0) \qquad W_{3,5}$

$CO_2(11^10) + CO$
$CO_2(03^10) + CO$ $\rightleftarrows CO_2(00^00) + CO\ (V=1) \qquad \begin{matrix} W_{5,1,2} \\ W_{5,2} \end{matrix}$

$CO\ (V=0) + N_2 \rightleftarrows CO\ (V=1) + N_2\ (V=0) \qquad W_{4,5}$

$CO\ (V=0) + NO \rightleftarrows CO\ (V=1) + NO\ (V+0) \qquad W_{5,6}$

$CO\ (V=0) + H_2O \rightleftarrows CO\ (V=1) + H_2O\ (000) \qquad W_{5,8}$

$CO\ (V=0) + M \rightleftarrows CO\ (V=1) + M \qquad W_{5,0}^M$

$NO\ (V=0) + O_2\ (V+1) \rightleftarrows NO\ (V=1) + O_2\ (V=0) \qquad W_{6,7}$

$NO\ (V=0) + M \rightleftarrows NO\ (V=1) + M \qquad W_{8,0}^M$

$NO\ (V=0) + H_2O(010) \rightleftarrows NO\ (V=1) + H_2O(000) \qquad W_{6,8}$

$H_2\ (V=0) + M \rightleftarrows H_2\ (V=1) + M \qquad W_{11,0}^M$

$$H_2(V=0) + H_2O(001) \rightleftarrows H_2(V=1) + H_2O(000) \quad W_{11,10}$$

$$H_2(V=0) + OH(V=1) \rightleftarrows H_2(V=1) + OH(V=0) \quad W_{11,12}$$

$$OH(V=0) + H_2O(100) \rightleftarrows OH(V=1) + H_2O(000) \quad W_{12,9}$$

$$OH(V=0) + M \rightleftarrows OH(V=1) + M \quad W_{12,0}$$

$$OH(V=0) + H_2O(001) \rightleftarrows OH(V=1) + H_2O(000) \quad W_{12,10}$$

$$H_2O(100) + M \rightleftarrows H_2O(001) + M \quad W^M_{10,9}$$

$$H_2O(020) + M \rightleftarrows H_2O(100) + M \quad W^M_{9,8}$$

$$CO_2(10^01) + H_2O(00^00) \rightleftarrows H_2O(001) + CO_2(00^00) \quad W_{10,13}$$

$$CO_2(10^01) + H_2O(000) \rightleftarrows H_2O(100) + CO_2(00^00) \quad W_{0,13}$$

$$H_2O(000) + M \rightleftarrows H_2O(001) + M \quad W^M_{10,0}$$

$$H_2O(000) + M \rightleftarrows H_2O(100) + M \quad W_{9,0}$$

where W_{pK} is the rate of the intemolecular energy exchange between the p and K states of the colliding molecules; and W^M_{pq} is the rate of the intermolecular exchange between p and q states upon collision with the M partner.

To distinguish the M partners (M = CO_2, N_2, CO, NO, O_2, H_2O, OH, H_2, H, O), it is necessary to assign to each partner a value of i = 1, 2, 3, 4, 5, 6, 7, 8, 9, 10.

The p and q states can adopt values from 0 to 12, where p = 1-3 corresponds to the symmetric deformation and asymmetric modes of CO_2 molecule vibration; p = 4-7 corresponds to the vibrations of N_2, CO, NO, O_2 molecules; p = 8-10 corresponds to the deformation, antisymmetric, and symmetric modes of H_2O molecule vibration; and p = 11, 12 corresponds to H_2 and OH molecules vibrations. CO_2 molecule combination states 11^10 and 10^01 correspond to q = 1.2 and 1.3, respectively. With the exchange channels in a H_2O-H_2-OH system and the data of Levin and Starik (1980a) and Blauer and Nickerson (1974), it is possible to show that the above system can be described by three kinetic equations.

Since the rates $W_{10,9}$, $W_{11,10}$, $W_{11,12}$, $W_{12,9}$, and $W_{12,10}$ are approximately 10-100 times higher than the rates $W_{10,0}^M$, $W_{9,0}^M$, $W_{11,0}^M$, $W_{12,0}^M$ (Blauer and Nickerson 1973; Levin and Starik 1980) and $W_{9,8}^M$, the quasistationary distribution between the H_2O molecule symmetric and asymmetric vibrations types and also between the H_2 and OH molecules vibrations is considered to be established. Calculations of Levin and Starik (1980a) indicate that the temperatures T_9, T_{10}, T_{11}, and T_{12} are equal in supersonic nozzle flows over a wide range of the flow parameters and conditions.

Thus it is possible to describe all processes of the vibrational quantum changes in H_2O, H_2, OH gas mixtures with two-temperature (T_8 and T_9) model. For the nonresonant transfer the rates $W_{10,13}$, $W_{9,13}$ are assumed to be negligible. The above assumption is correct because the concentrations of H_2O, H_2, OH molecules in a mixture are small. (For the compositions considered in our paper $\xi_{H_2} \leq \xi_{OH} \leq \xi_{H_2O} \leq 0.06$, where ξ_i is the molecular fraction of the i-th component of the mixture.) The kinetics equations system was established in the form presented by Biryukov (1975) and assumes the existence of the full resonance between (100) and (020) CO_2 molecule states.

To calculate the distribution of the medium nonequilibrium parameters, the gasdynamic and vibrational kinetics equations were integrated. The numerical integration details and calculation method are presented by Britan and Starik (1980) and Levin and Starik (1980a). The dependence of W_{pq}^M on temerature was predicted from recent results (Genich et al. 1979; Kulagin 1979; Blauer and Nickerson 1974; Levin and Starik 1980; Center 1973). The calculations were based on the inviscid non-heat-conducting gas approximation with a stationary, one-dimensional flow.

The results of Britan et al. (1980) indicate that the above approximation provides a satisfactory accuracy in calculating inverse flow characteristics behind the flat nozzle edge if the nozzle is profiled for a continuum flow with uniform parameters in the outlet section. The dissociated mixture is assumed to be at equilibrium at the nozzle inlet, and the mixture in the supersonic flow is assumed to be frozen. The calculation for a mixture with ξ_{CO} and $\xi_{NO} < 0.04$ assumed that $T_5=T_4$ and $T_6=T_7$.

The T_0 values were calculated according to Eq. (1) with P_K measured in the explosive chamber and with P_0 taken as the initial condition behind the nozzle inlet. It is considered that the real combustion product composition corresponded to the calculated one, and both the possible deviation from the calculated composition and uncontrollable impurities slightly influenced the gain. For calculations of inversion characteristics of acetylene combustion product flow, the above mathematical model was used. The calculated

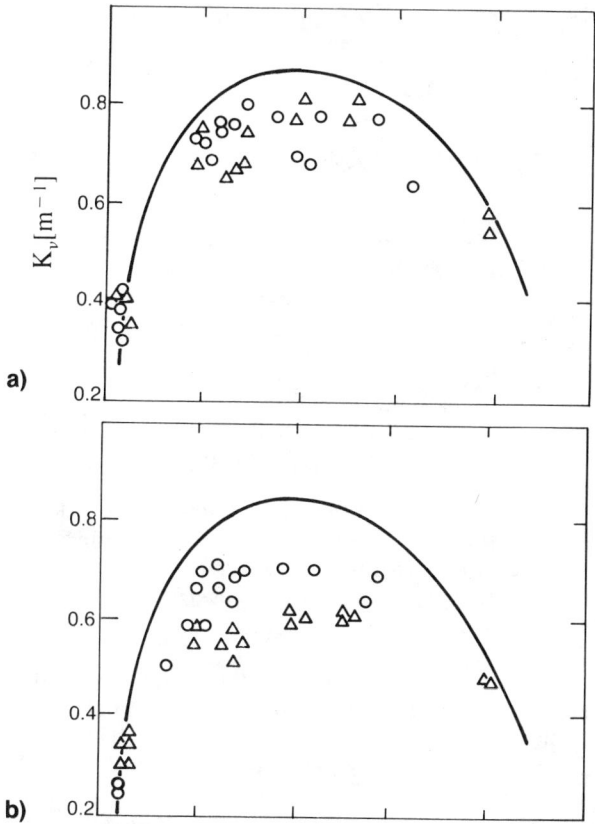

Fig. 3 Effect of stagnation pressure P_0 on the gain K_ν. ——— - calculated values of K_ν; o - observed K_ν in nozzle grid plane; Δ - observed K_ν in the single nozzle. Observations at the nozzle grid plane: a) for 4% C_2H_2-air mixture; c) for 6.54% C_2H_2-air mixture. Observations at 60 mm downstream of the nozzle grid plane: b) for 4% C_2H_2-air mixture; d) for 6.54% C_2H_2-air mixture.

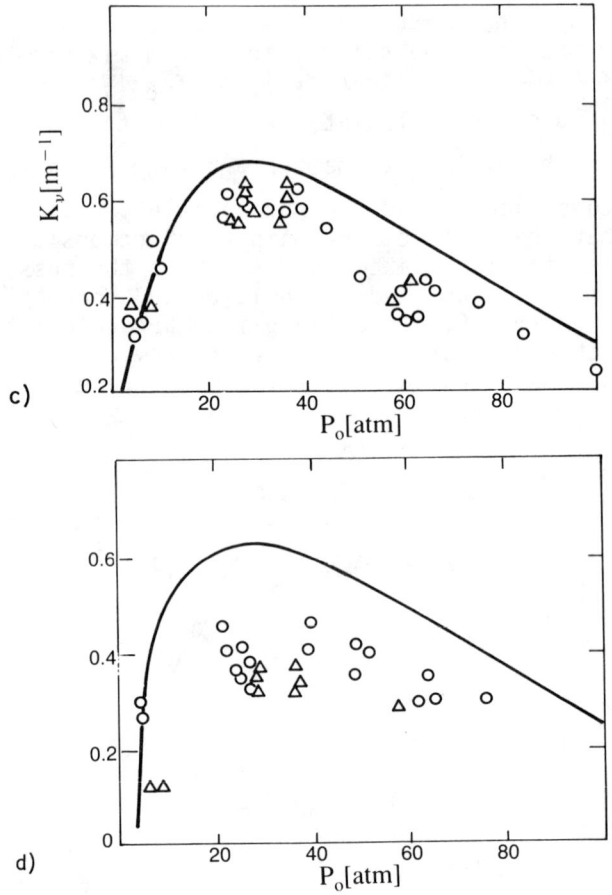

Fig. 3 (cont.) Effect of stagnation pressure P_0 on the gain K_ν. —— calculated values of K_ν; o - observed K_ν in nozzle grid plane; Δ - observed K_ν in the single nozzle. Observations at the nozzle grid plane: c) for 6.54% C_2H_2-air mixture. Observations at 60 mm downstream of the nozzle grid plane: d) for 6.54% C_2H_2-air mixture.

gain was compared with experimental ones measured behind the nozzle grid exit. This comparison allowed analysis of the use of this model and the prediction of the high-temperature GDL energetic characteristics.

Results

Two runs of gain measurements were carried out with acetylene-air mixtures having different compositions. In each experiment, the gain was simultaneously measured in two sections behind the grid exit: near the exit and at a distance of 60 mm from the downflow. The K_ν measurements were carried out at two positions of the grid also relative to the probe laser beam direction (Fig. 1b). The gain measurement scheme is shown in the grid symmetry plane (Fig. 1a). Measurement of the gain in the plane normal to grid symmetry determines the gain in the mononozzle formed by two nozzle vanes. The results of the first run (4% C_2H_2 + air) are presented in Figs. 3a and 3b. Here o and Δ are the experimental data on the gain measured along the nozzle grid and in the plane normal to the grid (in the mononozzle), respectively.

The experimental gain values are self-consistent and agree with the calculation results obtained at different positions of the grid near the grid exit (Fig. 3a). The maximum values of $K_\nu = 0.8$ m^{-1} are attained at $P_0 = 50$ atm, but over the 20-55 atm range K_ν varied slightly and remained in the 0.7-0.8 m^{-1} range. Gain of 0.6-0.7 m^{-1} range were observed by Ivanov (1978) for a 5.5% C_2H_2-air mixture ($T_0 = 1800$ K) over the $P_0 = 4-16$ atm pressure range. Satisfactory agreement between calculation and experiment made it possible to analyze the variation of K_ν with a pressure. Analyses of nonequilibrium parameter distribution calculated along the nozzle revealed that increases in P_0 increases the upper (00^01) laser level deactivation and the excitation of the V=1 state population N_2 molecule caused by a fast quasiresonant V-V' process.

At the same time, the CO_2 molecule deformation vibrations are, in fact, in equilibrirum with the translational degrees of freedom $T_2=T$ at $P_0 \geq 30$ atm. The decrease in upper laser level population is compensated by an increase in the density of the inverse molecules (gain is maximum) over the P_0 range from 30 to 60 atm. With a further increase of the stagnation pressure, together with a decrease of the level population, the spectral line of the radiation is greatly increased because of molecular collisions; and the gain value is reduced. At $P_0 < 30$ atm and $T_2 > T$, the quasistationary distribution in the nozzle between the CO_2 asymmetric mode and N_2 vibrations does not reach equilibrium as it had at the higher P_0 values. With the increase of stagnation pressure from 6 to 30 atm, the lower laser level population begins to decrease, which leads, together with a growth of the density of the inverse molecule, to larger K_ν in spite of decreasing T_2 and decreasing upper laser level population. If K_ν distribution maximum is achieved at $P_0 > 30$ atm in the nozzle, the values rise at $P_0 = 6$ atm in the flow behind the edge, and in any case the gain measured near the nozzle is not maximum. The $K_\nu(P_0)$ measurements and calculations for the section at a distance of 60 mm from the nozzle exit along the flow are presented in Fig. 3b. The experimental values are lower than the calculated ones.

The maximum divergence takes place at $P_0 = 30$-50 atm and reaches 20%. The divergence between theory and experiment probably are the consequence of the real flow structure connected with the presence of the shocks and wakes behind the nozzle vane exits (the angle of truncation at the wall is equal to 11 deg) and other details not taken into account in the calculations. Divergence of the experimental K_ν values obtained in the mononozzle and in the grid plane is apparently attributed to the strong effects of the boundary-layer increase and to a possible flow separation from the parallel channel walls, as compared to the effects caused by the finite angle of truncation at the nozzle wall.

To investigate the inverse medium having greater vibrational energy store than in the previous case, the gain measurements were conducted in the combustion products of the acetylene-air mixture containing a greater percentage of C_2H_2. Combustion of the mixture (6.54% C_2H_2 + air) produced a temperature $T_0 \geq 2500$ K. The gain K_ν as a function of the P_0 dependence obtained from calculations and experiments, is presented in Figs. 3c and 3d for a 6.54% C_2H_2-air mixture. Data of measurements conducted near the grid (Fig. 3c) are in good accord with those calculated over the whole range of stagnation pressure P_0 from 5 to 100 atm. A maximum K_ν value is achieved at $P_0 = 30$ atm. This fact is explained, first, by a greater content of CO_2 and H_2O and a smaller amount of N_2 in combustion products than in the previous cases, and, second, by increased gas temperature in each section of the nozzle. The above factors lead, first, to more rapid deactivation (in comparison with the previous cases) of CO_2 and N_2 vibrational modes, and, second, to the reduction of density. The maximum K_ν is here somewhat smaller than for the previously discussed mixture and is equal to 0.65 m^{-1} (in comparison with 0.82 m^{-1} for a mixture with 4% C_2H_2). The experimental values of gain obtained in the second section (Fig. 3d) are lower than the calculated values. The maximum of divergence amounts to 40% over the $P_0 < 20 \div 40$ atm range. The increase in the above divergence, in comparison with that obtained for a 4% C_2H_2-air mixture, is explained by acceleration of the relaxation process behind the front of oblique shock waves at increased stagnation temperatures.

Recently, the use of nitrous oxide as oxidizer in GDL's has become a subject of great interest (Losev 1977; Cassady 1979; Anderson 1979). The energy yield of the hydrocarbon combustion in N_2O is considerably greater than in air. Higher temperatures ($T_0 > 3000$ K) reached with nitrous oxide results in a significant increased in energy stored in the vibrational levels of the product molecules. As a consequence, experiments were made with a 6.54% C_2H_2 + 33.3% N_2O + 60% N_2 mixture at P_0 = 5-10 atm. In comparison to the products of 6.54% C_2H_2-air combustion, the combustion products of this mixture (see Table 1) contained about one half the concentration of CO_2 and more N_2 and CO. The deactivation of N_2 and CO vibrational modes occurs for this mixture more slowly than for mixtures produced by C_2H_2

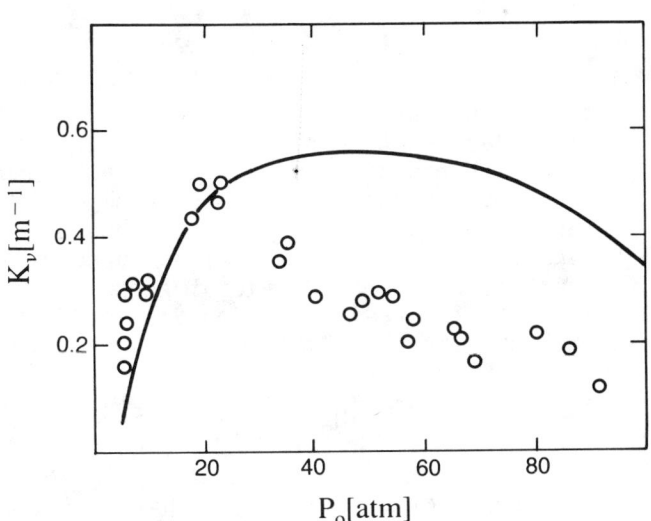

Fig. 4 Effect of the stagnation pressure P_0 on the gain K_ν for 6.67% C_2H_2 33.33% N_2O-60% N_2 mixture. Observations at the nozzle grid edge ——— calculated values of K_ν.

combustion in air. It is caused by the smaller amount of CO_2 and H_2O and greater N_2 and by reduced number density at increased T_0. These factors offset the increased rates of relaxation processes at higher temperature. Comparison of calculated and observed gains shows (Fig. 4) that the calculation method embodies a correct idea of the kinetics of processes for this system. Calculated results are in good agreement with experimental data up to $P_0 = 25$ atm.

The divergence between observed and calculated K_ν at $P_0 > 30$ atm may be explained by the following factors. The processes of ignition and combustion of acetylene differ appreciably with N_2O than air. In most of the experiments, at $P_h > 5$ atm, the diaphragm break occurs before the maximum pressures of explosion products are achieved. The presence of unreacted N_2O in a mixture may increase the deactivation rate of the CO_2 (00^01) upper laser level because the energy level of N_2O (00^01) to those of CO_2 (00^01) and N_2 $(V=1)$. The energy defect is $\Delta\varepsilon = 178$ and 152 K, respectively. The rates of V-V' processes between these states are very significant (10^{-12} cm^3/s), and the relaxation processes in N_2O are considerably faster than in CO_2 (Kulagin 1979; Cassady 1979).

Fuel Efficiency

To compare the efficiency of different fuels, it is necessary to consider the conditions at the nozzle grid exit which control the distribution of energy in the vibrational modes of CO_2 and N_2. In general terms, the objective is to convert the maximum portion of energy into coherent laser radiation before the collisional deactivation degrades this energy into translational and rotational energy of mixture molecules. The specific radiation energy stored per unit gas mass at the inversion transition is (Anderson 1976)

$$E_m^n = (R/\mu)\theta_{mn} [\varepsilon_3 \xi_{CO_2} + \varepsilon_4 \xi_{N_2} - (\varepsilon_3' \xi_{CO_2} + \varepsilon_4' \xi_{N_2}) K_\nu = 0] \quad (2)$$

where $\varepsilon_j = r_j y_j/(1-y_j)$ $(j = 3,4)$; r_j is the multiplicity of the j-th mode degeneration; θ_{mn} is the laser frequency in temperature units; and the term $(\varepsilon_3' \xi_{CO_2} + \varepsilon_4' \xi_{N_2})_{K_\nu=0}$ is the

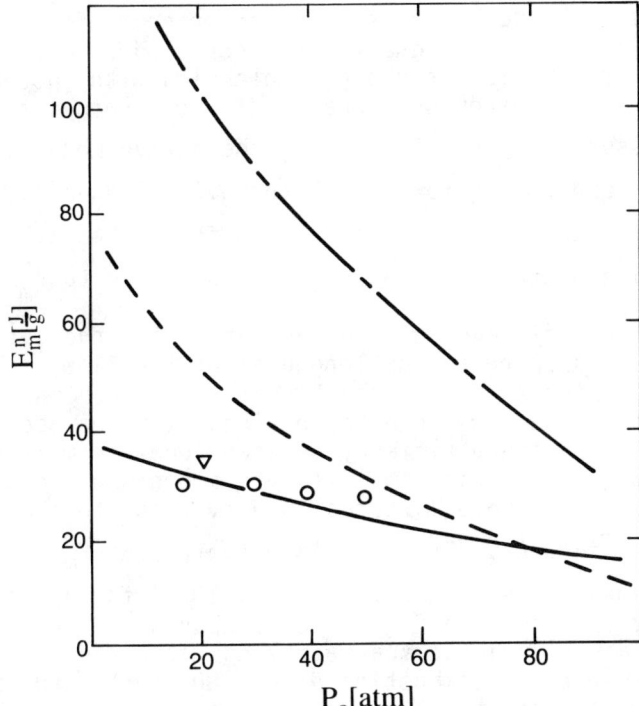

Fig. 5 Effect of the stagnation pressure P_0 on the specific radiation energy E_m^n; —— 4% C_2H_2- air mixture; ---- 6.54% C_2H_2-air mixture —·—·— 6.67% C_2H_2 - 33.33% N_2O - 60% N_2 mixture; calculated values of E_m^n for benzene-air mixture o - Ktalkherman (1979) and Δ - Hill et al. (1978).

average number of quanta in the CO_2 asymmetric vibrations and in N_2 vibration in a zero gain medium. The expression for $\varepsilon_3' = y_3'/(1-y_3')$ can be obtained by setting the expression for the gain calculation equal to zero (Losev 1977). As a result:

$$y_3' = y_2' \frac{B_{10°0}}{B_{00°1}} \exp\left[\frac{B_{00°1}j(j-1) - B_{10°0}j(j+1)}{T}\right]$$

For $\varepsilon_4' = y_4'/(1-y_4')$, it is assumed that $y_3' = y_4' \exp[(\theta_3-\theta_4)/T]$. The predicted values of E_m^n (shown in Fig.

5) decrease with increasing stagnation pressure for each mixture. This result is due to the increased rate of vibrational relaxation of the gas molecules with increased P_0. The maximum values of stored radiation energy, achieved at low pressures (P_0 = 6-10 atm) of the active medium, are equal to 35 and 70 J/g for the 4% and 6.54% C_2H_2-air mixtures, respectively, and to 125 J/g for the acetylene nitrous oxide mixtures. These values are close to E_m^n for mixing lasers.

When the efficiency of the resonator with the optimum parameters (transparency and length along the flow) is sufficiently high ($\eta \sim 0.7$), the highest values of the specific radiation power can be obtained with the above mixtures. As for the efficiency of acetylene-air mixture, it should be noted that with small ($\sim$ 4%) amounts of acetylene, the efficiency of these mixtures is very close to those of benzene-air mixtures. The calculated values of E_m^n obtained by Ktalkherman (1979) and Hill et al. (1978) for a benzene-air mixture are shown in Fig. 5 for comparison. Thus the results obtained indicate that acetylene as a fuel can be successfully used in combustion driven GDL's. Mixtures of C_2H_2-N_2-N_2O have greater promise as an active medium because of the large value of the specific radiation energy $E_m^n \sim 120$ (J/g) with gains of 0.4-0.5 m^{-1}.

Conclusions

The calculation flow model appears to accurately represent the kinetics of vibrational energy exchange in the supersonic flow of the acetylene combustion products over the whole range of the experimental parameters T_0 = 1800-3100 K, P_0 = 5-100 atm. The calculated values of the gain are in agreement with the experimental ones only near the nozzle grid exit. Downstream of the nozzle exit, the real flow structure differs markedly from that of the model because of a large angle ($\sim$ 11 deg) of truncation at the wall. As a consequence the discrepancies between predicted and observed gains are appreciable. The deviation is 20-40% at a distance of 60 mm from the nozzle exit.

Acknowledgments

The authors would like to express their gratitude to M. S. Jijoev for his help in conducting these experiments and for useful discussion of the results, and to V. N. Makarov and Y. V. Tunik for carrying out calculations in designing the nozzle grid.

References

Anderson, J. D. (1976) <u>Gasdynamic Lasers: An Introduction</u>. Academic Press, New York.

Biryukov, A. S. (1975) Kinetics of physical processes in gasdynamic lasers. <u>Trans. Phys. Inst. Acad. Sci. USSR</u> 83, 1386.

Blauer, J. A. and Nickerson, G. R. (1974) A survey of vibrational relaxation rate data for processes important to CO_2-N_2-H_2O infrared plume radiation. AIAA Paper 74-536, Palo Alto, Calif.

Britan, A. B., Losev, S. A., and Shatalov, O. P. (1974) The effect of additions on the mixture gain in the carbon dioxide gas dynamic laser. <u>Sov. J. Quantum Electron</u>. 1 (12), 2620.

Britan, A. B. and Starik, A. M. (1980) Study of vibrationally nonequilibrium flow CO_2-N_2-H_2-H_2O mixture in the wedge nozzle. <u>J. Appl. Mech. Tech. Phys. USSR</u> 4, 27.

Britan, A. B., Starik, A. M., and Khailov, V. M. (1980) Experimental study of vibrationally nonequilibrium flow of gas in the profiling nozzle. <u>Fluid Dyn. USSR</u>, 1, 203.

Cassady, P. E., Pindroh, A. L., and Newton, I. F. (1979) Performance potential of advanced GDL concepts. <u>AIAA J</u>. 17, 845.

Center, R. E. (1973) Vibrational relaxation of CO_2 by O atoms. <u>J. Chem. Phys</u>. 59, 3523.

Egorov, V. V. and Komarov, V. N. (1975) Calculation of population inversion and gain in relaxing gases mixture flowing through the nozzle. <u>Trans. CAGI</u>, 1959, 35.

El-patanov, A. I. and Srizhevskii, N. N. (1968) Combustion products of acetylene near the lower explosive limit. <u>Sov. J. Phys. Chem</u>. 152, 1294.

Evtyukhin, N. V., Genich, A. P., and Manelis, G. B.(1978) Model of active media composition for the gas dynamic CO_2-laser in combustion. <u>Combust. Explos. Shock Waves USSR</u> 4, 36.

Genich, A. P., Evtyukhin, N. V., and Kulikov, S. V. (1979) Gain calculations of multicomponents active media in combustion-driven CO_2 gas dynamic laser. <u>J. Appl. Mech. Tech. Phys. USSR</u> 1, 34.

Hill, G. J., Juel, N. T., and Jones, A. T. (1977) Airbreathing gasdynamic lasers. J. Energy 1, 125.

Ivanov, V. N. (1978) Study of gasdynamic laser characteristics on combustion products hydrocarbon-air mixtures. Ph.D. Thesis, Institute of Mechanical Problems, USSR Academy of Sciences, Moscow, USSR.

Kozlov, G. I., Ivanov, V. N. and Selezneva, I. K. (1979) About gain and power of the GDL on $CO_2-N_2-CO-H_2O-H_2$ mixture. Combust. Explos. Shock Waves USSR 15, 88.

Ktalkherman, M. G. and Malkov, V. M. (1979) Effects of determining parameters on a gain in GDL on combustion products. The GDL Working Process Study (edited by V. K. Baev). Institute for Theoretical and Applied Mechanics, Novosibirsk, USSR.

Kulagin, Y. A. (1979) Active media for gas dynamic lasers. Trans. Phys. Inst. Acad. Sci. USSR 107, 110.

Levin, V. A. and Starik, A. M. (1980a) Vibrational energy-exchange in $H_2-O-H_2-O_2$ mixtures by fast cooling in supersonic nozzle. Fluid Dyn. USSR 2, 102.

Levin, V. A. and Starik, A. M. (1980b) On some methods of creation of the inverse population in molecule vibrational levels. A. Nonequilibrium Flow of a Gas with a Physical-- Chemical Transformation (edited by G. G. Chernyi and V. A. Levin) pp. 4-25. Moscow University, Moscow, USSR.

Losev, S. A. (1977) Gasdynamic Lasers. Izd. Nauka, Moscow, USSR.

Losev, S. A. and Makarov, V. N. (1976) Multifactor optimization of the gas dynamic CO_2-laser. II. Optimization of the radiated power. Sov. J. Quantum Electron. 3, 960.

Makarov, V. N. and Tunik, Y. V. (1978) Determination of optimal nozzle parameters for the gas dynamic laser. J. Appl. Mech. Tech. Phys. USSR 19, 23.

Odintsov, A. I., Fedoseev, A. M., and Bakanov, D. G. (1976) Gas dynamic laser with heating of active media by electric arc pulse. Sov. J. Quantum Electron. 1, 2620.

Smekhov, G. D. and Fotiev, V. A. (1978) About calculation of equilibrium composition of the high-temperature gas. USSR J. Comp. Math. Phys. 18, 1284.

Volkov, A. Y. and Demin, A. I. (1979) Data analysis of vibrational relaxation rates in $CO_2-N_2-H_2O$ mixtures and optimization CO_2 gas dynamic laser. Trans. Phys. Inst. Acad. Sci. USSR 113, 150.

Influence of Flow Structure on Optical Gain in Gasdynamic Lasers

M.G. Ktalkherman,* V.M. Malkov,† and N.A. Ruban‡
Academy of Sciences, Novosibirsk, USSR

Abstract

The aerodynamics of small-sized nozzles having a sharp cornered throat and profiles of different shapes are investigated. The difference of flow parameters from calculated ones for all nozzles are shown. High losses of stagnation pressure and the pressure of shock waves in the stream have been observed. The data on the real flow structure effect on the gain measurements in GDL have been obtained. A correlation for the gain for the exit section of nozzles is shown to fit the data.

Introduction

Both the selection of fuel and aerodynamic design are very important in the construction of high flow-rate driven gasdynamic lasers (GDL). For these systems the working fluid is exhausted to the atmosphere, and the laser work efficiency, to a considerable extent, is determined by the channel aerodynamics.

Losev (1977) and Anderson (1976) report numerous studies on the effects of the plenum chamber pressure P_f, the temperature T_f, the mixture composition, and the nozzle geometry on small signal gain g_0. Various nozzle designs were used in these investigations. The nozzle influences the course of gas mixture relaxation processes and consequently the working fluid in the laser cavity. The

---
Presented at the 8th ICOGER, Minsk, USSR, Aug. 23-26, 1981. Copyright © American Institute of Aeronautics and Astronautics, Inc., 1983. All rights reserved.

*Senior Research Scientist, Institute of Theoretical and Applied Mechanics.
†Research Scientist, Institute of Theoretical and Applied Mechanics.
‡Research Engineer, Institute of Theoretical and Applied Mechanics.

flow homogeneity is disturbed more or less by boundary layers, wakes, and shock waves in flow around trailing edges of the nozzles and walls of the laser cavity. These disturbances depend on the nozzle type and structural features.

Attention has not been paid to the influence of real flow structure on GDL gain. The discrepancies in the results of different investigations obtained under similar experimental conditions may be due to this oversight. The actual aerodynamics of the nozzle must be recognized if gain measurements are to be correctly interpreted and compared with calculations (Ktalkherman, Levin et al. 1979). Also, the evaluation of nozzle designs for GDLs with real pressure sources requires an understanding of real flow structures.

In this work, the aerodynamics of the sharp cornered nozzles is investigated experimentally, and the influence of shock waves on the gain of a combustion driven CO_2 GDL is shown. The combustion products of liquid hydrocarbons served as the GDL working fluid. The use of these fuels for the combustion driven GDL is technologically promising. Single plane nozzles were used in the experiments, and the shock waves were initiated by their supersonic profiles. Numerical analysis of shock wave effects on the inverse medium characteristics of the CO_2 laser has been performed by Kozlov and Stupitsky (1975), Levin and Tunik (1976) and Egorov et al (1975); and experimental data on g_0 change behind the shock waves and rarefaction waves have been reported by Soloukhin and Fomin (1976).

Installation and Measurements

The present experimental investigation consists of two parts. First, visualization of the shock structure for various nozzle designs, was performed in cold air with a separate apparatus, and second, basic parameters of the stream were measured. Analysis of these results made it possible to localize the cross section in which the small-signal gain measurements were made. Thus the gain variation may be compared with the changes of the stream characteristics.

In the aerodynamic investigations the nozzles were smoothly joined to a rectangular constant cross-section channel 20 or 10 mm high, 80 mm wide, and 90-100 mm long. The nozzle profiles were similar to those used in the GDL. The optical glass or steel inserts with drain holes along the axis served as sidewalls for static pressure measurements. The static pressure distribution along the shaped nozzle wall was also recorded.

OPTICAL GAIN IN GASDYNAMIC LASERS

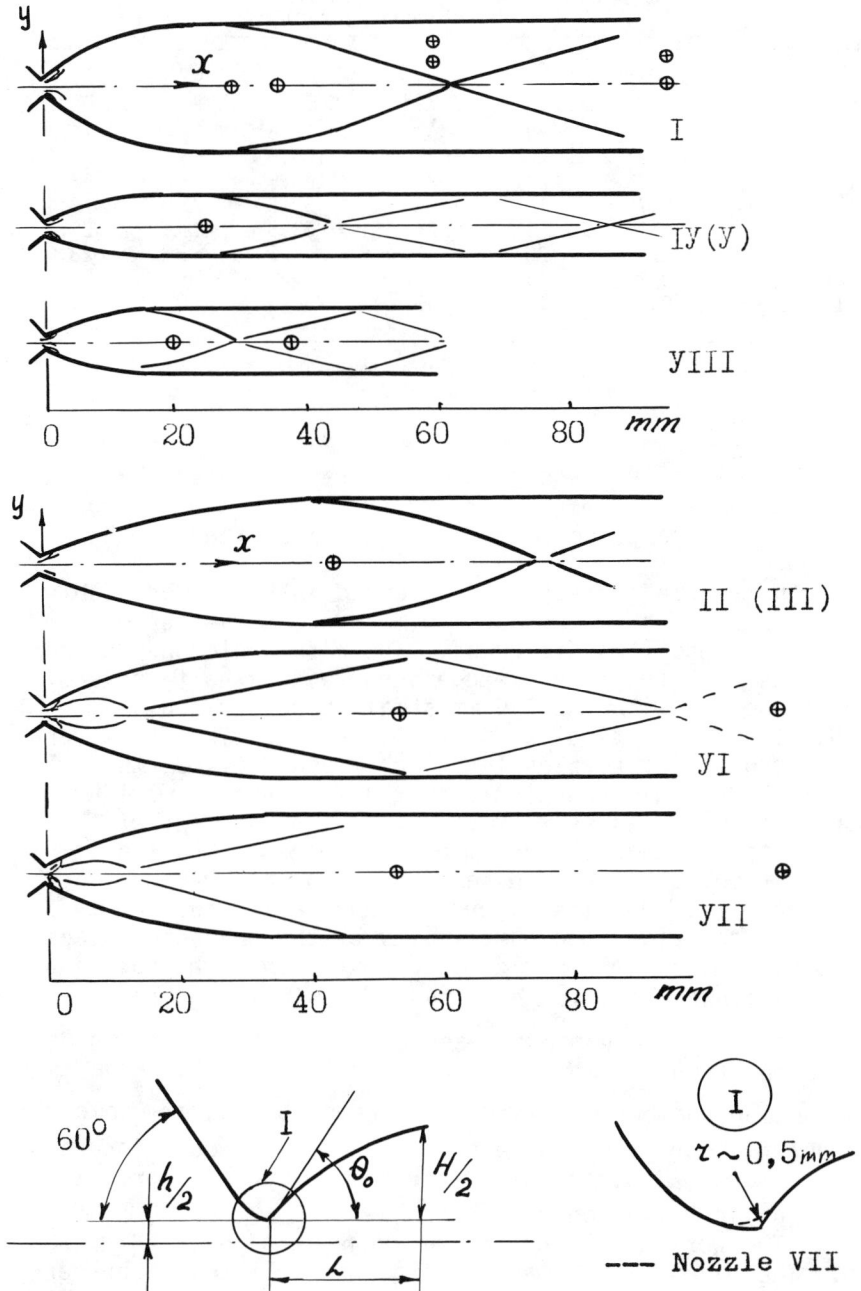

Fig. 1 Flow structure and nozzle geometry.

Table 1 Nozzle configurations

Nozzle	h, mm	H, mm	D, deg	R,[a] mm	L,[b] mm	H/h
I	0.7	20	42	37.5	25	28.6
II	0.7	20	25	106.5	45	28.6
III	0.38	20	25	106.5	45	52.7
IV	0.7	10	25	53.5	22.5	14.3
V	0.35	10	25	53.5	22.5	28.6
VI	0.7	20	41.5	...	56	28.6
VII	0.7	20	...	...	56	28.6
VIII	0.38	10	37	25	15	26.4

[a] R is the shape radius (for the radius nozzle).
[b] L is the length of the nozzle.

The pitot (P_0') and static (P) pressure distributions along the channel height were measured with a pitot tube and static-pressure probe. The flow was visualized by the shadow method.

The nozzle geometry is presented in Fig. 1 and Table 1. The nozzle sizes were chosen to make one parameter of the supersonic profiles different. Their subsonic and transonic profiles were the same: wedge input with a half-angle of 60 deg connected with the 0.5 mm straight part via 0.5 mm radius in the critical cross section (Fig. 1). The supersonic profile of nozzles I through V and VIII was the arc of a circle R (radial nozzles), so that the nozzle wall was parallel to the axis of symmetry. The attachment of the nozzle to the channel was smooth. These nozzles are of interest because of the ease with which they may be manufactured. Nozzles VI and VII have a minimum length profile, as calculated numerically by the characteristics method. The radial nozzles are shorter than the nozzle of "minimum length," which have the same H/h and angle θ_0.

Gain coefficient was measured in the test section of the combustion driven CO_2 GDL. The laser contained a combustion chamber with a burner, removable water-cooled copper plane nozzles, a test section with windows for gain measurements, a supersonic diffuser, and an ejector. The limiting installation parameters were $T_f \sim 2000$ K, $P_f \sim 40$ atm. The working channel had a cross section of 20 x 150 mm. The gaseous fuel was supplied directly to the burner; the liquid fuel was burnt after preliminary evaporation to increase the combustion efficiency.

The gain was measured by the translucence method, which employed a differential circuit. Two radiation detectors were connected to different arms of the bridge circuit. The

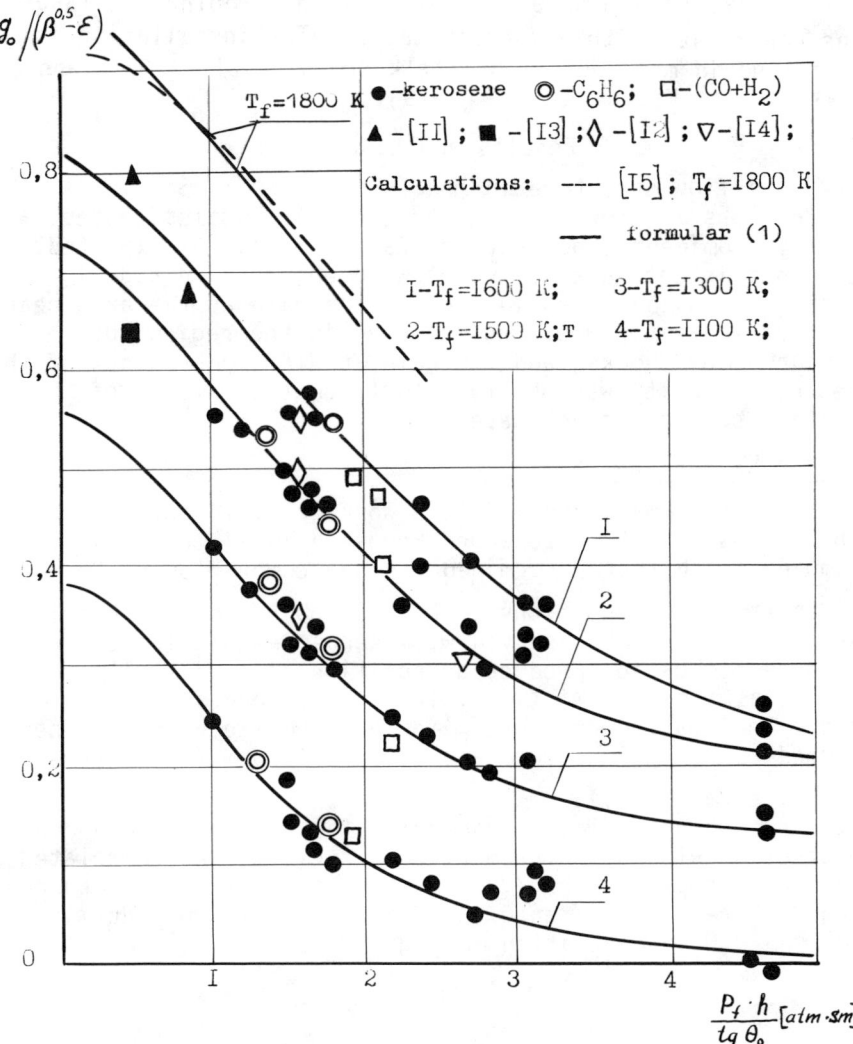

Fig. 2 GDL nozzle exit plane gain (———): Correlation of this work, Eq. (1); (---): Anderson (1972). (See Table 2.)

Table 2 Legend for Figure 2 GDL nozzle exit plane gain.

Curve	Temp., K	Kerosene	C_6H_6	$CO + H_2$	Hill (1977)	Meinzer (1971)	Vamos (1974)	Evtyukin (1978)
A	1100	●	○	◻	—	—	—	—
B	1300	●	○	◻	—	—	◊	—
C	1500	●	○	◻	—	◼	◊	▽
D	1600	●	○	◻	△	—	—	—
E	1800	—	—	—	—	—	—	—

working transition of a power-stabilized probing CO_2 laser was controlled with a monochromator. The installation is described in more detail by Ktalkherman et al. (1977) and Ktalkherman, Malkov et al. (1979).

Results and Discussion

The flow structure in models observed in Schlieren photographs is shown in Fig. 1. For all nozzles tested, a fairly prominent shock pattern was evident. As expected, the most intensive shock waves were observed in radial nozzles. As indicated in Fig. 1, the gain at the axis near the nozzle exit planes was measured in the region not disturbed by shocks, and the gain at different points of the resonator cavity was observed to trace the effects of various strength shock waves.

Nozzle Exit Plane Gain

The gain was observed in a working fluid produced by the combustion of benzene and kerosene in air and was compared with that determined for the combustion of $CO+H_2$ under the same conditions. The gain observed with gasoline as a fuel and that observed with kerosene coincide exactly because combustion product compositions are virtually identical for the two fuels with equal amounts of excess air. The measurements of g_0 for each nozzle have been performed at T_f = 900-1700 K and P_f = 13-35 atm (Ktalkherman, Malkov et al. 1979).

As shown in Fig. 2, the gains for all sharp cornered nozzles and for all aforementioned fuels may be correlated in terms of parameters Z, β, ε (for the collision contour of CO_2). Here $Z = P_f h / \tan \theta_0$, and β and ε are the functions of CO_2 and H_2O concentration in a gas mixture:

$$\beta = \xi_{CO_2} / \xi_{H_2O}$$

$$\varepsilon = \frac{2 \ln[0.1(\xi_{CO_2} + \xi_{H_2O})]}{[1 + \exp(10-\xi_{CO_2})] \exp(0.6 \xi_{H_2O}^{-2})}$$

where ξ_{CO_2} and ξ_{H_2O} are in percent.

The correlation of g_0 (at nozzle exit planes) for different shape nozzles is a consequence of the rapid freezing of the vibrational energy and the population of the upper laser level N_{001} near the critical section (at a distance of about 2h). N_{001} and the vibrational energy change slightly along the nozzle and therefore do not depend on the nozzle shape. The gain coefficient g_0 is proportional to N_{001} because $N_{100} \ll N_{001}$ for a sufficient nozzle expansion. N_{001} is determined by the flow regime near the throat (P_f, T_f, h, θ_0) and by the gas mixture composition.

The gain measured at the nozzle exit planes was correlated by the empirical formula:

$$\frac{g_0}{\beta-\epsilon} = \Delta_1 + \Delta_2 \frac{T_f}{200} - (\exp \frac{T_f}{100} - 19 \times \frac{\beta^{0.06}}{Z^{0.08}}) - 5.5$$

$$\Delta_1 = \frac{0.72}{Z^2+2} \quad \Delta_2 = 0.18 \left(\frac{2}{Z^{0.6}+1} + \frac{Z-1}{\exp 1.4Z}\right)$$

The agreement of the gain predicted with the empirical formula with that observed is good for $T_f < 1800$ K, $Z = 0.5-5$ ($P_f = 13-35$ atm), $\beta = 0.4-9.5$ (ξ_{CO_2} and ξ_{H_2O} ranged from 4% to 20% and from 1.2% to 10%, respectively). The results of comparable calculations and experimental data of other works (Hill et al. 1977; Vamos 1974; Meinzer 1971; Evtyukhin et al. 1978; Anderson 1972) are given in Fig. 2. These data are in good agreement with the correlation.

The population inversion in the test gas mixture increases with the reduction of h and increase of θ_0. However, in this case the nozzle performance is degrading although the flowfield does not change qualitatively over the variation of the above parameters and expansion.

Aerodynamic Measurements

The static pressure profile P/P_f dependence on x, the distance from the critical cross section, at the side-($y = 0$) and shaped walls of models I, III, and VII, as well as the profiles of static and pitot (P_0') pressure measured by the static pressure and pitot probes, are shown in Fig. 3. They have been performed for all nozzles and for the plenum chamber pressures $P_f = 21, 31, 41$ atm. The results did not depend on P_f.

Pressure profiles are determined by the character of shock waves. Thus the growth of $P(x)$ at the shaped wall outlet of nozzles I and III is a consequence of the shock wave departure from the wall at this point. The smooth course of the pressure curves is determined by the interaction of shock waves with the wall boundary layers. The measurements indicate that the flow is not one dimensional, especially in radial nozzles. Nozzle VII exhibits the most uniform profiles of flow parameters. For

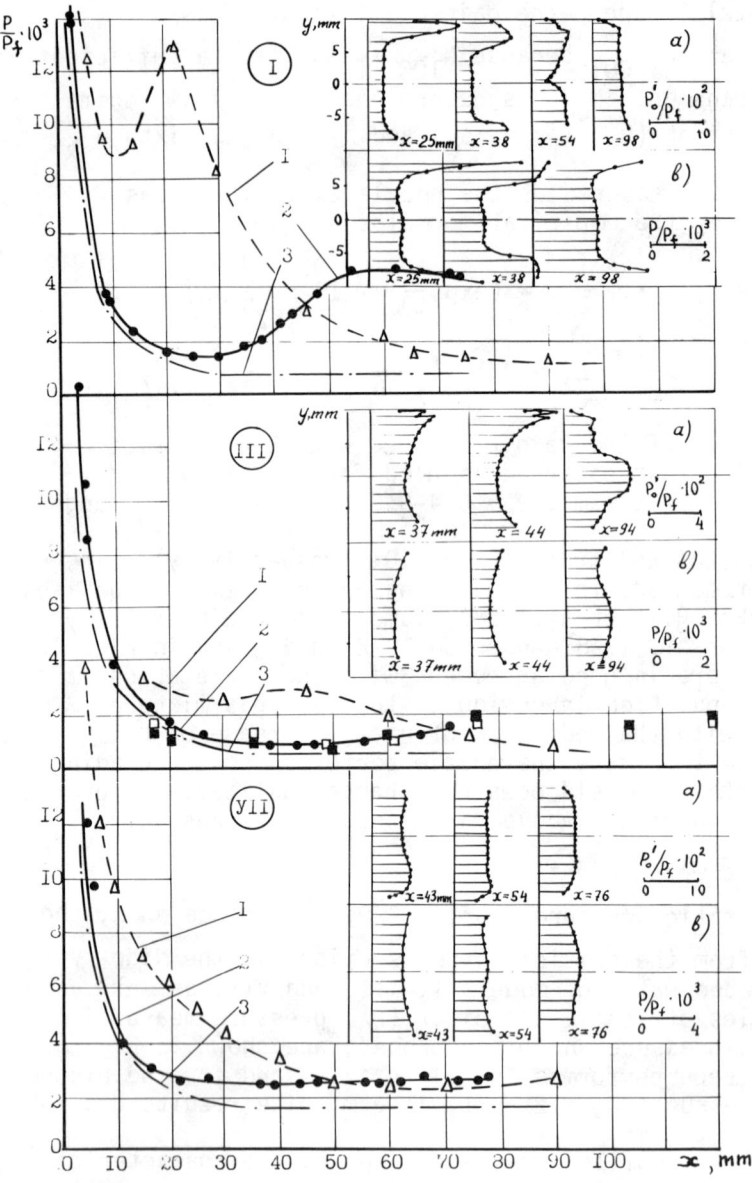

Fig. 3. Characterization of aerodynamic flow cold model study. 1) △: shape wall, 2) ●: sidewall, 3) (—●—): one-dimensional calculation, 4) x: method of characteristics calculation; roman numerals refer to GDL channel pressure distribution. $T_f(K) = 260$ ▫ and $= 1050$ ◼. Cross-section profiles: a) P_0'/P_f, b) P/P_f.

instance, the values of P(x) on the side- and shaped walls practically coincide if the distance x > 50 mm. But in this case as well (i.e., for minimum length nozzles) the observed gasdynamic parameters differ from the calculated ones.

The ratio P/P_0' was used to estimate the local M values. Section parameters were averaged. The mean M values, the stagnation flow pressure $\bar{P}_{0_2}$ and the coefficient of stagnation pressure recovery $\sigma = \bar{P}_0/P_f$ were estimated. The value of σ is low for radial nozzles: $\sigma = 0.53$ for nozzle II exit and $\bar{M} = 4.2$ (the calculated $M_c = 5.17$). The stagnation pressure losses increase with increasing θ_0 and decreasing h. The coefficient $\sigma = 0.\underline{45}$ and $\bar{M} = 4.1$ for nozzle I exit ($M_c = 5.17$), and $\sigma = 0.35$, $\bar{M} = 4.4$ ($M_c = 6.1$) for nozzle exit III. The losses increase downstream along the channel. Further investigation showed that the total pressure losses for a system of nozzle plus diffuser depended very little on the form of pressure shape at the nozzle exit plane.

Apparently, "minimum length" nozzles have better performances. The measured average flow parameters for the exit plane of profile VII are $\sigma = 0.6$, $\bar{M} = 4.6$ ($M_c = 5.17$). But in this case as well, the stagnation pressure losses increase with decreasing h, namely, the nozzle with h = 0.4 (calculated $M_c = 6.1$) has $\sigma = 0.44$ and $\bar{M} = 4.6$. The difference between theoretical and experimental values increases. The shock waves in the "minimum length" nozzle also appear in a stream, their intensity being much lower in comparison with the radial nozzle.

While the nozzle shape determines shock initiation in the radial nozzles, for "minimum length" nozzles the shock generation is attributed to the flow features behind a sharp cornered throat. Pirumov (1974) has predicted that the positive pressure gradient occurs in the supersonic part of such nozzles. The flow mode behind the sharp corner throat was investigated experimentally in Chephanov (1978). It was demonstrated that the flow did not turn immediately behind the break of the profile because of real gas properties, but instead attained a velocity component directed to the wall. This initiated the shock and stagnation pressure losses. More detailed results of aerodynamic measurements for the nozzles of interest are given by Ktalkherman et al. (1980).

Flow Structure Effect on the Gain

While the exit sections of various profile nozzles with identical parameter Z may produce nearly the same gain, the aerodynamic performance may be different. Pressure losses along the gas-flow passage determine the minimum starting

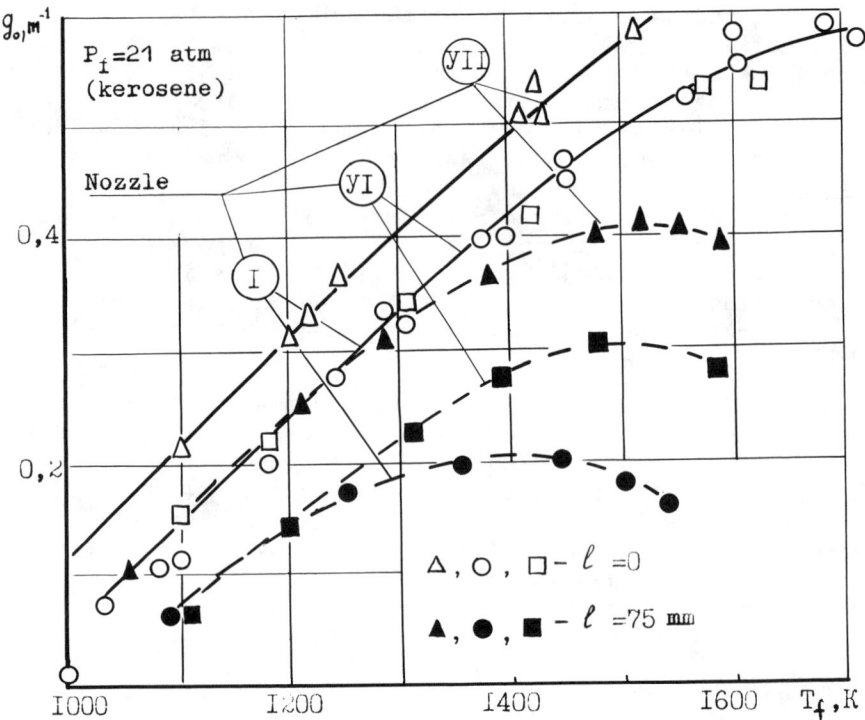

Fig. 4 GDL gain verification along channel. Nozzle I: ○ = 0 mm, ● = 75 mm. Nozzle VI: ☐ = 0 mm, ■ = 75 mm. Nozzle VII: △ = 0 mm, ▲ = 75 mm. P_f = 21 atm; fuel = kerosene.

pressure of the gasdynamic laser. In other words, the use of nozzles with poorer aerodynamic performance leads to an increased starting pressure in the real GDL and therefore to deterioration of flow inversion.

Furthermore, inversion behavior in the resonator cavity behind nozzles with the same h, H/h, and θ_0, but with different profile shapes, may differ. Figure 4 shows g_0 measurements for three nozzles: I, VI, and VII (see Table 1). The functions $g_0(T_f)$ and given here for the exit section and for the cross section at a distance of 75 mm. Gain measurements were performed both before and after the shock interception as is shown in Fig. 1. Profiles VI and VII were identical except the sharp cornered edge rounded to give r ∼ 0.5 mm (Fig. 1) for nozzle VII. This resulted in lower shock strength in the throat region and in smoothing of parameter profiles in the channel cross sections.

The difference in the shock wave strength strongly affected g_0 measurements, especially in the cross section l

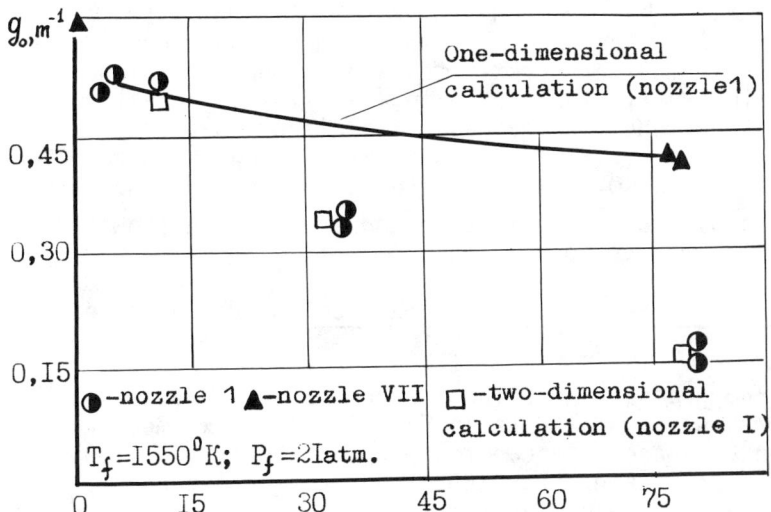

Fig. 5 Comparison of predicted GDL gains with observations. (———): One-dimensional calculation for nozzle I. ◐ nozzle I, experimental; ▲ nozzle VII, gain; ▭ two-dimensional calculations for nozzle I. T_f = 1550 K; P_f = 21 atm.

= 75 mm. The most pronounced decrease of g_0 was observed for nozzle I, whose shocks were strongest. For nozzle profile VII with a rounded edge in the throat (the weakest shock) g_0 measurements were greater than those of both nozzle I and VI, and here the gain increase was also observed in the nozzle exit section.

It is interesting to note that one-dimensional calculations (Anderson and Harris 1972) indicates that the rounding of sharp edges must result in a decrease of g_0. Undoubtedly, the observed effect of a g_0 increase is connected with attenuation of shock waves in the throat region.

Hill et al. (1977) and Meinzer (1971) reported the presence of shock waves in sharp cornered nozzles but observed that their influence on the obtained results was negligible. The present work points to the necessity of a comprehensive allowance for the actual nozzle aerodynamics in the interpretation of the gain measurements. This point is of particular relevance to the reverse problem studies in which kinetic rate constants are to be extracted from g_0 measurements.

The shock effects on inversion properties have been examined in more detail by (Ktalkerman, Levin et al. (1979), who compared experimental results with the two-dimensional calculations by Levin and Tunik (1976). While the gain profile along the channel downstream of the rounded edge nozzle VII may be approximated with a one-dimensional model, the gain profile for a relaxing gas flow in radial nozzles with strong shocks requires a two-dimensional model, as is shown in Fig. 5. The gain decrease behind the shock (for the collision CO_2 contour) results mainly from the rise of the translational temperature and, hence, the growth of the lower laser level population.

Extension of the Results of the Model Wind Tunnel Test to Real GDL Operating Conditions

The pressure distribution on GDL channel walls are compared with model data In Fig. 3. The P(x) measurements in the GDL were also carried out in hot air (T_f = 1050 K) and the results of "hot" GDL and "cool" wind-tunnel tests are in agreement because at the low-pressure and -temperature levels of these experiments the ideal gas law adequately describes the gas behavior.

The experiments also indicate the absence of condensation in "cool" wind-tunnel tests. Daum and Gyarmathy (1968) showed that the supercooling of several tens of degrees may be achieved in the wind-tunnel nozzles depending on the flow expansion rate. The expansion rate is higher for the GDL nozzles than for ordinary wind-tunnel nozzles. At higher temperatures, a discrepancy in the pressure distributions may appear because the boundary-layer thickness δ will increase severalfold. Measurements of δ on a "cool" flow indicate that δ is large for the investigated profiles. The displacement thickness of boundary layer δ^* is equal to 0.85 mm for nozzles VI and VII.

Summary

The aerodynamics of small-size GDL nozzles (the "minimum length" and radial nozzles) were investigated. The nozzles had the following dimensions: a throat between 0.35 and 0.7 mm and expansions 14.3, 28.6, and 50. Shocks were observed in the downstream flow for all the nozzles. The difference between the observed and calculated flowfield distributions is determined by the nozzle expansion. High stagnation pressure losses $\sigma = P_0/P_f$ were observed. For instance, the average Mach number is equal to 4.6 and σ = 0.6 for the "minimum length" nozzle at M_c = 5.17.

The gain was measured in combustion products of gasoline, kerosene, and benzene with air in the nozzle exit sections and at some points of the resonator cavity. Experiments were performed for temperatures of 900-1700 K and pressures of 13-35 atm. The data on the gain for the exit planes were correlated by an empirical formula. This work shows that a comprehension of the actual aerodynamics of nozzles is necessary for the correct interpretation of the gain results, correct comparison with calculations, and appropriate choice of the nozzle device for a GDL.

References

Anderson, J. D. and Harris, E. L. (1972) Modern advances in the physics of gasdynamic laser. AIAA Paper 72-143. 10th Aerospace Sciences Meeting, San Diego, Calif.

Anderson, J. D. (1976) Gasdynamic laser. An Introduction. Academic Press, New York.

Chephanov, V. M. (1978) Techenie gaza v gorle sverhzvukovogo sopla. In: Gazodinamika letatel'nyh apparatov. Vypusk 1, Kazan'. (in Russian).

Daum, F. L. and Gyarmathy, G. (1968) Condensation of air and nitrogen in hypersonic wind tunnels. AIAA J. 6, 458.

Egorov, B. V., Komarov, V. N. and Sayapin, G. N. (1975) Ob izmenenii koefficienta usilenia v udarnom sloe pri obtekanii zatyplenych tel potokom s inversnoi naselenostiu. Prikl. Mekh. Tekh. Fiz. 15, 13.

Evtyukhin, A. V., Genich, A. P., Yadanov, L. A., and Manelis, G. B. (1978) Koefficienty usilenia rabochih sred v CO_2-GDL na produktah sgorania. Kvantovaya Elektron. Moscow 5, 1013.

Hill, R. J., Jewell, N. T., Jones, A. T., and Price, R. B. (1977) Airbreathing gasdynamic lasers. J. Energy 1, 125.

Kozlov, G. I. and Stupitsky, E. A. (1975) Izmenenia koefficienta usilenia v udarnoi volne, rasprostraniausheisia po inversnoi srede. Tekh. Fiz. 45, 359.

Ktalkherman, M. G., Malkov, V. M., Petukhov, A. V., and Kharitonova, Y. I. (1977) Koefficienty usilenia v gasodinamicheskom lasere na producktah gorenia benzola. Kvantovaya Elektron. Moscow 4, 173.

Ktalkherman, M. G., Malkov, V. M., Shevyrin, A. Y., and Sheitelman, G. Y. (1979) Vlianie parametrov tormozhenia, razmera i kontura sopla na koefficient usilenia v GDL na produktah sgorani zhidkih topliv. Fiz. Goreniya Vzryva 15, 64.

Ktalkherman, M. G., Levin, V. A., Malkov, V. M., and Tunik, Y. V. (1979) Pole techenia i koefficienty usilenia v rezonatornoi polosti GDL na produktah gorenia kerosina. Dvumernyi rashet i sravnenie s eksperimentom. Fiz. Goreniya Vzryva 15, 84.

Ktalkherman, M. G., Malkov, V. M., and Ruban, N. A. (1980) Eksperimental'noe issledovanie techenia v soplah gazodinamicheskih laserov. Izv. Akad. Nauk SSSR Mekh. Zhid. Gaza 15, 178.

Levin, V. A., and Tunik, Y. V. (1976) Dvizhenie relaksiruyushei smesi gazov v dvumernyh soplah. Izv. Akad. Nauk SSSR Mekh. Zhid. Gaza 11, 118.

Losev, S. A. (1977) Gazodinamicheskiye Lasery. Nauka, Moscow, USSR.

Meinzer, R. A. (1971) Experimental gasdynamic laser studies. AIAA Paper 71-125. 9th Aerospace Sciences Meeting, New York, NY.

Pirumov, U. G. (1974) Nekotoryje Primeneniya Setochnykh Metodov v Gazodinomike, No. 6, MGU, Moskwa.

Soloukhin, R. I. and Fomin, N. A. (1976) Izmenenie inversii v potoke s gazodinamicheskimi vozmusheniami. Dokl. Akad. Nauk SSSR 228, 596.

Vamos, I. S. (1974) Small-signal gain measurements in high aeration nozzle shock tunnel GDL. AIAA Paper 74-177. 12th Aerospace Sciences Meeting, Washington, D.C.

Operation of Arc-Heated Gasdynamic CO_2 Laser at 16.4-18.6 μm

A.I. Demin,* E.M. Kidriavtsev,† and A.Y. Volkov*
Academy of Sciences, Moscow, USSR
and
D.G. Bakanov,‡ A.I. Fedoseev,‡ and A.I. Odintsov§
Moscow State University, Moscow, USSR
and
V.F. Sharkov^π
Kurchatov Institute of Atomic Energy, Moscow, USSR

Abstract

This paper outlines the laser action on the transitions between the bending and symmetric modes of CO_2 and on the inside of ν_2 mode transitions in quasicontinuous and continuous wave gasdynamic lasers with electric arc heating of a test mixture.

Introduction

Vedeneev et al. (1978) and Konyukhov and Faizulaev (1978) have suggested that gasdynamic lasers (GDL) may operate on the transitions between levels of coupled modes of CO_2. Vedeneev and co-workers demonstrated laser action on the transition 03^10-10^00 in shock-tube-driven CO_2-Ar GDL and presented a theoretical model of its operation. The process of vibrational relaxation was interpreted in terms of total energy of coupled modes (ν_1 and ν_2). In the following theoretical and experimental investigations (Vendeneev et al. 1979; Brunne et al. 1981; Vedeneev et al. 1981a, 1981b), the influence of stagnation parameters (initial temperature T_0 and pressure p_0) and mixture composition was considered.

Paper presented at the 8th ICOGER, Minsk, USSR, Aug. 23-25, 1981. Copyright © 1983 by A. I. Demin. Published by the American Institute of Aeronautics and Astronautics, Inc. with permission.
*Research Scientist, Lebedev Institute of Physics.
†Senior Research Scientist, Lebedev Institute of Physics.
‡Research Scientist, Department of Physics.
§Senior Research Scientist, Department of Physics.
^πSenior Research Scientist.

The expected power output was calculated by Volkov et al. (1979) and Brunne et al. (1980). It was shown that CO_2-Ar GDL (λ = 18.4 µm) may have advantages in comparison with CO_2-N_2 GDL (λ = 10.6 µm), the main ones being the higher efficiency and the possibility of application of the mixtures under lower stagnation temperature.

Coupled-mode CO_2 GDL is also of interest in the spectral range extension of infrared (i.r.) lasers. Calculations by Vedeneev et al. (1978) and Vedeneev et al. (1979) show that population inversions may be reached in a number of vibrational-rotational transitions (02^00-01^10, 03^10-02^20, 04^20-03^30, 04^20-11^10, 04^00-11^10, 03^10-10^00, 04^00-00^01) in the spectral range 16.2-50.2 µm. The scheme of low-lying vibrational levels of a CO_2 molecule and the transitions mentioned above are shown in Fig. 1. The transition 03^10-10^00 (λ = 18.4 µm) proved to be most suitable for lasing (Vedeneev et al. 1978).

Until recently the only experimental apparatus for the CO_2-Ar GDL investigations was the shock tube with a short optical path ($\sim$ 9 cm), and a praticable system for exploiting the CO2-Ar GDL population inversion had not been developed.

This paper reports a study of the arc-heated CO_2 GDL operating on the transitions 03^10-10^00 (λ = 18.4 µm), 02^00-01^10, and 03^10-02^20 (λ = 16.4-17.2 µm). Descriptions of both quasicontinuous and continuous wave (cw) operations are presented.

Quasicontinuous Wave Laser

The quasi-cw device was used for the investigation of 10.6 µm CO_2 GDL (Odintsov et al. 1976). To obtain lasing on coupled-mode transitions, only the parameters of the optical reasonator and nozzle of the device were changed. The sketch of the equipment is shown in Fig. 2. The gas was heated by pulsed electric discharge (time duration $\sim 10^{-4}$ s) initiated by the electric explosion of a thin wire. Pressure in the stagnation chamber (2) was monitored with a calibrated piezoelectric transducer (4). The supersonic flow was formed in a flat contoured nozzle 4 cm long (throat height h* = 0.01 cm, expansion ratio A/A* = 150). The nozzle was connected to the constant cross-section channel

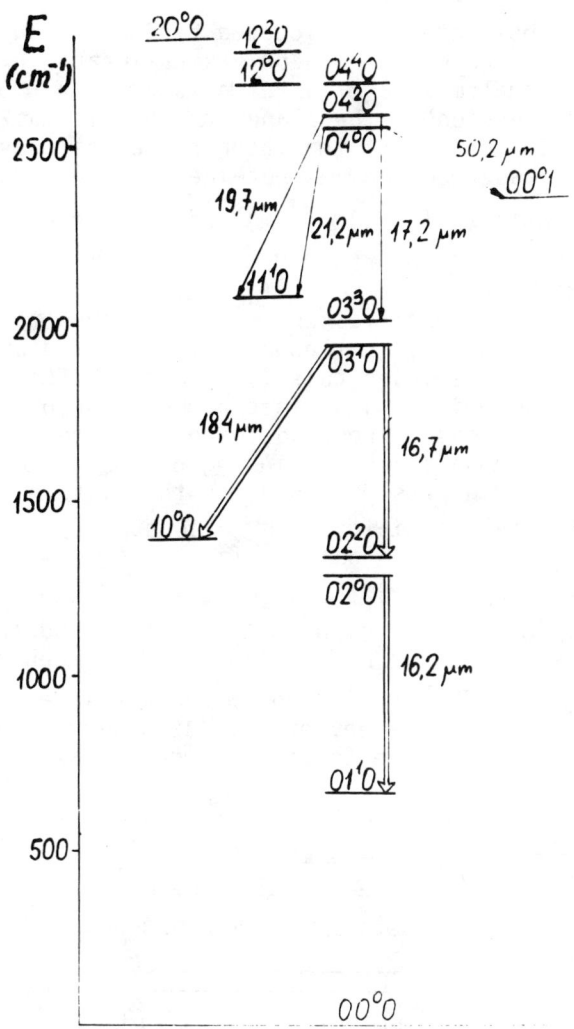

Fig. 1 Low-lying vibrational CO_2 levels. Arrows show possible laser transitions. Wide arrows, laser transitions obtained in the present work.

linked to a vacuum chamber. The optical axis was located 3.5 cm from the throat. Cavity length across the flow was about 40 cm. The resonator consisted of two spherical gold-coated mirrors with central holes (D = 0.05 cm) for the extraction of laser radiation.

Laser wavelengths were measured with the help of the Abert-scheme IR monochromator containing grating (50 grooves/mm). Laser radiation was recorded by a semiconductor bolometer (12) located behind the outlet slit (0.5 mm wide). The monochromator was calibrated over the high diffractional maxima of a He-Ne laser ($\lambda = 0.6328$ μm). The linear dispersion at the plane of the slit was equal to 0.04 μm/mm. The energy of the laser pulse was measured by a calorimeter. Laser action was obtained on a Q branch of $03^10\text{-}10^00$ transition ($\lambda = 18.4$ μm) and on P branches of the transitions $02^00\text{-}01^10$ and $03^10\text{-}02^20$ (spectral range 16.4-17.2 μm).

Figure 3 shows oscilloscope traces of the stagnation plenum pressure and 18.4 μm laser pulse recorded by the Ge-Zn detector. The laser pulse, in time, shifts from the pressure maximum point. This fact proves that optimal generation conditions correspond to lower temperatures and pressures than those at the beginning of the cycle. Maximum power output (~ 3 W) was obtained for the mixture CO_2:Ar = 1:2 at the initial conditions: $p_o = 10$ atm, $T_o = 1000$ K. The estimate of the power density inside the cavity, made by output power, resulted in $I = 3 \cdot 10^3$ W/cm^2.

The predicted laser gain at $\lambda = 18.4$ μm and the results of power measurements as a function of CO_2 concentration are shown in Fig. 4. The predictions were calculated by the method described by Vedeneev et al. (1979): one-dimensional gasdynamic equations were solved with the relaxation

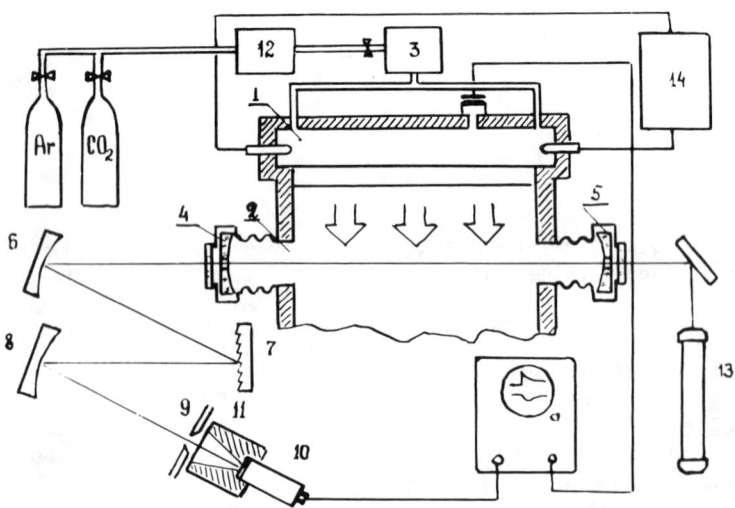

Fig. 2 Experimental setup for the investigations of quasi-cw gasdynamic CO_2 laser on coupled modes transitions.

ARC HEATED GASDYNAMIC CO$_2$ LASER

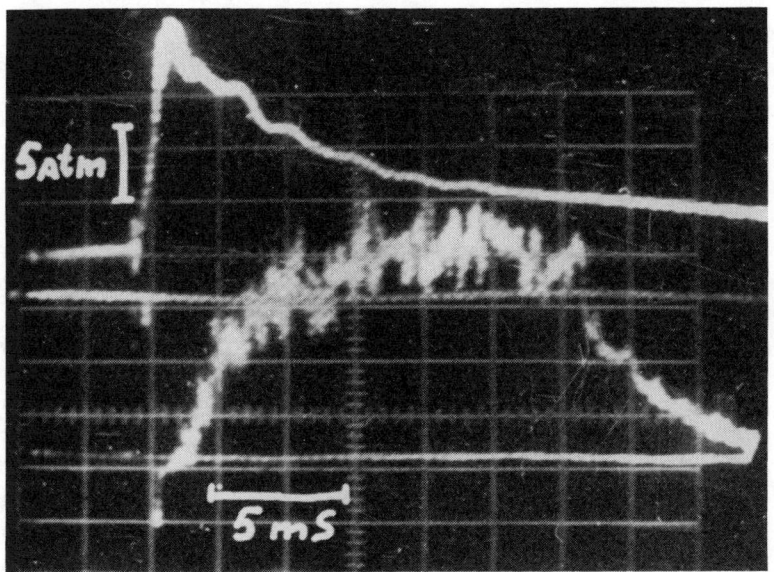

Fig. 3 Oscillograms of the stagnation pressure and the laser pulse.

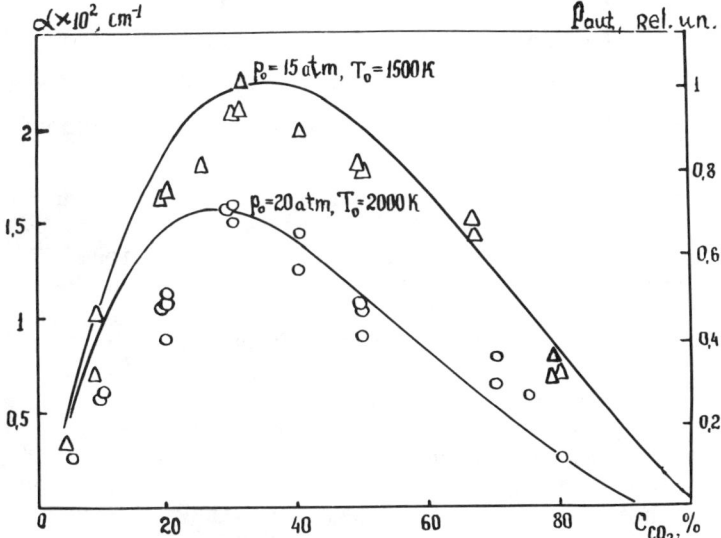

Fig. 4 Gain (calculated curves) and laser energy (experimental points) vs CO$_2$ concentration. Initial conditions: 1) T_0 = 2000 K, p_0 = 20 atm; 2) T_0 = 1500 K, p_0 = 15 atm.

equation for the total energy of coupled modes of CO_2. Figure 4 shows that the maximum power coincides with the gain maximum, which confirms the validity of the model. The gain and output decrease at large CO_2 concentrations can be attributed to the decrease of gas cooling due to the reduction of the specific heat ratio, as well as to the relative decrease in V-T relaxation processes.

The output power achieved in the laser is not an accurate indication of its energy potential, because the resonator and nozzle parameters were not optimized. In the experiments with a CO_2-N_2-He mixture at approximately optimum initial conditions, the output power at 10.6 μm was somewhat less than 3 W. From this fact one can infer that both types of GDL are comparable from the point of view of their output power.

At this optical path length laser action was observed on other transitions of the CO_2 molecule. Preliminary calculations based on the one-dimensional model (Brunne et al. 1980) showed that considerable gain could be obtained not only on the transition 03^10-10^00, but also at the P-branch of the transition 02^00, at the P-, Q-, and R-branches of the transition 03^10-02^20 and at the P- and Q-branches of the transition 04^20-03^30 (spectral range 16.2-17.5 μm) under typical initial conditions (T_0 = 1000 K, p_0 = 7.5 atm, CO_2:Ar = 1:9). The calculated gain spectrum is presented in Fig. 5a.

As is shown in Fig. 5a, the maximum gain for each of the three vibrational transitons must occur for P-branches, a manifestation of a partial inversion. However, the calculated spectrum given in Fig. 5a does not account for some peculiarities which are important for inversion formation. The presence of impurities in the working fluid (e.g., water vapor), which effectively decreases the energy of the coupled modes (Vedeneev et al. 1981a), has a substantial effect on the vibrational (T_v) and translational (T) temperatures during the gas expansion. Figure 5b shows the spectrum of the CO_2-Ar GDL gain under the same conditions as in Fig. 5a, but with H_2O added to the working fluid as an impurity (γ_{H_2O} = 0.01%). Even an insignificant quantity of H_2O affects T_v and T in such a way that a noticeable transformation of the gain spectrum takes place. Without H_2O (Fig. 5a), the maximum gain corresponds

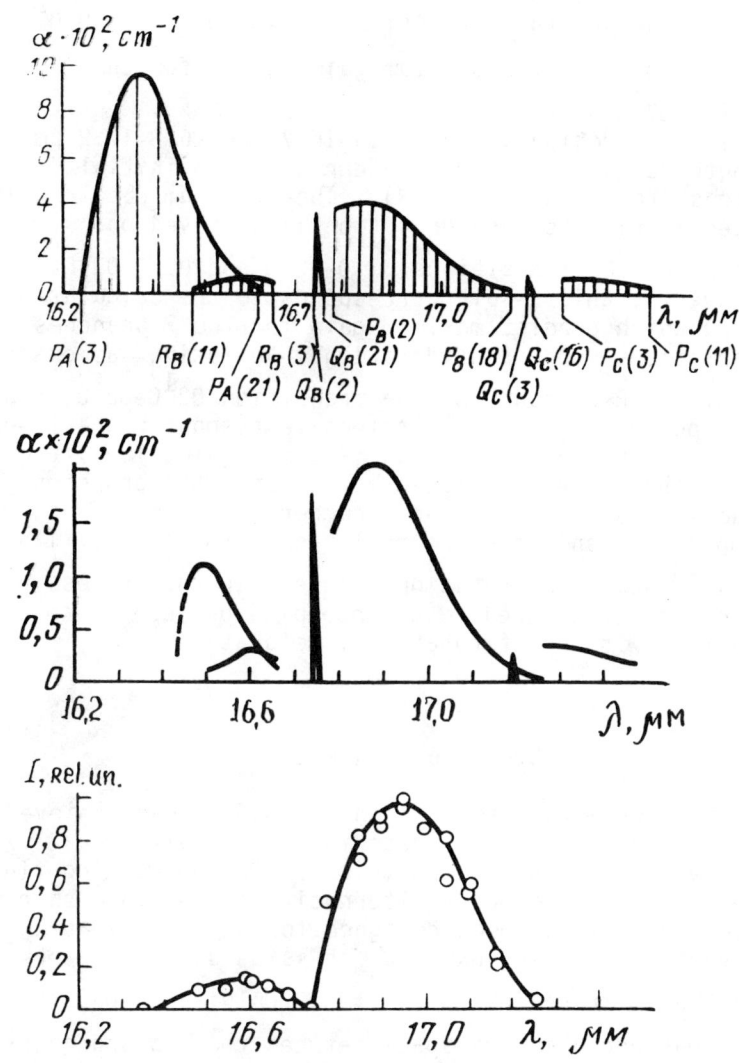

Fig. 5 Calculated values of the gain (a,b) and experimental values of the laser power (c) vs the wavelength. $P_A(J)$, P-branch of $02^00 - 01^10$ transition, P_B, Q_B, $R_B(J)$, P-, Q- and R-branches of $04^20 - 03^30$ transition. Calculation and experiment correspond to the following conditions: mixture $CO_2:Ar = 1:9$, $T_0 = 1000$ K, $p_0 = 7.5$ atm, throat height $h^* = 0.1$ mm, distance from the throat $X = 200\ h^*$. (a) $\gamma_{H_2O} = 0$, $T_v = 474$ K, $T = 37$ K, $p = 1.4 \cdot 10^{-3}$ atm. (b) $\gamma_{H_2O} = 0.01\%$, $T_v = 507$ K, $T = 45$ K, $p = 1.6 \cdot 10^{-3}$ atm.

to the P-branch of the transition 02^00-01^10. With H_2O impurity (Fig. 5b), the maximum gain occurs for the transition 03^10-02^20.

The laser action in the 16.4-16.7 and 16.8-17.2 μm wavelength range was recorded under the same initial conditions (total power ∿ 2 W). The spectrum obtained is presented in Fig. 5c. Laser action is achieved on P-branches of the transitions 03^10-02^20 and 02^00-01^10. To a certain extent this result corresponds to the calculations (Fig. 5b) which predict maximum gain for the P-branches of the transitions under consideration. No lasing was observed at the Q- and R-branches of the transition 03^10-02^20, since in the experiments with a nonselective resonator the laser action began at the branch with maxixmum gain, i.e., the P-branch, thus prevented laser action at other branches. Laser action was observed for a number of lines. Thus, for low temperature and pressure realized in the cavity (T = 60 K, p = 10^{-3} atm), the rotational levels in the process of lasing can be considered to be uncoupled because of the relatively low speed of rotational relaxation.

Continuous Wave Laser

Continuous wave laser action at 18.4 μm was achieved with the experimental setup described in detail by Goriatchev et al. (1980). In this cw GDL the working fluid was heated by a three-phase alternative current arc source. While this system had been designed for an investigation of the conventional $CO_2-N_2-H_2O$ GDL, it is used in this work with another mixture (CO_2:Ar = 1:9). Only argon was heated; cold CO_2 was added in the space between the arc source exit and nozzle set. The initial temperature T_o and pressure p_o range from 1250 to 240 K and from 4.5 to 9.5 atm, respectively. The expansion of the working fluid occurred through the grid, which consisted of 38 profile nozzles with a 0.014-cm throat height and an A/A* = 40 expansion ratio. In these experiments the resonator was placed immediately after the nozzle exits; but laser action was observed even after the resonator was moved 10 cm downstream. The gas pressure in the resonator section was varied from 0.5 to 2.5 Torr. The resonator parameters were approximately the same as those of the quasi-CW GDL described above.

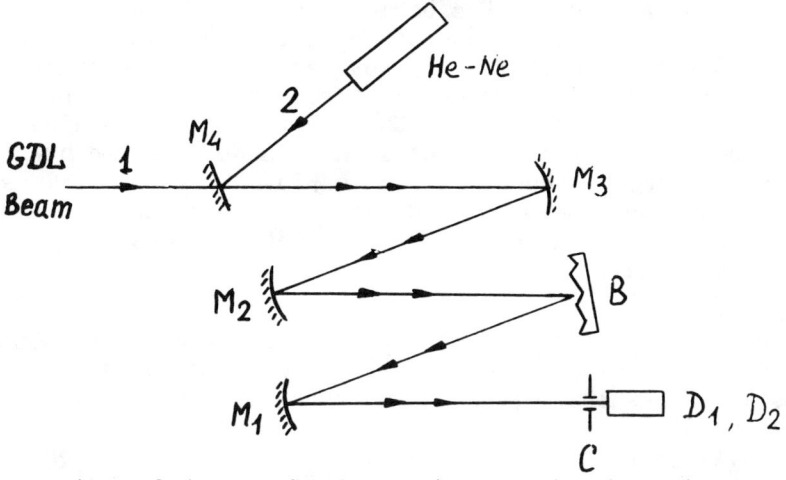

Fig. 6 Scheme of the scanning IR-monochromator for the registration of GDL spectrum.

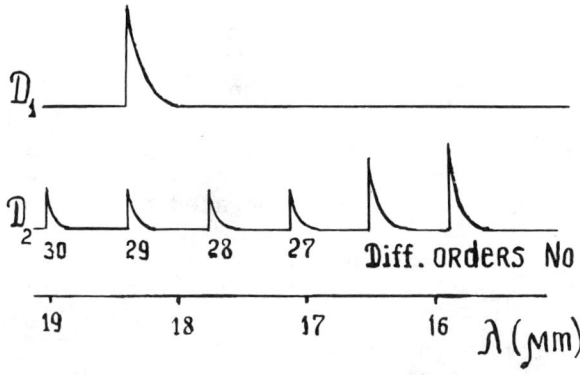

Fig. 7 Oscillograms obtained by scanning IR-monochromator. D_1 - signal from Ge-Zn detector; D_2 - signal from photodiode (for frequency calibration).

The cw GDL spectrum was recorded with the scanning i.r. monochromator shown in Fig. 6. The monochromator was assembled from diffraction grating (B) and three spherical mirrors (M_1-M_3). The additional mirror, M_4, was placed below the beam (1) and served to introduce the radiation of the He-Ne laser (λ = 0.6328 μm), which was used to calibrate the wavelength range. Infrared radiation from the GDL resonator and the He-Ne laser signal was recorded by the photoresistor Ge-Zn (D_1) and the photodiode (D_2), respectively, which were located behind the slit in the cross section normal to the plane of design. Spectrum scanning with a speed of about 200 m/s was realized by the rotation of the grating. The frequency of rotation (about 1 Hz) allowed a few records of the GDL spectrum to be obtained from one run (3-4 s). One of such records obtained by the low-speed oscilloscope is given in Fig. 7. The positions of the peaks, corresponding to definite diffractional orders of the He-Ne laser radiation gave information about the wavelength scale.

Only radiation with $\lambda = 18.4$ μm (± 0.1 μm) corresponding to the lines of the Q-branch of the transition $03^10\text{-}10^00$ was recorded in a wide range of initial temperatures and pressures. As the resonator employed was not optimized, these results should be regarded as a demonstration of the feasibility of laser action at 18.4 μm. The experiments led to an estimate of the energy abilities of a cw 18.4-μm GDL. Specific power of approximately 30 kW/kg was estimated by accounting for the portion of energy extracted from the resonator. The laser gain was estimated, by modifying resonator losses, to be about 0.3 m^{-1}.

Conclusion

1) Quasi-cw and cw laser action with $\lambda = 18.4$ μm and quasi-cw action in the spectral range 16.4-17.2 μm were observed in a gasdynamic laser operated on CO_2 coupled-mode transitions.

2) Neither the possible impurities due to the arc source nor the gas flow nonuniformity at the nozzles outlet lead to destruction of the population inversion between CO_2 coupled-mode levels in the above devices.

3) High levels of specific power $P_g \sim 30$ kW/kg and intercavtity power density $I \sim 3$ kW/cm^2 estimated for $\lambda = 18.4$ μm make it conceivable that the 18.4-μm CO_2 GDL can be comparable to the 10.6-μm CO_2 GDL in its power output.

4) Lasing on CO_2 coupled-mode transitions was obtained on devices which were designed for the investigation of conventional CO_2 GDL ($\lambda = 10.6$ μm).

References

Brunne, M., Zielinski, A., Milewski, J., Volkov, A. Y., Demin, A. I., and Kudriavtsev, E. M. (1980) Analytical modelling of the influence of the nozzle geometry and stagnation conditions on parameters of a CW CO_2 18.4 m gasdynamic laser. <u>J. Phys. Paris Colloq.</u> 41(11), 183-188.

Brunne, M., Zielinski, A., Milewski, J., Volkov, A. Y., Demin, A. I., and Kudriavtsev, E. M. (1981) Simplified calculations for prediction of parameters of 18.4 m CO2 continuous wave gasdynamic laser. J. Appl. Phys. 52, 74-86.

Goriatchev, S. B. et al. (1980) Experimental setup C2P. Gasdynamic CO_2 laser with gas heating by plasma source. Preprint 3320/7, pp. 1-39, Kurchatov Institute of Atomic Energy, Moscow, USSR.

Konyukhov, V. K. and Faizulaev, V. N. (1978) Possibility of building a gasdynamic laser utilizing transitions between levels of paired CO_2 modes. Sov. J. Quantum Electron. 8, 1473-1474.

Odintsov, A. I., Fedoseev, A. I., and Bakanov, D. G. (1976) A gasdynamic laser with working substance heated by pulsed arc discharge. Pis'ma Zh. Tech. Phys. 2, 145-149.

Steverding, S. B. (1979) On rotational equilibrium in infrared laser cavities. J. Appl. Phys., 50,5994-5995.

Vendeneev, A. A., Volkov, A. Y., Demin, A. I., Logunov, A. N., Kudriavtsev, E. M., and Sobolev, N. N. (1978) Thermally pumped gasdynamic CO_2 laser based on deformation and symmetric modes. Sov. Tech. Phys. Lett. 4, 275-276.

Vedeneev, A. A., Volkov, A. Y., Gomenuk, Y. V., Demin, A. I., Kudriavtsev, E. M., and Poluian, V. P. (1979) Theoretical and experimental investigation of heat pumped CO_2-Ar (Xe) GDL with 18.4 μm wavelength. Preprint 120, pp. 1-50, Lebedev Institute of Physics, USSR Academy of Sciences, Moscow, USSR.

Vedeneev, A. A., Volkov, A. Y., Demin, A. I., and Kudriavtsev, E. M. (1981a) The influence of water, hydrogen and helium on CO_2 vibrational levels population at strong nonequilibrium conditions of supersonic cooling. Preprint 26, pp. 1-24, Lebedev Institute of Physics, USSR Academy of Sciences, Moscow, USSR.

Vedeneev, A. A., Volkov, A. Y., Demin, A. I., Kudriavtsev, E. M., Stanco, J., Milewski, J., and Brunne, M. (1981b) Shock-tube CO_2 gasdynamic laser operating on the $(03^10)-(10^00)$ transiton at 18.4 μm. Appl. Phys. Lett. 38,199-201.

Volkov, A. Y., Demin, A. I., Kudriavtsev, E. M., Brunne, M., and Milewski, J. (1979) Estimation of parameters of CW 18.4 μm CO_2/Ar gasdynamic laser. Proceedings of the 2nd International Symposium on Gas Flow and Chemical Lasers (edited by J. F. Wendt) pp. 249-252. Hemisphere Publishing Co., Washington, D.C.

Author Index for Volume 88

Al Fakir, M.S. 167
Altman, R.L. 273
Azov, H. 182
Baev, V.K. 369
Bakanov, D.G. 425
Bellet, J.C. 119
Bilger, R.W. 81
Borisov, A.A. 239
Bousgarbies, J.L. 105
Britan, A.B. 391
Brun, R. 293
Cambray, P. 119
Champion, M. 119
Chauveau, Y. 119
Clarke, J.F. 3
Demin, A.I. 425
Doroshenko, V.M. 319
Dourieu, M.F. 293
Duran, G. 293
Efremov, V.L. 228
Escudie, D. 147
Eubank, C.S. 252
Fedoseev, A.I. 425
Galant, S. 182
Gardiner, W.C. Jr. 252
Gel'fand, B.E. 239
Gengembre, E. 119
Golovichev, V.I. 369
Guenoche, H. 369
Hashimoto, T. 37
Joulain, P. 167, 182
Khmelevskii, A.N. 391
Khomik, S.V. 239
Kidin, N.I. 305
Kidriavtsev, E.M. 425
Kiskin, A.B. 208
Klimov, A.M. 133
Kolesnikov, A.B. 228
Kolesnikov, B.Y. 228
Korobeinichev, O.P. 197
Ktalkherman, M.G. 411
Kudryavtsev, N.N. 319
Kuibida, L.V. 197
Lankford, D.W. 336
Levin, V.A. 391
Librovich, V.B. 305
Losev, S.A. 391
Lugovskoi, V.V. 391
Malkow, V.M. 411
McIntosh, A.C. 3
Most, J.M. 167
Nérault, J. 105
Niemitz, K.J. 252
Novikov, S.S. 319
Odintsov, A.I. 425
Ohta, Y. 38
Paranthoen, P. 147
Philippi, P.C. 293
Rapagnani, N.L. 336
Roberts, J.P. 305
Ruban, N.A. 411
Sedes, C. 369
Sharkov, V.F. 425
Simmie, J.M. 252
Simonenko, V.N. 208
Smekhov, G.D. 391
Souil, J.M. 182
Starik, A.M. 391
Stårner, S.H. 81
Takahashi, H. 38
Takeno, T. 57
Timofeev, E.I. 239
Tosello, R. 293
Trinite, M. 147
Tsyganov, S.A. 239
Umarbekov, N.S. 228
Volkov, A.Y. 425
Vuillermoz, M.L. 252
Warnatz, J. 305
Yamasaki, S. 57
Zarko, V.E. 208, 220
Zellner, R. 252
Zyryanov, V.Y. 220